Modeling and Remote Sensing of Forests Ecosystem

Modeling and Remote Sensing of Forests Ecosystem

Guest Editors

Jianping Wu
Zhongbing Chang
Xin Xiong

Basel • Beijing • Wuhan • Barcelona • Belgrade • Novi Sad • Cluj • Manchester

Guest Editors

Jianping Wu
Guangdong Provincial Key Lab of Remote Sensing and Geographical Information System
Guangzhou
China

Zhongbing Chang
Key Laboratory of Natural Resources Monitoring in Tropical and Subtropical Area of South China
Surveying and Mapping Institute Lands and Resource Department of Guangdong Province
Guangzhou
China

Xin Xiong
Jiangxi Provincial Key Laboratory of Carbon Neutrality and Ecosystem Carbon Sink
Lushan Botanical Garden
Jiangxi Province and Chinese Academy of Sciences
Jiujiang
China

Editorial Office
MDPI AG
Grosspeteranlage 5
4052 Basel, Switzerland

This is a reprint of the Special Issue, published open access by the journal *Forests* (ISSN 1999-4907), freely accessible at: https://www.mdpi.com/journal/forests/special_issues/H6GFYH11VG.

For citation purposes, cite each article independently as indicated on the article page online and as indicated below:

Lastname, A.A.; Lastname, B.B. Article Title. *Journal Name* **Year**, *Volume Number*, Page Range.

ISBN 978-3-7258-3329-0 (Hbk)
ISBN 978-3-7258-3330-6 (PDF)
https://doi.org/10.3390/books978-3-7258-3330-6

Cover image courtesy of Jianping Wu

© 2025 by the authors. Articles in this book are Open Access and distributed under the Creative Commons Attribution (CC BY) license. The book as a whole is distributed by MDPI under the terms and conditions of the Creative Commons Attribution-NonCommercial-NoDerivs (CC BY-NC-ND) license (https://creativecommons.org/licenses/by-nc-nd/4.0/).

Contents

About the Editors vii

Preface ix

Zhongbing Chang, Xin Xiong, Shuo Zhang and Jianping Wu
Modeling and Remote Sensing of the Forest Ecosystem
Reprinted from: *Forests* **2025**, *16*, 101, https://doi.org/10.3390/f16010101 1

Yichen Luo, Shuhua Qi, Kaitao Liao, Shaoyu Zhang, Bisong Hu and Ye Tian
Mapping the Forest Height by Fusion of ICESat-2 and Multi-Source Remote Sensing Imagery and Topographic Information: A Case Study in Jiangxi Province, China
Reprinted from: *Forests* **2023**, *14*, 454, https://doi.org/10.3390/f14030454 4

Xavier Gallagher-Duval, Olivier R. van Lier and Richard A. Fournier
Estimating Stem Diameter Distributions with Airborne Laser Scanning Metrics and Derived Canopy Surface Texture Metrics
Reprinted from: *Forests* **2023**, *14*, 287, https://doi.org/10.3390/f14020287 22

Lina Liu, Yaqiu Liu, Yunlei Lv and Xiang Li
A Novel Approach for Simultaneous Localization and Dense Mapping Based on Binocular Vision in Forest Ecological Environment
Reprinted from: *Forests* **2024**, *15*, 147, https://doi.org/10.3390/f15010147 40

Chang Liu, Wenzhi Du, Honglin Cao, Chunyu Shen and Lei Ma
Aboveground Biomass and Endogenous Hormones in Sub-Tropical Forest Fragments
Reprinted from: *Forests* **2023**, *14*, 661, https://doi.org/10.3390/f14040661 70

Jiaming Wang, Han Xu, Qingsong Yang, Yuying Li, Mingfei Ji, Yepu Li, et al.
Topographic Variation in Ecosystem Multifunctionality in an Old-Growth Subtropical Forest
Reprinted from: *Forests* **2024**, *15*, 1032, https://doi.org/10.3390/f15061032 81

Lei Li, Guangxing Ji, Qingsong Li, Jincai Zhang, Huishan Gao, Mengya Jia, et al.
Spatiotemporal Evolution and Prediction of Ecosystem Carbon Storage in the Yiluo River Basin Based on the PLUS-InVEST Model
Reprinted from: *Forests* **2023**, *14*, 2442, https://doi.org/10.3390/f14122442 96

Han Ren, Chaonan Chen, Yanhong Li, Wenbo Zhu, Lijuan Zhang, Liyuan Wang, et al.
Response of Vegetation Coverage to Climate Changes in the Qinling-Daba Mountains of China
Reprinted from: *Forests* **2023**, *14*, 425, https://doi.org/10.3390/f14020425 110

Yi Yang, Lei Yao, Xuecheng Fu, Ruihua Shen, Xu Wang and Yingying Liu
Spatial and Temporal Variations of Vegetation Phenology and Its Response to Land Surface Temperature in the Yangtze River Delta Urban Agglomeration
Reprinted from: *Forests* **2024**, *15*, 1363, https://doi.org/10.3390/f15081363 139

Vladimir Tabunshchik, Roman Gorbunov, Tatiana Gorbunova and Mariia Safonova
Vegetation Dynamics of Sub-Mediterranean Low-Mountain Landscapes under Climate Change (on the Example of Southeastern Crimea)
Reprinted from: *Forests* **2023**, *14*, 1969, https://doi.org/10.3390/f14101969 168

Yuxin Wang, Zhipei Liu, Baowei Qian, Zongyu He and Guangxing Ji
Quantitatively Computing the Influence of Vegetation Changes on Surface Discharge in the Middle-Upper Reaches of the Huaihe River, China
Reprinted from: *Forests* **2022**, *13*, 2000, https://doi.org/10.3390/f13122000 191

Mengya Jia, Shixiong Hu, Xuyue Hu and Yuannan Long
Response Mechanism of Annual Streamflow Decline to Vegetation Growth and Climate Change in the Han River Basin, China
Reprinted from: *Forests* **2023**, *14*, 2132, https://doi.org/10.3390/f14112132 . . . . **202**

Guangxing Ji, Shuaijun Yue, Jincai Zhang, Junchang Huang, Yulong Guo and Weiqiang Chen
Assessing the Impact of Vegetation Variation, Climate and Human Factors on the Streamflow Variation of Yarlung Zangbo River with the Corrected Budyko Equation
Reprinted from: *Forests* **2023**, *14*, 1312, https://doi.org/10.3390/f14071312 . . . . **214**

About the Editors

Jianping Wu

Jianping Wu received his Ph.D. from the University of Chinese Academy of Science, Beijing, China. He currently serves as an Associate Researcher at the Guangzhou Institute of Geography, Guangdong Academy of Sciences, Guangzhou, China. He is primarily engaged in ecological and geographical scientific research, focusing on the response mechanisms of forest ecosystems to climate change and urban ecological environments. Dr. Wu has published over 40 peer-reviewed papers in the field of global ecology, ecosystem ecology, urban ecosystems, and forest ecosystems. He serves on the editorial board of *Forests*. Dr. Wu is the project leader for several significant research projects, including the Natural Science Foundation of Guangdong Province (grant number 2024A1515030190), the Science and Technology Program of Guangzhou (grant number 2024A04J3347), the Young Talent Project of GDAS (grant number 2023GDASQNRC-0217), and the GDAS Project of Science and Technology Development (grant number 2024GDASZH-2024010102).

Zhongbing Chang

Zhongbing Chang received his PhD from the South China Botanical Garden, Chinese Academy of Sciences. He is an engineer at the Surveying and Mapping Institute Lands and Resource Department of Guangdong Province. He developed a methodology to estimate the distribution and dynamics of forest aboveground carbon stock in China by integrating optical vegetation indices and microwave VOD remote sensing products. Dr. Chang is the project leader for several significant research projects, including the Science and Technology Program of Guangzhou (2023A04J0927), the Key Laboratory of Surveying and Mapping Science and Geospatial Information Technology of MNR, CASM (20230505), the Key Laboratory of Vegetation Restoration and Management of Degraded Ecosystems, Chinese Academy of Sciences (VRMDE2306), and the Science and Technology Program of Guangdong Province (2021B1212100003).

Xin Xiong

Xin Xiong received his Ph.D. from the University of Chinese Academy of Science, Beijing, China. He is a postdoctoral fellow at South China Botanical Garden, Chinese Academy of Sciences, Guangzhou, China. He works at the Lushan Botanical Garden, Jiangxi Province, and the Chinese Academy of Sciences, Jiujiang, China. His research interests include global change ecology and forest ecology, with a focus on carbon cycling processes and mechanisms in subtropical forest ecosystems. He is the group leader of the ecosystem ecology research group (associate researcher) and is a Master's supervisor at Nanchang University and Anhui Agricultural University. He is also an Executive Director of the Jiangxi Botanical Society. Dr. Xiong is the project leader or principal investigator for several significant research projects, including the Natural Science Foundation of Guangdong Province (grant number 2022A1515110403), the Science and Technology Program of Jiangxi Province (grant numbers 20232BAB215009, 20242BCC32138), and the Jiujiang Municipal Science and Technology Program (grant number S2024KXJJ0001).

Preface

As the largest carbon pool in terrestrial ecosystems, forests play a pivotal role in regulating the Earth's climate, influencing both the global carbon cycle and climate change mitigation strategies. Forests not only act as significant carbon sinks, absorbing and storing carbon dioxide from the atmosphere, but they also provide a wide range of other ecosystem services, including biodiversity conservation, water regulation, and soil protection. As such, understanding the dynamics of forest ecosystems is critical in addressing the global challenges posed by climate change.

In recent decades, satellite and remote sensing technologies have emerged as powerful tools for monitoring forest changes across large spatial scales. These advancements offer unprecedented opportunities for the real-time tracking of forest health, structure, composition, and carbon fluxes, all of which are crucial for developing effective forest management strategies. By integrating various remote sensing sources, including optical imagery, synthetic aperture radar (SAR), light detection and ranging (LiDAR), and microwave sensors, researchers can now capture a more comprehensive and nuanced picture of forest ecosystems. These technologies enable the detection of forest cover changes, the assessment of forest biomass and carbon storage, and the monitoring of environmental stressors such as droughts, pests, and deforestation.

In this reprint, titled *"Modeling and Remote Sensing of Forest Ecosystem,"* we are pleased to present a collection of 12 insightful articles that represent the forefront of forest ecosystem research. These contributions highlight the valuable insights into how remote sensing technologies and computational models are being used to enhance our understanding of forest dynamics. These articles cover a broad spectrum of topics, including the development of new remote sensing techniques, the application of advanced modeling frameworks to simulate forest processes, and the integration of multi-source data to improve the accuracy and reliability of forest monitoring systems. The articles also address some of the most pressing challenges in forest ecosystem research, including the need for high-resolution, long-term data to track forest changes over time, the complexities of modeling forest carbon dynamics under varying environmental conditions, and the integration of remote sensing technologies into forest management practices. By bridging the gap between data collection, modeling, and on-the-ground forest management, these articles provide valuable insights into how remote sensing and modeling can inform decision-making processes and contribute to sustainable forest management practices.

We hope that this collection will inspire continued innovation in remote sensing and modeling techniques and encourage further research into the role of forests in the context of global environmental change. By advancing our ability to monitor, analyze, and manage forest ecosystems, we can work towards more effective strategies for conserving and restoring forests, mitigating climate change, and ensuring the long-term sustainability of these vital ecosystems.

Jianping Wu, Zhongbing Chang, and Xin Xiong
Guest Editors

Editorial

Modeling and Remote Sensing of the Forest Ecosystem

Zhongbing Chang [1], Xin Xiong [2], Shuo Zhang [3] and Jianping Wu [4,*]

1. Key Laboratory of Natural Resources Monitoring in Tropical and Subtropical Area of South China, Surveying and Mapping Institute Lands and Resource Department of Guangdong Province, Guangzhou 510663, China
2. Jiangxi Provincial Key Laboratory of Carbon Neutrality and Ecosystem Carbon Sink, Lushan Botanical Garden, Jiangxi Province and Chinese Academy of Sciences, Jiujiang 332900, China
3. Zhaoqing Municipal Bureau of Forestry, Center for Zhaoqing High-Level Talent Development, Zhaoqing 526040, China
4. Guangdong Provincial Key Lab of Remote Sensing and Geographical Information System, Guangdong Open Laboratory of Geospatial Information Technology and Application, Guangzhou Institute of Geography, Guangdong Academy of Sciences, Guangzhou 510070, China
* Correspondence: wujianping@gdas.ac.cn

Forests cover around one-third of the global land surface, store about half of the terrestrial carbon, and are the dominant contributors to terrestrial net primary production. As the largest carbon pool of terrestrial ecosystems, the forest ecosystem plays a critical role in both the global carbon cycle and climate change mitigation. Time-series monitoring is essential for understanding forest ecosystem processes and forest response to anthropogenic activities and climate change. In recent decades, satellite records have offered the potential to monitor forest changes by combining diverse remote sensing sources including optical, synthetic aperture radar (SAR), light detection and ranging (LiDAR), and microwave sensors. Remote sensing data from different sources and with various land surface process models could provide better spatial coverage with high resolution, and are available for long-term time series, which can enable the effective global mapping and monitoring of forest trends. In light of these advantages, we organized this Special Issue, "Modeling and Remote Sensing of the Forest Ecosystem". This Special Issue covers potential topics including the response of forest dynamics to anthropogenic activities and climate change; time-series change detection and trend analyses of forest ecosystems; the impacts of climate extremes (e.g., drought to wetness) on the forest ecosystem; the monitoring of forest biomass and carbon dynamics; the mapping of forest structure parameters.

In this Special Issue, we are pleased to present a collection of 12 insightful articles that highlight the latest advancements in forest ecosystem research. Among these, five articles focus on the inversion of forest structural parameters and biomass, as well as an analysis of forest ecosystem functions [1–6]. These studies employ advanced modeling techniques and remote sensing data to better understand the complex relationships between forest structure, biomass, and ecosystem processes. Several articles integrate remote sensing data, such as ICESat-2, airborne laser scanning, and UAV-based imagery, to map forest attributes like canopy height, stem diameter, and biomass distribution. Others focus on understanding how forest fragmentation, topography, and land-use changes affect ecosystem functions like carbon storage, biomass, and biodiversity. Modeling techniques, including machine learning and the InVEST model, are widely applied to predict future scenarios and optimize forest management strategies. Three articles focus on the impact of climate change on vegetation dynamics across different regions [7–9]. They emphasize the role of climate variables such as temperature, precipitation, and sunshine duration in shaping vegetation coverage and phenology. The research highlights how vegetation responds to changes in land surface temperature, urbanization, and other environmental factors, with varying patterns of growth, biomass, and seasonality. Three articles focus on the impact of vegetation changes on streamflow variation and hydrological processes in different regions of China [10–12]. They use advanced methods like the Budyko equation and elastic coefficient analysis to quantify how vegetation, climate, and human activities influence surface runoff and streamflow. The results indicate that vegetation growth has a generally positive effect on streamflow reduction, with varying contributions across regions. The studies highlight the importance of understanding the interplay between climate, vegetation, and hydrology for water resource management and ecological restoration efforts in different river basins.

The contributions in this issue provide valuable insights into how remote sensing technologies and computational models can be leveraged to enhance our understanding of forest dynamics. These articles not only discuss the technical methodologies used to derive forest structural parameters but also explore the implications of these findings for forest management, biodiversity conservation, and climate change mitigation. Together, these articles offer a comprehensive perspective of the role of remote sensing and modeling in advancing forest ecosystem research, addressing key challenges in monitoring, analyzing, and managing forest ecosystems in the face of global environmental changes. We hope that this collection will inspire further research and foster the development of innovative solutions for sustainable forest management.

Author Contributions: Conceptualization, J.W.; investigation, X.X.; data curation, Z.C.; writing—original draft preparation, Z.C.; writing—review and editing, S.Z. and X.X.; visualization, J.W. All authors have read and agreed to the published version of the manuscript.

Conflicts of Interest: The authors declare no conflicts of interest.

References

1. Luo, Y.; Qi, S.; Liao, K.; Zhang, S.; Hu, B.; Tian, Y. Mapping the Forest Height by Fusion of ICESat-2 and Multi-Source Remote Sensing Imagery and Topographic Information: A Case Study in Jiangxi Province, China. *Forests* **2023**, *14*, 454. [CrossRef]
2. Gallagher-Duval, X.; van Lier, O.R.; Fournier, R.A. Estimating Stem Diameter Distributions with Airborne Laser Scanning Metrics and Derived Canopy Surface Texture Metrics. *Forests* **2023**, *14*, 287. [CrossRef]
3. Liu, L.; Liu, Y.; Lv, Y.; Li, X. A Novel Approach for Simultaneous Localization and Dense Mapping Based on Binocular Vision in Forest Ecological Environment. *Forests* **2024**, *15*, 147. [CrossRef]
4. Liu, C.; Du, W.; Cao, H.; Shen, C.; Ma, L. Aboveground Biomass and Endogenous Hormones in Sub-Tropical Forest Fragments. *Forests* **2023**, *14*, 661. [CrossRef]
5. Wang, J.; Xu, H.; Yang, Q.; Li, Y.; Ji, M.; Li, Y.; Chang, Z.; Qin, Y.; Yu, Q.; Wang, X. Topographic Variation in Ecosystem Multifunctionality in an Old-Growth Subtropical Forest. *Forests* **2024**, *15*, 1032. [CrossRef]
6. Li, L.; Ji, G.; Li, Q.; Zhang, J.; Gao, H.; Jia, M.; Li, M.; Li, G. Spatiotemporal Evolution and Prediction of Ecosystem Carbon Storage in the Yiluo River Basin Based on the PLUS-InVEST Model. *Forests* **2023**, *14*, 2442. [CrossRef]
7. Ren, H.; Chen, C.; Li, Y.; Zhu, W.; Zhang, L.; Wang, L.; Zhu, L. Response of Vegetation Coverage to Climate Changes in the Qinling-Daba Mountains of China. *Forests* **2023**, *14*, 425. [CrossRef]
8. Yang, Y.; Yao, L.; Fu, X.; Shen, R.; Wang, X.; Liu, Y. Spatial and Temporal Variations of Vegetation Phenology and Its Response to Land Surface Temperature in the Yangtze River Delta Urban Agglomeration. *Forests* **2024**, *15*, 1363. [CrossRef]
9. Tabunshchik, V.; Gorbunov, R.; Gorbunova, T.; Safonova, M. Vegetation Dynamics of Sub-Mediterranean Low-Mountain Landscapes under Climate Change (on the Example of Southeastern Crimea). *Forests* **2023**, *14*, 1969. [CrossRef]
10. Wang, Y.; Liu, Z.; Qian, B.; He, Z.; Ji, G. Quantitatively Computing the Influence of Vegetation Changes on Surface Discharge in the Middle-Upper Reaches of the Huaihe River, China. *Forests* **2022**, *13*, 2000. [CrossRef]

11. Jia, M.; Hu, S.; Hu, X.; Long, Y. Response Mechanism of Annual Streamflow Decline to Vegetation Growth and Climate Change in the Han River Basin, China. *Forests* **2023**, *14*, 2132. [CrossRef]
12. Ji, G.; Yue, S.; Zhang, J.; Huang, J.; Guo, Y.; Chen, W. Assessing the Impact of Vegetation Variation, Climate and Human Factors on the Streamflow Variation of Yarlung Zangbo River with the Corrected Budyko Equation. *Forests* **2023**, *14*, 1312. [CrossRef]

Disclaimer/Publisher's Note: The statements, opinions and data contained in all publications are solely those of the individual author(s) and contributor(s) and not of MDPI and/or the editor(s). MDPI and/or the editor(s) disclaim responsibility for any injury to people or property resulting from any ideas, methods, instructions or products referred to in the content.

Article

Mapping the Forest Height by Fusion of ICESat-2 and Multi-Source Remote Sensing Imagery and Topographic Information: A Case Study in Jiangxi Province, China

Yichen Luo [1,2,†], Shuhua Qi [1,*,†], Kaitao Liao [1,2], Shaoyu Zhang [1], Bisong Hu [1] and Ye Tian [1]

[1] Key Laboratory of Poyang Lake Wetland and Watershed Research (Ministry of Education), School of Geography and Environment, Jiangxi Normal University, Nanchang 330022, China
[2] Jiangxi Academy of Water Science and Engineering, Nanchang 330029, China
* Correspondence: qishuhua11@jxnu.edu.cn
† These authors contributed equally to this work.

Abstract: Forest canopy height is defined as the distance between the highest point of the tree canopy and the ground, which is considered to be a key factor in calculating above-ground biomass, leaf area index, and carbon stock. Large-scale forest canopy height monitoring can provide scientific information on deforestation and forest degradation to policymakers. The Ice, Cloud, and Land Elevation Satellite-2 (ICESat-2) was launched in 2018, with the Advanced Topographic Laser Altimeter System (ATLAS) instrument taking on the task of mapping and transmitting data as a photon-counting LiDAR, which offers an opportunity to obtain global forest canopy height. To generate a high-resolution forest canopy height map of Jiangxi Province, we integrated ICESat-2 and multi-source remote sensing imagery, including Sentinel-1, Sentinel-2, the Shuttle Radar Topography Mission, and forest age data of Jiangxi Province. Meanwhile, we develop four canopy height extrapolation models by random forest (RF), Support Vector Machine (SVM), K-nearest neighbor (KNN), Gradient Boosting Decision Tree (GBDT) to link canopy height in ICESat-2, and spatial feature information in multi-source remote sensing imagery. The results show that: (1) Forest canopy height is moderately correlated with forest age, making it a potential predictor for forest canopy height mapping. (2) Compared with GBDT, SVM, and KNN, RF showed the best predictive performance with a coefficient of determination (R^2) of 0.61 and a root mean square error (RMSE) of 5.29 m. (3) Elevation, slope, and the red-edge band (band 5) derived from Sentinel-2 were significantly dependent variables in the canopy height extrapolation model. Apart from that, Forest age was one of the variables that the RF moderately relied on. In contrast, backscatter coefficients and texture features derived from Sentinel-1 were not sensitive to canopy height. (4) There is a significant correlation between forest canopy height predicted by RF and forest canopy height measured by field measurements (R^2 = 0.69, RMSE = 4.02 m). In a nutshell, the results indicate that the method utilized in this work can reliably map the spatial distribution of forest canopy height at high resolution.

Keywords: forest canopy height; ICESat-2; Sentinel-1; Sentinel-2; topographic information; forest age; machine learning

Citation: Luo, Y.; Qi, S.; Liao, K.; Zhang, S.; Hu, B.; Tian, Y. Mapping the Forest Height by Fusion of ICESat-2 and Multi-Source Remote Sensing Imagery and Topographic Information: A Case Study in Jiangxi Province, China. *Forests* **2023**, *14*, 454. https://doi.org/10.3390/f14030454

Academic Editor: Helmi Zulhaidi Mohd Shafri

Received: 9 December 2022
Revised: 18 February 2023
Accepted: 20 February 2023
Published: 22 February 2023

Copyright: © 2023 by the authors. Licensee MDPI, Basel, Switzerland. This article is an open access article distributed under the terms and conditions of the Creative Commons Attribution (CC BY) license (https://creativecommons.org/licenses/by/4.0/).

1. Introduction

Forest ecosystems, as an integral component of terrestrial ecosystems, play a vital role in global climate change and carbon sinks [1]. Forest canopy height is defined as the distance between the highest point of the tree canopy and the ground, which is used to calculate above-ground biomass, leaf area index, and carbon stocks [2–4]. In addition, large-scale forest canopy height monitoring can provide scientific information regarding deforestation and forest degradation to policymakers [5,6].

Remote sensing technology is regarded as a useful technique for global and regional forest canopy height mapping [7,8]. Traditional optical remote sensing for forest canopy height mapping is based on reflectance information, which has limitations such as quick signal saturation and sensitivity to meteorological conditions, such as clouds and rain [9,10]. In contrast, a microwave radar can operate in any weather condition, regardless of clouds or rain. However, in areas with varied terrain, microwave radar emissions can be influenced by signal saturation, which dramatically reduced the accuracy of canopy height estimates [11]. Light Detection and Ranging (LiDAR), in contrast to the preceding two types of technology, measures targets by producing pulsed lasers that can penetrate dense forest surfaces to acquire information regarding both the understory and surface. As a result, LiDAR is regarded as one of the most precise methods for the measurement of forest structure parameters [6,12].

The Geoscience Laser Altimeter System (GLAS) onboard the Ice, Cloud, and Land Elevation Satellite (ICESat) was operated by NASA in 2010, which was the first satellite based LiDAR to be used to observe ice cap and forest structure parameters [13]. A series of researchers have undertaken research regarding the integration of the GLAS and other optical data for large-scale forest canopy height mapping [14–16]. For example, in 2010, Lefsky [15] mapped the first global 500 m resolution forest canopy height map by combining GLAS and MODIS (NASA). Simard et al. [16] used random forest (RF) to map 1 km of global resolution forest canopy height based on the GLAS and seven remote sensing images. In addition, in 2016, Wang et al. [17] used the GLAS and 13 auxiliary variables to globally map forest canopy height at a 500 m resolution using the balanced random forest algorithm. However, the GLAS has a low spot density and a big spot size, resulting in low accuracy and resolution in forest canopy height mapping products [18]. In 2018, the Ice, Cloud, and Land Elevation Satellite-2 (ICESat-2) was launched by NASA at Vandenberg Air Force Base, which carried the Advanced Topographic Laser Altimeter System (ATLAS) [19,20]. The ATLAS instrument employs multi-beam, micropulse, and photon-counting LiDAR technology. Multiple beams can map more information at the same time compared with a single beam. Meanwhile, the micropulse increases the emission frequency and thus the spot density, which can greatly improve measurement accuracy [19]. The ATLAS instrument is not primarily charged with studying vegetation, but the ATL08 data packages for both land and vegetation provide more opportunities for large-scale forest investigations. On the other hand, ATL08 data can only offer a footprint rather than continuous forest canopy height information at regional and global scales [20–22]. Furthermore, the Global Ecosystem Dynamics Investigation (GEDI) is another high-resolution laser ranging system operated by NASA [23]. The GEDI is designed to observe the forest areas between 51.6° N and S. Meanwhile, the basic metric of the GEDI are wave-forms related to the density profile of canopy height for 25 m footprints [24]. To date, the GEDI has been widely used for global and regional forest canopy height mapping. In 2019, Potapov et al. [25] integrated the GEDI and Landsat data to create a global, 30 m resolution map of forest canopy height.

To overcome the spatial discontinuity of LiDAR measurements [21,26], several studies have shown that integrating multi-source continuous remote sensing data with LiDAR data could extrapolate forest canopy height from the footprint scale to the regional scale, resulting in wall-to-wall forest canopy height maps [14,16,18,27]. Furthermore, it was found that the large-scale spatial feature information provided by optical remote sensing data such as Landsat (NASA), Sentinel (ESA), and MODIS correlated well with the vegetation structure information acquired from LiDAR [6,7,16,21]. Wu et al. [28] constructed an ecological zoning random forest algorithm based on ATLAS and Landsat for 30 m resolution forest height mapping in China. Compared with the optical satellite data, Sentinel-2's red-edge bands were shown to provide more accurate information regarding vegetation growth [29]. Zhang et al. [30] fused ATLAS, Landsat, and Sentinel to map boreal forest canopy height via RF. In contrast, this study demonstrated that the red-edge band (band 5) and the NDRE derived from band 5 in Sentinel-2 are significantly dependent features in RF. In addition, Sentinel-1 SAR data, topographic information, textural features, and climatic data were used to map global and regional forest canopy height maps [1,7,18]. In the Himalayan Foothills of India, Nandy et al. [1] mapped forest canopy height by integrating ICESat-2 and Sentinel-1 data using the RF algorithm. Meanwhile, Sothe et al. [31] mapped forest canopy height in Canada by combining GEDI and ICESat-2 with PALSAR and Sentinel in 2020, which demonstrated the potential of the L-band for forest canopy height prediction.

Machine learning models, as statistical predictive models, have many advantages, including lower computational complexity, less need for parameter tuning, stronger classification and regression capabilities, and higher performance in integrating multi-source data [5,32]. Machine learning models such as RF, the Support Vector Machine (SVM), K-nearest neighbor (KNN), Gradient Boosting Decision Tree (GBDT), and Mixed Logistic Regression (MLR) have been frequently employed for forest canopy height extrapolation [5,6,21,27,33]. For example, Zhu et al. [14] generated a forest canopy height map with a spatial resolution of 30 m, utilizing Multiple Altimeter Beam Experimental Lidar (MABEL) and Landsat 8 OLI. Jiang et al. [6] successfully developed a stacking algorithm consisting of MLR, RF, SVM, and KNN to map the forest canopy height in northern China. Compared with the single algorithm, it was shown that the stacking algorithm effectively improved the extrapolation accuracy. Xi et al. [5] developed machine learning models for different forest types by integrating ICESat-2, Sentinel, and topographic information. This study demonstrated the better performance of machine learning models developed based on different forest types relative to the whole forest.

Although numerous studies have successfully mapped continuous forest canopy height by fusing satellite-based LiDAR data and remotely sensed images, few have considered the relationship between forest-related indicators and canopy height when conducting forest canopy height mapping [34]. Previous research has demonstrated that forest age and forest canopy height are closely related [35]. Several methods for combining forest canopy height information to map forest age have been proposed [36–38]. Furthermore, no relevant studies have explored the potential of forest age in forest canopy height mapping. In addition, the spatial and temporal inconsistency of remote sensing and validation datasets is another challenge in mapping regional forest canopy height [39].

The goal of this research was to create a high-resolution forest canopy height map for Jiangxi Province in southeastern China by utilizing machine learning models that incorporate ICESat-2, Sentinel, and forest age, and then to assess the forest canopy height map using time-consistent validation data. To achieve this goal, four specific objectives were proposed: (1) investigating the relationship between forest age and forest canopy height at the sample scale and providing insight into the key drivers of canopy height extrapolation models in the region, (2) developing four forest canopy height extrapolation models based on machine learning models to explore significant predictors that relate to forest canopy height, (3) generating a Jiangxi Province forest canopy height map based on the best-performing model, and (4) validating the forest canopy height map using field measurements and comparing it with existing forest canopy height maps.

2. Materials and Methods

2.1. Study Area

The study was conducted in Jiangxi Province, which is located in southeastern China and has an area of about 166,900 km^2 (Figure 1). The research area's topography is complex and diversified, dominated by hills and mountains, and it features wide basins and valleys with an average elevation above sea level of 245.6 m. Jiangxi Province is abundant in water resources, with a thick network of rivers connecting to the Yangtze River via Poyang Lake, China's largest freshwater lake. The climate is a humid subtropical monsoon climate with an annual temperature range of 16.3 to 19.5 °C. Furthermore, the region has a frost-free period of 240–307 days and an annual precipitation average of 1341–1943 mm.

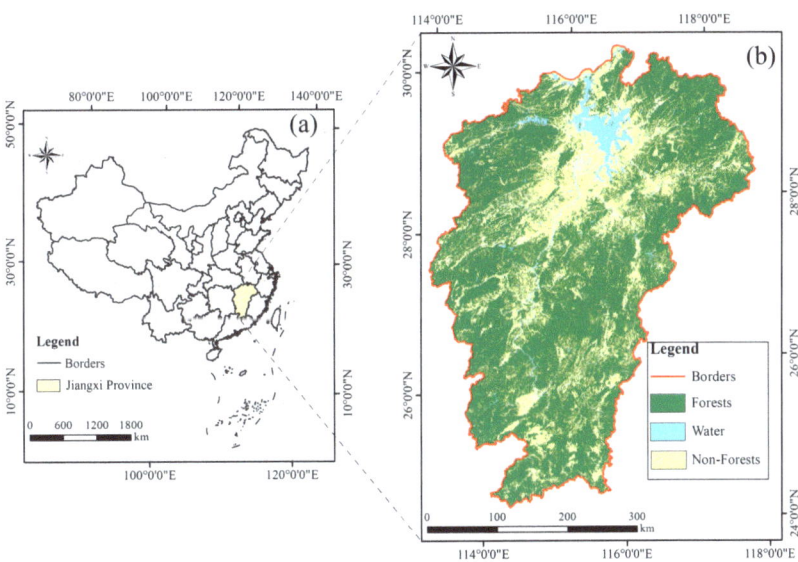

Figure 1. (**a**) Location of the study area in China. (**b**) Land cover, including forest, water, and non-forest, in the study area.

Jiangxi Province is rich in forest resources, with a forest coverage rate of 63.35%, ranking second in China. Broad-leaved evergreen forests are the zonal vegetation of Jiangxi Province. Due to the region's broad geographical span from north to south, the southern section is influenced by the southern subtropical climate and flora components, and the vegetation features exhibit some southern subtropical characteristics. In the north, there are more tropical flora components, and some warm temperate flora components are increasingly mixed. The dominant tree species in the region are Masson's pine (*Pinus massoniana Lamb*), Chinese fir (*Cunninghamia lanceolata*), and slash pine (*Pinus elliottii*).

2.2. Data Acquisition and Processing

In this study, the following data were used to accomplish the objectives: (1) ATL08 data collected using ICESat-2, (2) SAR data collected using Sentinel-1, (3) optical image data collected using Sentinel-2, (4) topographic data collected from the Shuttle Radar Topography Mission (SRTM), (5) forest age data in Jiangxi Province calculated using a time-series change monitoring algorithm, (6) the Land Use Cover and Change product collected using GlobeLand30, and (7) validation data provided by the 7th Jiangxi Province Forest Resources Second Class Survey. Canopy height information from the ICESat-2 point and 33 predictors (Table 1) corresponding to the geographical location were obtained from the data sources listed above and were used in subsequent machine learning modeling.

2.2.1. ATL08 Data

The ATLAS instrument uses green (532 nm) laser light and single-photon sensitive detection to measure the time of flight and, subsequently, surface height along each of its six beams [19]. ATLAS data are distributed as Hierarchical Data Format Version 5 (HDF-5) through the National Snow and Ice Data Center (NSIDC, https://nsidc.org/data/icesat-2 (accessed on 10 January 2022)), which is divided into 4 levels (ATL01-ATL21, without ATL05) [40]. ATL08 is a product that was developed in response to one of ICESat-2's scientific tasks: measuring vegetation canopy height as a basis for estimating large-scale biomass and biomass change, and it provides us with several parameters such as surface elevation, absolute canopy height, relative canopy height, and so on. Each metric is derived from photons classified as ground or canopy within a 100-m segment. The primary canopy height parameter in the data product, h_{canopy}, is defined as 98% canopy height in 100 m segments [20]. In addition, as a parameter used to describe the terrain elevation, $h\_te\_best\_fit$ is defined as the best-fit terrain elevation at the mid-point location of each 100 m segment [20].

The ATL08 data (580876 ICESat-2 points) from 2020 were utilized in this study to ensure the chronological validity of the verified data and the coverage of the research area. As noise points are always present, we used several methods to filter out potentially erroneous ICESat-2 points. ICESat-2 points with h_{canopy} less than 2 m or greater than 50 m were filtered out based on forest structure information in the research area. Furthermore, SRTM was utilized as terrain calibration data, and if the discrepancy between the $h\_te\_best\_fit$ and SRTM was greater than 20 m, it was assumed that the point was impacted by clouds or the atmosphere [18]. At the same time, strong beam data recorded at night was used in the study as it was shown to have the highest performance in canopy height retrieval [41]. In summary, 5777 filtered ICESat-2 points (Figure 2a) based on the method described above were used for subsequent machine modeling.

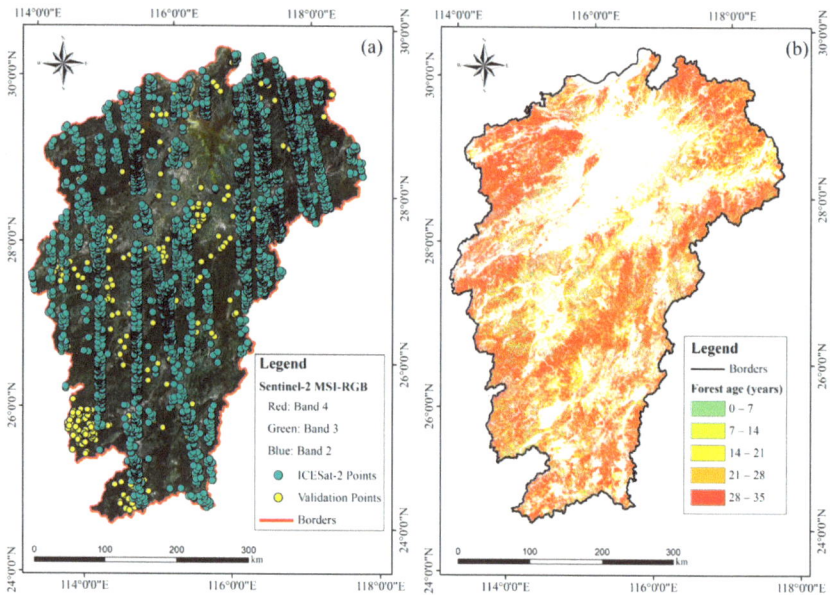

Figure 2. (**a**) Distribution of ICESat-2 points and field measurements in Sentinel-2 RGB composite images. (**b**) Forest age of Jiangxi Province.

2.2.2. Forest Age

Forest age data were used in this study as key input data in models to estimate forest canopy height, which were created by the contributing author last year. The data showed that the secondary forest age ranged from 0 to 34 years annually, and the stable forest with no changes in the time series was defined as having an age of 35. It was mapped using multiple change detection algorithms with dense Landsat time series and inventory data, which presents the first map of forest age (Figure 2b) in 2020 at a 30 m resolution in Jiangxi Province. Meanwhile, we used 160 field measurements to validate the forest age map for Jiangxi Province. The variable from the forest age product was reliable to establish the relationship with forest canopy height due to the high accuracy of the coefficient of determination (R^2) = 0.87; root mean square error (RMSE) = 3.17 years.

2.2.3. Sentinel Data

Sentinel-1 is a polar-orbiting binary constellation with the ability to provide daily, day/night, and all-weather medium-resolution observations for the continuation and improvement of SAR operational services and applications, which can provide single (HH and VV) or dual polarization modes (HH + HV and VV + VH) [42]. Sentinel-2 provides high-resolution multispectral image acquisition on a global scale through the Multi-Spectral Instrument (MSI), which carries 13 spectral bands from the visible and near-infrared to the short-wave infrared [43]. In addition, the red-edge band obtained using the MSI is widely used to monitor vegetation because it is sensitive to vegetation growth [44].

In this study, Sentinel data regarding the vegetation growing season (2020) were acquired and processed in Google Earth Engine (GEE, https://earthengine.google.com/ (accessed on 28 April 2022)). The backscatter coefficients, VV and VH, of Sentinel-1 were obtained from Sentinel-1 SAR GRD products. Sentinel-2 images from "Level-2A" data were cloud-masked using the s2cloudless algorithm [45]. Considering the spatial consistency and availability of data, both Sentinel-1 and Sentinel-2 data (Figure 2a) were resampled to a 30 m resolution and processed using a median image composition. Additionally, composite images were used to generate eighteen texture features and three vegetation indices as the base image. In general, features contained in the gray-level co-occurrence matrix (GLCM) in Table 1 are used to characterize the texture; the features include Mean, Variance, Homogeneity, Contrast, Dissimilarity, Entroy, Angular Second Moment, and Correlation [46]. Texture features are calculated using a GLCM with third-order kernels based on the backward scattering coefficient. The three normalized (NDRE_BAND5, NDRE_BAND6, and NDRE_BAND7) red-edge indices are calculated based on the three red-edge bands and the infrared band [47]. The formula for NDRE is as follows:

$$\text{NDRE} = \frac{NIR - RedEdge}{NIR + RedEdge} \quad (1)$$

where NIR represents the near-infrared band (band 8), and $RedEdge$ represents the three red-edge bands of band 5, 6, and 7 in Sentinel-2.

Table 1. Predictors obtained by multi-source data.

Data Source	Predictors	Description	References
Forest Age	FOREST_AGE	30 m-resolution forest age map in Jiangxi Province	-
Sentinel-1	VV	VV band extracted from Sentinel-1	[42]
	VH	VH band extracted from Sentinel-1	
	VV_MEAN	Mean Value calculated by GLCM based on VV	
	VV_VAR	Variance calculated by GLCM based on VV	
	VV_HOM	Homogeneity calculated by GLCM based on VV	
	VV_CON	Contrast calculated by GLCM based on VV	
	VV_DISS	Dissimilarity calculated by GLCM based on VV	
	VV_ENT	Entropy calculated by GLCM based on VV	
	VV_SEC	Angular Second Moment calculated by GLCM based on VV	
	VV_COR	Correlation calculated by GLCM based on VV	[46]
	VH_MEAN	Mean Value calculated by GLCM based on VH	
	VH_VAR	Variance calculated by GLCM based on VH	
	VH_HOM	Homogeneity calculated by GLCM based on VH	
	VH_CON	Contrast calculated by GLCM based on VH	
	VH_DISS	Dissimilarity calculated by GLCM based on VH	
	VH_ENT	Entropy calculated by GLCM based on VH	
	VH_SEC	Angular Second Moment calculated by GLCM based on VH	
	VH_COR	Correlation calculated by GLCM based on VH	

Table 1. Cont.

Data Source	Predictors	Description	References
Sentinel-2	S2_BAND2	Blue band extracted from Sentinel-2	[43]
	S2_BAND3	Green band extracted from Sentinel-2	
	S2_BAND4	Red band extracted from Sentinel-2	
	S2_BAND5	Vegetation Red Edge band extracted from Sentinel-2 (705 nm)	
	S2_BAND6	Vegetation Red Edge band extracted from Sentinel-2 (740 nm)	
	S2_BAND7	Vegetation Red Edge band extracted from Sentinel-2 (782 nm)	
	S2_BAND8	NIR band extracted from Sentinel-2	
	S2_BANDA	Narrow NIR band extracted from Sentinel-2	
	NDRE_BAND5	Normalized difference red-edge vegetation index based on S2_BAND5 and S2_BAND8	[47]
	NDRE_BAND6	Normalized difference red-edge vegetation index based on S2_BAND6 and S2_BAND8	
	NDRE_BAND7	Normalized difference red-edge vegetation index based on S2_BAND7 and S2_BAND8	
SRTM	ELEVATION	Elevation extracted from SRTM	[48]
	SLOPE	Slope extracted from SRTM	
	ASPECT	Aspect extracted from SRTM	

2.2.4. Auxiliary Data

The SRTM uses one-way interferometry to acquire topographic data. The SRTM collects radar data for 80% of the Earth's surface between 60° N and 56° S, which is used to construct a global digital elevation model. In this study, the SRTM was called and processed using GEE to obtain the elevation, slope, and aspect of the study area [48].

GlobeLand30 2020, a 30 m resolution global land cover data product, is the latest version developed in China. It includes 10 land cover classifications, including forests. The product has an overall accuracy of 85.72% [49]. GlobeLand30 2020 was used for the study area in this study, which effectively helped us to distinguish between forested and non-forested areas.

The 7th Jiangxi Province Forest Resources Second Class Survey was completed in Jiangxi Province in 2019–2020, which established a database of forest resources in Jiangxi Province consisting of irregular blocks [6]. Each block consists of the same forest species and contains geographic coordinates, forest type, forest canopy height, forest age, etc. Meanwhile, the forest canopy height represents the average height of the forest canopy within the whole block. In this study, 380 field measurement points (Figure 2a) converted from the center of gravity of blocks were used to validate the forest canopy height maps generated by the machine learning model. The coordinates of the field measurement points were used as a reference, and cell values at locations were extracted from one raster and recorded in the attribute table of the point.

Previous research has generated forest canopy height maps of China at 30 m resolutions using various methodologies [18,50]. Zhu [18] produced a 30 m resolution forest altitude map of China (R^2 = 0.38; RMSE = 2.67 m) using a random forest model in collaboration with data from satellite-based LiDAR (GEDI and ICESat-2) and optical imagery (Landsat-8 and Sentinel). Meanwhile, Liu et al. [50] developed a deep learning interpolation model by integrating the GEDI and ICESat-2 to map another 30 m resolution forest canopy height map of the Chinese region (R^2 = 0.6; RMSE = 4.88 m). In addition, the Global Forest Canopy Height (GFCH) mapped based on the GEDI and Landsat by Potapov et al. [25] in 2019 is also open access (https://glad.umd.edu/dataset/gedi/ (accessed on 11 February 2023)). We gathered the forest canopy height in Jiangxi Province from three data [18,25,50] using the Jiangxi Province boundary mask to assess the correctness of the forest canopy height map in this study.

2.3. Methods

In this study, correlation analysis was used to understand the relationship between canopy height and predictors, and in addition, machine learning models were used to link canopy height and predictors and perform continuous forest canopy height mapping. Meanwhile, SPSS Statistics software (version 25) was used for correlation analysis, and R (version 4.2.1) was used for machine learning modeling. A workflow (Figure 3) was created to summarize the main steps of the implemented methodology in the study.

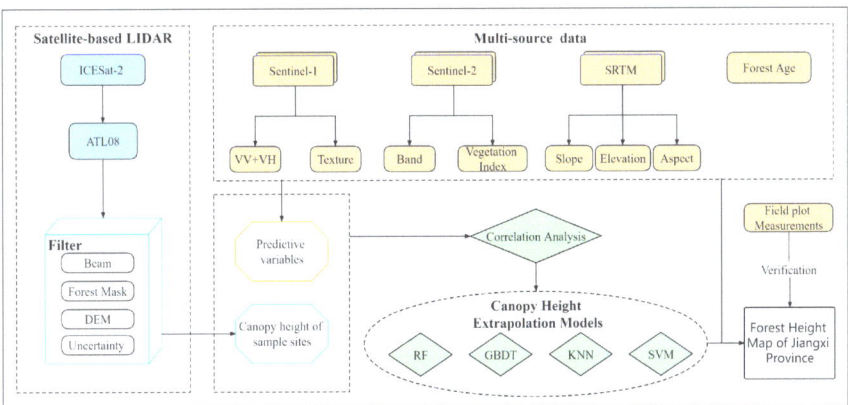

Figure 3. Workflow of integration of ICESat-2 and multi-source data for forest canopy height mapping in Jiangxi Province.

2.3.1. Correlation Analysis

Correlation analysis is the examination of two or more variables that may be correlated, and it often uses the correlation coefficient to determine the degree of correlation between two variables. The correlation coefficient is between −1 and 1. The more connected the variables are, the closer it gets to the extreme value [51]. In this study, the Spearman correlation coefficient (r_s) was utilized to link the forest canopy height and the relevant predictors at ICESat-2 sites based on the distribution features of the data. To reduce the dimensionality of the predictors, a p-value < 0.01 and an $|r_s| > 0.2$ were used for model development. The formula of r_s is as follows:

$$r_s = 1 - \frac{6\sum d_i^2}{n(n^2 - 1)} \qquad (2)$$

where r_s represents the Spearman correlation coefficient; d_i represents the difference in the place value of the data pair; and n represents the number of observation samples.

2.3.2. Model Development and Hyperparameter Optimization

Four machine learning models (RF, GBDT, SVM, and KNN) were utilized to construct canopy height extrapolation models for the proper estimation of forest canopy height in Jiangxi Province. RF is a bagging integration algorithm that builds a robust model by integrating multiple decision trees and has better generalization performance and less possible overfitting compared with a single decision tree [52]. GBDT, similar to RF, is an integrated model based on decision trees that uses a boosting algorithm to concatenate the decision trees and eventually approximate the true value by minimizing the model's bias [53]. The basic principle of an SVM is to discover a hyperplane that minimizes the distance between it and all of the data. Although the SVM technique is best suited for classification issues, it can also perform well in regression [6]. The final method, KNN, solves the regression problem by looking for comparable K samples based on spectral distance and making predictions using the mean of the samples [54].

In this study, 80% of the samples (n = 4621) served as a training set, while 20% of the data (n = 1156) served as a testing set. The RandomForest (version 4.7.1) package was used for the RF algorithm [52], the gbm (version 2.1.8.1) package was used for the GBDT algorithm [53], the e1071 (version 1.7.11) package was used for the SVM algorithm [55], and the Caret (version 6.0.92) package was used for the KNN algorithm [56]. To improve the model's performance, the parameters were modified by running the "train" function in the Caret package several times. The hyperparameter optimization approach was developed using a grid search and 10-fold cross-validation [56]. The Ntree and Mtry parameters had to be input to the RF algorithm. The "randomforest" function was also used to test the value of the Ntree. After determining the optimal Ntree, a hyperparameter optimization approach was utilized to determine the best Mtry. The GBDT algorithm required the input of four parameters: Trees, Depth, Minobsinnodehe, and Shrinkage, which were divided into two groups for the implementation of the hyperparameter optimization algorithm. The linear function was used as the kernel function in the SVM algorithm, and the only parameter that needed to be changed was the cost value (C). A range was established based on various changes to determine the best-performing cost value. Simultaneously, for the KNN algorithm, the Euclidean distance was set to the spectral distance, and the optimal number of neighborhood points (K) was found using a hyperparametric optimization algorithm. All the parameters are shown in Table 2, which presents the parameters and descriptions of them in the machine learning models. The RMSE was used as an evaluation tool for hyperparametric optimization algorithms, and the parameters with the lowest RMSE were used to develop the canopy height extrapolation model.

Table 2. Parameters and descriptions in machine learning models.

Model	Parameter	Description
RF	Ntree	Number of decision trees
	Mtry	Number of features in each decision tree
SVM	Cost	Penalty factor
KNN	K	Number of neighboring points
GBDT	Trees	Number of iterative regression trees
	Depth	Maximum depth of decision tree
	Minobsinnode	The minimum number of observations in the terminal nodes of the decision tree
	Shrinkage	Learning rate or step reduction

2.3.3. Continuous Forest Canopy Height Mapping and Validation

Once all of the parameters were optimized, four canopy height extrapolation models were created, with the canopy height in the training set serving as the dependent variable and other remote sensing data serving as the predictors. The canopy heights of the testing set were assessed using R^2 and RMSE. Furthermore, the best-performing model was used to map the forest canopy height in Jiangxi Province. Finally, all the forest canopy height maps were evaluated using field measurements. The formulas for R^2 and RMSE are as follows:

$$R^2 = 1 - \frac{\sum_{i=1}^{n}(x_i - y_i)^2}{\sum_{i=1}^{n}(y_i - \bar{y})^2} \quad (3)$$

$$\text{RMSE} = \sqrt{\frac{1}{n} \cdot \sum_{i=1}^{n}(x_i - y_i)^2} \quad (4)$$

where n represents the number of observations; x_i represents the i-th observation; y_i represents the i-th predicted value; $\bar{y}$ represents the average of the predicted values.

3. Results

3.1. Correlation between Canopy Height and Predictors

For each ICESat-2 point, the Spearman correlation coefficient was used to investigate the relationship between canopy height and 33 predictors. The statistical results (Table 3) showed that 26 variables were significantly correlated with canopy height, and 7 variables were not significantly correlated. Among the 26 significantly correlated variables, the r_s ranged from 0.08 to 0.62.

Table 3. r_s of predictors and canopy height at sample scale.

Predictors	r_s	Predictors	r_s	Predictors	r_s
FOREST_AGE	0.415 **	VV_COR	0.007	NDRE_BAND5	0.454 **
VH_DISS	0.313 **	VV_DISS	0.326 **	NDRE_BAND6	0.361 **
VH_ENT	0.316 **	VV_HOM	−0.316 **	NDRE_BAND7	0.007
VH_CON	0.310 **	VV_MEAN	0.147 **	S2_BAND2	−0.278 **
VH_COR	0.011	VV_SEC	−0.327 **	S2_BAND3	−0.409 **
VH_HOM	−0.302 **	VV_VAR	0.326 **	S2_BAND4	−0.411 **
VH_MEAM	0.191 **	SLOPE	0.548 **	S2_BAND5	−0.436 **
VH_SEC	−0.315 **	ELEVATION	0.620 **	S2_BAND6	−0.078 **
VH_VAR	0.309 **	ASPECT	0.031 *	S2_BAND7	0.018
VV_ENT	0.327 **	VV	0.163 **	S2_BAND8	0.016
VV_CON	0.322 **	VH	0.128 **	S2_BAND8A	0.034 *

− represents negative correlation, * represents p-value < 0.05, ** represents p-value < 0.01.

It is worth noting that elevation (r_s = 0.62) and slope (r_s = 0.55) were strongly connected with canopy height among all the statistically correlated variables. Sentinel-2 characteristics such as NDRE_BAND5 (r_s = 0.45) and S2_BAND5 (r_s = 0.44) were also moderately linked with canopy height. Sentinel-1-derived information was less correlated (r_s from 0.13 to 0.32), reflecting the limited penetration of Sentinel-1's C-band into the forest [27]. Furthermore, forest age showed a link with canopy height (r_s = 0.42), implying that it may contribute to canopy height extrapolation modeling. The correlation analysis and variable selection method were utilized to choose 21 variables (Figure 4) for models development.

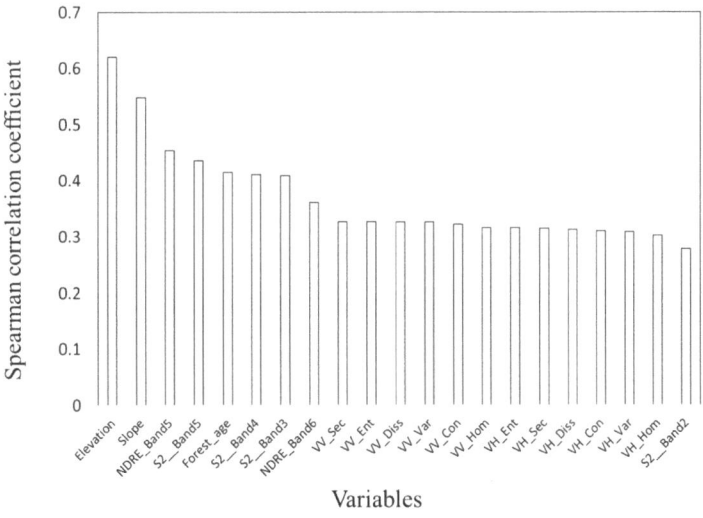

Figure 4. Correlation analysis of the 21 characteristic variables used in the model with canopy height.

3.2. Hyperparameter Optimization for Canopy Height Extrapolation Models

Ntree was initially set to 300 for the RF model, and the OOB error was steady after Ntree exceeded 150, whereupon it was set to 150 (Figure 5a). Furthermore, in the hyperparametric method, the ideal Ntree was used to obtain the Mtry, and Figure 5b shows that the RMSE was lowest when Mtry was 9. In contrast to RF, in this study, the KNN and SVM models required the development of a parameter. When K was set to 16 (Figure 5c) in KNN and cost was set to 0.3 (Figure 5d) in the SVM, the smallest RMSE was attained. Searching the mesh for all parameters at the same time in a model with several parameters may have resulted in computational complexity. Hence, for a four-parameter GBDT model, as shown in Figure 5e,f, the four parameters were divided into two groups and searched in the grid, and the best parameters were shown to be a combination of the two searches.

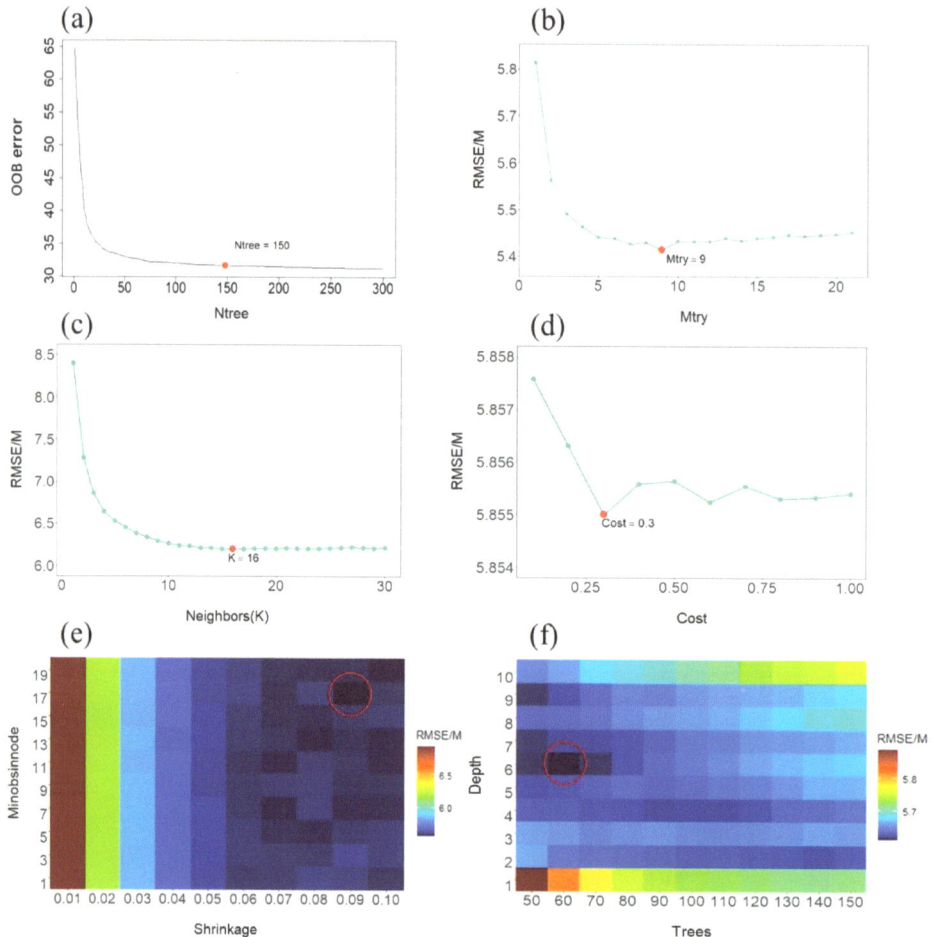

Figure 5. Parameters in the machine learning model selected based on cross-validation. (**a**) Ntree in the RF, (**b**) Mtry in the RF, (**c**) K in the KNN, (**d**) cost in the SVM, (**e**) Minobsinnode and Shrinkage in the GBDT, (**f**) Depth and Trees in the GBDT.

3.3. Comparison of Canopy Height Extrapolation Models

Once the optimal parameters in the model were determined, four machine learning models were utilized to establish the relationship between canopy height and predictors for the ICESat-2 point. The models were validated using the testing set to further verify their correctness and compare their performance using R^2 and RMSE metrics. In Figure 6, the R^2 for all models ranged from 0.47 to 0.61, and the RMSE ranged from 5.29 to 6.18 m. The best-fitting model was determined to be RF (R^2 = 0.61 and RMSE = 5.29 m). Figure 6a,b indicate that the two integrated models obtained from the decision tree, GBDT and RF, produced relatively similar results. Additionally, the mean coefficient of determination (R^2 = 0.55) and mean root mean square error (RMSE = 5.64 m) were calculated, indicating that all four models could extrapolate the sample plot scale to the regional scale. Lastly, Figure 6 shows all the model-predicted and actual canopy height values were significantly correlated (*p*-value < 0.01).

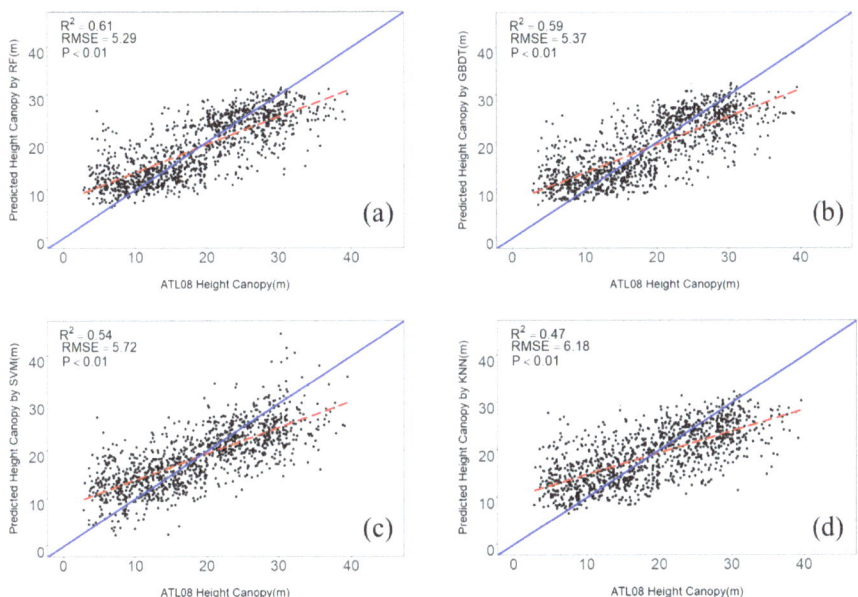

Figure 6. Comparison of accuracy of four machine learning models: (**a**) RF, (**b**) GBDT, (**c**) SVM, (**d**) KNN.

In Table 3, the Spearman coefficient was used to measure the linear relationship between the canopy height and predictors. However, the canopy height and predictors are not generally linearly related in environmentally complex forests [6]. To further explore the key drivers in the extrapolation process of forest canopy height, the importance evaluation in the random forest model (Figure 7) was used to analyze the degree of contribution of the variables in the canopy height extrapolation model. The results of Figure 7 show that the random forest model significantly depended on elevation, slope, S2_BAND5, and NDRE_BAND5, with elevation being the variable that contributed the most to the model. Meanwhile, Figure 7 suggests the backward scattering coefficient and texture features derived from Sentinel-1 were shown to have relatively low contributions to the model.

3.4. Mapping Wall-to-Wall Map of Forest Canopy Height

Continuous spatial feature information and canopy height extrapolation models provide the possibility of mapping forest canopy height in wall-to-wall areas. In Figure 8a, a 30-m resolution forest canopy height map for Jiangxi Province was generated based on the RF model and 21 consecutive remote sensing images. The heights of the predictions (Figure 8a) in the area ranged from 4.88 to 35.55 m, with an average height of 23.8 m. In general, the forest canopy height was higher in the mountains and lower in the plains and hills. Figure 8b shows that forest heights recorded in the field ranged from 2.7 m to 31.1 m, while samples were utilized to validate the forest canopy height map. The validation (Figure 8c) reveals that the canopy height extrapolation model we created could effectively map the forest canopy height in the research area ($R^2 = 0.69$; RMSE = 4.02 m).

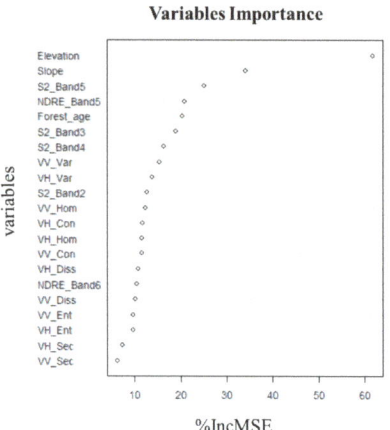

Figure 7. Importance assessment based on %IncMse in the RF model.

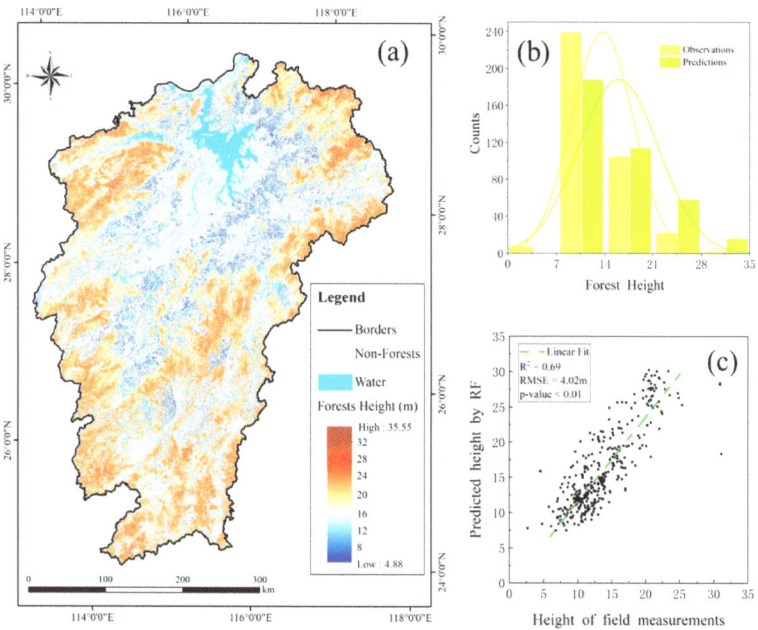

Figure 8. (a) A 30 m resolution forest canopy height map of Jiangxi Province based on the RF model. (b) The counts of field-measured height and predicted tree height. (c) Validation of forest canopy height map produced using RF model.

3.5. Comparison of Forest Canopy Height Maps

In Figure 8a, the validation shows that our method can produce reliable forest canopy height maps. Nonetheless, for a more comprehensive analysis of the accuracy of the forest canopy height maps, one global [25] and two Chinese [18,50] forest canopy height maps were processed using Jiangxi masks and validated using the same field measurements. Figure 9 shows that the R^2 of the validation ranged from 0.24 to 0.39, while the RMSE ranged from 4.11 to 5.48 m. The results show that the forest canopy height map of the Jiangxi Province produced in this study had higher accuracy.

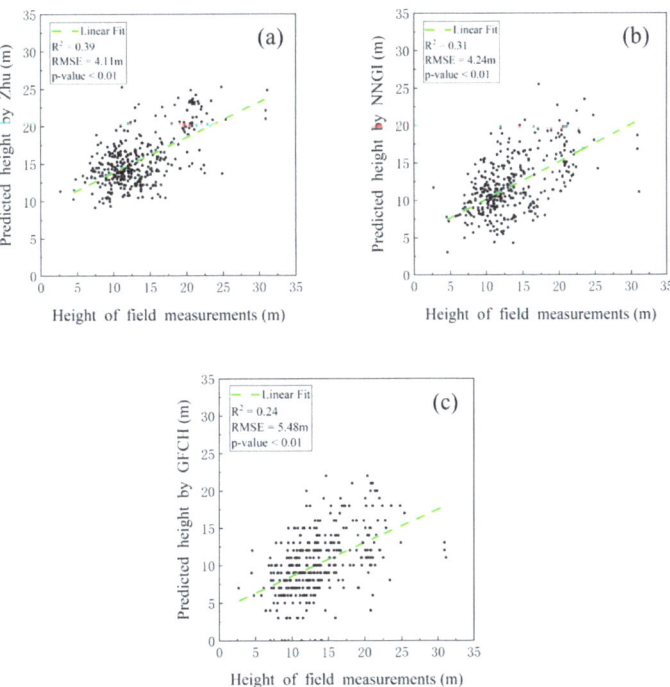

Figure 9. Validation of the forest height of Jiangxi Province generated by (**a**) Zhu [18], (**b**) NNGI [50], (**c**) GFCH [25].

4. Discussion

4.1. Key Drivers in the Extrapolation of Forest Canopy Height

Predictors derived from remote sensing data such as optical images can be efficiently integrated with ICESat-2 or MABEL modeling data to produce continuous forest canopy height maps [1,5–7,14]. However, few studies have discussed the role of other forest structures in forest canopy height mapping. Several studies have already shown a potential correlation between forest age and forest canopy height [36–38]. In this study, we explored continuous forest canopy height mapping using high-resolution forest age maps generated using a continuous change detection technique as predictors, along with other optical remote sensing data. The results in Table 3 indicate that there is a moderate correlation between forest age and forest canopy height at the sample scale.

Meanwhile, the importance ratings of the predictors (Figure 7) generated by the RF showed that forest age was one of the variables that the RF moderately relied on. These facts stated above imply that forest age has the potential to predict forest canopy height. From another perspective, the results in Figure 7 show that S2_BAND5 and NDRE_BAND5 are significantly dependent variables for forest canopy height prediction. Notably, this result has similarity with the results of Zhang et al. [30] and implies that the red-edge band can characterize vegetation growth to some extent, as demonstrated in previous studies [44,47]. In addition, Figure 7 shows that forest canopy height is also significantly affected by elevation and slope. One obvious reason for this is that changes in elevation and slope impact climate, temperature, and sunlight, all of which have an indirect effect on forest growth [57,58]. On the other hand, the impact of human activities on a forest will vary with elevation and slope.

4.2. Accuracy of Canopy Height Extrapolation Model and Forest Canopy Height

In Figure 6, all models overestimate low forests and underestimate forests with high heights. In fact, this bias appeared in previous studies that integrated ICESat-2 and different predictor variables using RF [5–7,18,28,30,31], KNN [6], SVM [6], GBDT [5], MLR [6], and a deep learning model [7] for continuous forest height mapping. However, the reason for this bias in the prediction of forest height maps via machine learning models is not clear. In addition, RMSE has been frequently used in the assessment and comparison of canopy height extrapolation models [5,6,31], and other metrics such as mean absolute error (MAE) and RMSE% should also be considered in subsequent studies to reduce uncertainty.

The results of Figures 8c and 9 reveal that the forest canopy height map generated in this study was more accurate. It is worth noting that the difference in accuracy between the current study and previous results may be attributable to a temporal mismatch between the validation data and ICESat-2 [28]. From another perspective, for global and national forest canopy height mapping, there are more complex elements to consider compared with regional studies. Both global and national data have achieved good accuracy [18,25,50,59]; however, part of the data may have lower accuracy due to geographical heterogeneity. Hence, forest canopy height mapping studies are important for forest resource management in the region.

4.3. Limitations and Prospects

In this study, 33 predictors were used to build a machine learning model to map wall-to-wall forest canopy height, including optical images and terrain features, as well as forest age, among others. Furthermore, a high-resolution forest canopy height map of the study area was mapped with a canopy height extrapolation model, which was developed using machine learning models. To investigate the potential of these variables in the estimation of forest canopy height, as seen in Figure 7, the feature variables in this study could be effectively linked to the canopy height in ICESat-2 points, with forest age performing particularly well as a new variable. Therefore, continuous forest canopy height maps can be used as predictors to map continuous forest structure parameters such as above-ground biomass and leaf area index in the future. Furthermore, due to latitudinal and longitudinal variances, environmental parameters such as precipitation and temperature should be incorporated as predictors for national or global forest canopy height mapping [50,58]. In terms of data, multicollinearity may have an impact on the ranking of importance in RF, and this issue should be taken into account in subsequent studies. In addition, several trials require additional consideration. ICESat-2 provides us with strip samples [19], which will leave many gaps that the GEDI [41] can fill in the study area. In terms of data, the number of vegetation signal photons in the ATL08 product is very low, and the vertical sampling error may underestimate the canopy height at the ICESat-2 point [20]. From another point of view, distinct forest species may have different canopy height extrapolation models [5]. Given the diverse biodiversity in the study area, canopy height extrapolation models for different tree species should be built once an appropriate categorization of the tree species is obtained.

5. Conclusions

In this study, we integrated ICESat-2, Sentinel, SRTM, and forest age data from Jiangxi Province, proposed a new method for variable selection, and developed four canopy height extrapolation models via RF, GBDT, KNN, and the SVM to link canopy height and spatial feature information. The model (RF) with the best-performance was used to map the forest canopy height in Jiangxi Province. Based on the results of this study, the following conclusions can be drawn: (1) Forest canopy height is moderately correlated with forest age, meaning it has the potential to be a predictor for forest canopy height mapping. (2) RF has the best-performance of the four canopy height extrapolation models, although it still underestimates higher forests and overestimates lower forests. Nonetheless, the use of multiple models can produce more robust results than the use of a single model. (3) Elevation, slope, and the red-edge band (band 5) derived from Sentinel-2 were significantly dependent variables in the canopy height extrapolation model. Apart from that, forest age was one of the variables that the RF moderately relied on. In contrast, backscatter coefficients and texture features derived from Sentinel-1 were not sensitive to canopy height. (4) There was a significant correlation between forest canopy height predicted using RF and forest canopy height measured using field measurements ($R^2 = 0.69$; RMSE = 4.02 m). In a nutshell, the results show that the method utilized in this work can reliably map the spatial distribution of forest canopy height.

Author Contributions: Conceptualization, Y.L., S.Q., K.L., B.H. and Y.T.; Methodology, Y.L. and B.H.; Formal analysis, Y.L. and K.L.; Software, Y.L.; Investigation, Y.L., S.Q. and K.L.; Resources, K.L. and S.Z.; Data curation, Y.L., S.Q. and S.Z.; Writing—original draft, Y.L.; Writing—review & editing, S.Q., S.Z., B.H. and Y.T.; Visualization, Y.L.; Project administration, S.Q.; Supervision, S.Q.; Funding acquisition, S.Q. and K.L. All authors have read and agreed to the published version of the manuscript.

Funding: The research was funded by the Water Conservancy Science and Technology Project of Jiangxi Province, China (No. 202124ZDKT25) and the National Natural Science Foundation of China (No. 41867012).

Institutional Review Board Statement: Not applicable.

Informed Consent Statement: Not applicable.

Data Availability Statement: The data presented in this study are available from the corresponding author by reasonable request.

Conflicts of Interest: The authors declare no conflict of interest.

References

1. Nandy, S.; Srinet, R.; Padalia, H. Mapping Forest Height and Aboveground Biomass by Integrating ICESat-2, Sentinel-1 and Sentinel-2 Data Using Random Forest Algorithm in Northwest Himalayan Foothills of India. *Geophys. Res. Lett.* **2021**, *48*, e2021GL093799. [CrossRef]
2. Asner, G.P.; Hughes, R.F.; Mascaro, J.; Uowolo, A.L.; Knapp, D.E.; Jacobson, J.; Kennedy-Bowdoin, T.; Clark, J.K. High-Resolution Carbon Mapping on the Million-Hectare Island of Hawaii. *Front. Ecol. Environ.* **2011**, *9*, 434–439. [CrossRef]
3. Asner, G.P.; Mascaro, J.; Muller-Landau, H.C.; Vieilledent, G.; Vaudry, R.; Rasamoelina, M.; Hall, J.S.; van Breugel, M. A Universal Airborne LiDAR Approach for Tropical Forest Carbon Mapping. *Oecologia* **2012**, *168*, 1147–1160. [CrossRef]
4. Zeng, Z.; Piao, S.; Li, L.Z.X.; Zhou, L.; Ciais, P.; Wang, T.; Li, Y.; Lian, X.; Wood, E.F.; Friedlingstein, P.; et al. Climate Mitigation from Vegetation Biophysical Feedbacks during the Past Three Decades. *Nat. Clim. Chang.* **2017**, *7*, 432–436. [CrossRef]
5. Xi, Z.; Xu, H.; Xing, Y.; Gong, W.; Chen, G.; Yang, S. Forest Canopy Height Mapping by Synergizing ICESat-2, Sentinel-1, Sentinel-2 and Topographic Information Based on Machine Learning Methods. *Remote Sens.* **2022**, *14*, 364. [CrossRef]
6. Jiang, F.; Zhao, F.; Ma, K.; Li, D.; Sun, H. Mapping the Forest Canopy Height in Northern China by Synergizing ICESat-2 with Sentinel-2 Using a Stacking Algorithm. *Remote Sens.* **2021**, *13*, 1535. [CrossRef]
7. Li, W.; Niu, Z.; Shang, R.; Qin, Y.; Wang, L.; Chen, H. High-Resolution Mapping of Forest Canopy Height Using Machine Learning by Coupling ICESat-2 LiDAR with Sentinel-1, Sentinel-2 and Landsat-8 Data. *Int. J. Appl. Earth Obs. Geoinf.* **2020**, *92*, 102163. [CrossRef]
8. Lu, D.; Chen, Q.; Wang, G.; Liu, L.; Li, G.; Moran, E. A Survey of Remote Sensing-Based Aboveground Biomass Estimation Methods in Forest Ecosystems. *Int. J. Digit. Earth* **2016**, *9*, 63–105. [CrossRef]
9. Chopping, M.; Moisen, G.G.; Su, L.; Laliberte, A.; Rango, A.; Martonchik, J.V.; Peters, D.P.C. Large Area Mapping of Southwestern Forest Crown Cover, Canopy Height, and Biomass Using the NASA Multiangle Imaging Spectro-Radiometer. *Remote Sens. Environ.* **2008**, *112*, 2051–2063. [CrossRef]
10. Balzter, H.; Luckman, A.; Skinner, L.; Rowland, C.; Dawson, T. Observations of Forest Stand Top Height and Mean Height from Interferometric SAR and LiDAR over a Conifer Plantation at Thetford Forest, UK. *Int. J. Remote Sens.* **2007**, *28*, 1173–1197. [CrossRef]
11. Wagner, W. Large-Scale Mapping of Boreal Forest in SIBERIA Using ERS Tandem Coherence and JERS Backscatter Data. *Remote Sens. Environ.* **2003**, *85*, 125–144. [CrossRef]

12. Narine, L.L.; Popescu, S.C.; Malambo, L. Synergy of ICESat-2 and Landsat for Mapping Forest Aboveground Biomass with Deep Learning. *Remote Sens.* **2019**, *11*, 1503. [CrossRef]
13. Zwally, H.J.; Schutz, B.; Abdalati, W.; Abshire, J.; Bentley, C.; Brenner, A.; Bufton, J.; Dezio, J.; Hancock, D.; Harding, D.; et al. ICESat's Laser Measurements of Polar Ice, Atmosphere, Ocean, and Land. *J. Geodyn.* **2002**, *34*, 405–445. [CrossRef]
14. Zhu, X.; Wang, C.; Nie, S.; Pan, F.; Xi, X.; Hu, Z. Mapping Forest Height Using Photon-Counting LiDAR Data and Landsat 8 OLI Data: A Case Study in Virginia and North Carolina, USA. *Ecol. Indic.* **2020**, *114*, 106287. [CrossRef]
15. Lefsky, M.A. A Global Forest Canopy Height Map from the Moderate Resolution Imaging Spectroradiometer and the Geoscience Laser Altimeter System: A GLOBAL FOREST CANOPY HEIGHT MAP. *Geophys. Res. Lett.* **2010**, *37*. [CrossRef]
16. Simard, M.; Pinto, N.; Fisher, J.B.; Baccini, A. Mapping Forest Canopy Height Globally with Spaceborne Lidar. *J. Geophys. Res.* **2011**, *116*, G04021. [CrossRef]
17. Wang, Y.; Li, G.; Ding, J.; Guo, Z.; Tang, S.; Wang, C.; Huang, Q.; Liu, R.; Chen, J.M. A Combined GLAS and MODIS Estimation of the Global Distribution of Mean Forest Canopy Height. *Remote Sens. Environ.* **2016**, *174*, 24–43. [CrossRef]
18. Zhu, X. Forest Height Retrieval of China with a Resolution of 30m Using ICESat-2 and GEDI Data. Ph.D. Thesis, Aerospace Information Research Institute, Chinese Academy of Sciences (CAS), Beijing, China, 2021.
19. Neumann, T.A.; Martino, A.J.; Markus, T.; Bae, S.; Bock, M.R.; Brenner, A.C.; Brunt, K.M.; Cavanaugh, J.; Fernandes, S.T.; Hancock, D.W.; et al. The Ice, Cloud, and Land Elevation Satellite—2 Mission: A Global Geolocated Photon Product Derived from the Advanced Topographic Laser Altimeter System. *Remote Sens. Environ.* **2019**, *233*, 111325. [CrossRef]
20. Neuenschwander, A.; Pitts, K. The ATL08 Land and Vegetation Product for the ICESat-2 Mission. *Remote Sens. Environ.* **2019**, *221*, 247–259. [CrossRef]
21. Narine, L.L.; Popescu, S.; Zhou, T.; Srinivasan, S.; Harbeck, K. Mapping Forest Aboveground Biomass with a Simulated ICESat-2 Vegetation Canopy Product and Landsat Data. *Ann. For. Res.* **2019**, *62*, 69–86. [CrossRef]
22. Narine, L.L.; Popescu, S.; Neuenschwander, A.; Zhou, T.; Srinivasan, S.; Harbeck, K. Estimating Aboveground Biomass and Forest Canopy Cover with Simulated ICESat-2 Data. *Remote Sens. Environ.* **2019**, *224*, 1–11. [CrossRef]
23. Dubayah, R.; Blair, J.B.; Goetz, S.; Fatoyinbo, L.; Hansen, M.; Healey, S.; Hofton, M.; Hurtt, G.; Kellner, J.; Luthcke, S.; et al. The Global Ecosystem Dynamics Investigation: High-Resolution Laser Ranging of the Earth's Forests and Topography. *Sci. Remote Sens.* **2020**, *1*, 100002. [CrossRef]
24. Patterson, P.L.; Healey, S.P.; Ståhl, G.; Saarela, S.; Holm, S.; Andersen, H.-E.; Dubayah, R.O.; Duncanson, L.; Hancock, S.; Armston, J.; et al. Statistical Properties of Hybrid Estimators Proposed for GEDI—NASA's Global Ecosystem Dynamics Investigation. *Environ. Res. Lett.* **2019**, *14*, 065007. [CrossRef]
25. Potapov, P.; Li, X.; Hernandez-Serna, A.; Tyukavina, A.; Hansen, M.C.; Kommareddy, A.; Pickens, A.; Turubanova, S.; Tang, H.; Silva, C.E.; et al. Mapping Global Forest Canopy Height through Integration of GEDI and Landsat Data. *Remote Sens. Environ.* **2021**, *253*, 112165. [CrossRef]
26. Hansen, M.C.; Potapov, P.V.; Goetz, S.J.; Turubanova, S.; Tyukavina, A.; Krylov, A.; Kommareddy, A.; Egorov, A. Mapping Tree Height Distributions in Sub-Saharan Africa Using Landsat 7 and 8 Data. *Remote Sens. Environ.* **2016**, *185*, 221–232. [CrossRef]
27. Huang, W.; Min, W.; Ding, J.; Liu, Y.; Hu, Y.; Ni, W.; Shen, H. Forest Height Mapping Using Inventory and Multi-Source Satellite Data over Hunan Province in Southern China. *For. Ecosyst.* **2022**, *9*, 100006. [CrossRef]
28. Wu, Z.; Shi, F. Mapping Forest Canopy Height at Large Scales Using ICESat-2 and Landsat: An Ecological Zoning Random Forest Approach. *IEEE Trans. Geosci. Remote Sens.* **2023**, *61*, 1–16. [CrossRef]
29. Vuolo, F.; Żółtak, M.; Pipitone, C.; Zappa, L.; Wenng, H.; Immitzer, M.; Weiss, M.; Baret, F.; Atzberger, C. Data Service Platform for Sentinel-2 Surface Reflectance and Value-Added Products: System Use and Examples. *Remote Sens.* **2016**, *8*, 938. [CrossRef]
30. Zhang, T.; Liu, D. Mapping 30 m Boreal Forest Heights Using Landsat and Sentinel Data Calibrated by ICESat-2. *Authorea*, 6 December 2021.
31. Sothe, C.; Gonsamo, A.; Lourenço, R.B.; Kurz, W.A.; Snider, J. Spatially Continuous Mapping of Forest Canopy Height in Canada by Combining GEDI and ICESat-2 with PALSAR and Sentinel. *Remote Sens.* **2022**, *14*, 5158. [CrossRef]
32. Liu, T.; Abd-Elrahman, A.; Morton, J.; Wilhelm, V.L. Comparing Fully Convolutional Networks, Random Forest, Support Vector Machine, and Patch-Based Deep Convolutional Neural Networks for Object-Based Wetland Mapping Using Images from Small Unmanned Aircraft System. *GIScience Remote Sens.* **2018**, *55*, 243–264. [CrossRef]
33. Neuenschwander, A.; Guenther, E.; White, J.C.; Duncanson, L.; Montesano, P. Validation of ICESat-2 Terrain and Canopy Heights in Boreal Forests. *Remote Sens. Environ.* **2020**, *251*, 112110. [CrossRef]
34. Zhu, X.; Wang, C.; Xi, X.; Nie, S.; Yang, X.; Li, D. Research progress of ICESat-2/ATLAS data processing and applications. *Infrared Laser Eng.* **2020**, *49*, 20200259. [CrossRef]
35. Zhang, C.; Ju, W.; Chen, J.M.; Li, D.; Wang, X.; Fan, W.; Li, M.; Zan, M. Mapping Forest Stand Age in China Using Remotely Sensed Forest Height and Observation Data: CHINA'S FOREST STAND AGE MAPPING. *J. Geophys. Res. Biogeosciences* **2014**, *119*, 1163–1179. [CrossRef]
36. Yu, Z.; Zhao, H.; Liu, S.; Zhou, G.; Fang, J.; Yu, G.; Tang, X.; Wang, W.; Yan, J.; Wang, G.; et al. Mapping Forest Type and Age in China's Plantations. *Sci. Total Environ.* **2020**, *744*, 140790. [CrossRef]
37. Zhang, Y.; Yao, Y.; Wang, X.; Liu, Y.; Piao, S. Mapping Spatial Distribution of Forest Age in China. *Earth Space Sci.* **2017**, *4*, 108–116. [CrossRef]

38. Racine, E.B.; Coops, N.C.; St-Onge, B.; Bégin, J. Estimating Forest Stand Age from LiDAR-Derived Predictors and Nearest Neighbor Imputation. *For. Sci.* **2014**, *60*, 128–136. [CrossRef]
39. García, M.; Saatchi, S.; Ustin, S.; Balzter, H. Modelling Forest Canopy Height by Integrating Airborne LiDAR Samples with Satellite Radar and Multispectral Imagery. *Int. J. Appl. Earth Obs. Geoinformation* **2018**, *66*, 159–173. [CrossRef]
40. Markus, T.; Neumann, T.; Martino, A.; Abdalati, W.; Brunt, K.; Csatho, B.; Farrell, S.; Fricker, H.; Gardner, A.; Harding, D.; et al. The Ice, Cloud, and Land Elevation Satellite-2 (ICESat-2): Science Requirements, Concept, and Implementation. *Remote Sens. Environ.* **2017**, *190*, 260–273. [CrossRef]
41. Liu, A.; Cheng, X.; Chen, Z. Performance Evaluation of GEDI and ICESat-2 Laser Altimeter Data for Terrain and Canopy Height Retrievals. *Remote Sens. Environ.* **2021**, *264*, 112571. [CrossRef]
42. Torres, R.; Snoeij, P.; Geudtner, D.; Bibby, D.; Davidson, M.; Attema, E.; Potin, P.; Rommen, B.; Floury, N.; Brown, M.; et al. GMES Sentinel-1 Mission. *Remote Sens. Environ.* **2012**, *120*, 9–24. [CrossRef]
43. Drusch, M.; Del Bello, U.; Carlier, S.; Colin, O.; Fernandez, V.; Gascon, F.; Hoersch, B.; Isola, C.; Laberinti, P.; . Sentinel-2: ESA's Optical High-Resolution Mission for GMES Operational Services. *Remote Sens. Environ.* **2012**, *120*, 25–36. [CrossRef]
44. Liu, Y.; Gong, W.; Xing, Y.; Hu, X.; Gong, J. Estimation of the Forest Stand Mean Height and Aboveground Biomass in Northeast China Using SAR Sentinel-1B, Multispectral Sentinel-2A, and DEM Imagery. *ISPRS J. Photogramm. Remote Sens.* **2019**, *151*, 277–289. [CrossRef]
45. Sanchez, A.H.; Picoli, M.C.A.; Camara, G.; Andrade, P.R.; Chaves, M.E.D.; Lechler, S.; Soares, A.R.; Marujo, R.F.B.; Simões, R.E.O.; Ferreira, K.R.; et al. Comparison of Cloud Cover Detection Algorithms on Sentinel–2 Images of the Amazon Tropical Forest. *Remote Sens.* **2020**, *12*, 1284. [CrossRef]
46. Haralick, R.M.; Shanmugam, K.; Dinstein, I. Textural Features for Image Classification. *IEEE Trans. Syst. Man Cybern.* **1973**, *SMC-3*, 610–621. [CrossRef]
47. Jorge, J.; Vallbé, M.; Soler, J.A. Detection of Irrigation Inhomogeneities in an Olive Grove Using the NDRE Vegetation Index Obtained from UAV Images. *Eur. J. Remote Sens.* **2019**, *52*, 169–177. [CrossRef]
48. Farr, T.G.; Rosen, P.A.; Caro, E.; Crippen, R.; Duren, R.; Hensley, S.; Kobrick, M.; Paller, M.; Rodriguez, E.; Roth, L.; et al. The Shuttle Radar Topography Mission. *Rev. Geophys.* **2007**, *45*, RG2004. [CrossRef]
49. Chen, J.; Chen, J. GlobeLand30: Operational Global Land Cover Mapping and Big-Data Analysis. *Sci. China Earth Sci.* **2018**, *61*, 1533–1534. [CrossRef]
50. Liu, X.; Su, Y.; Hu, T.; Yang, Q.; Liu, B.; Deng, Y.; Tang, H.; Tang, Z.; Fang, J.; Guo, Q. Neural Network Guided Interpolation for Mapping Canopy Height of China's Forests by Integrating GEDI and ICESat-2 Data. *Remote Sens. Environ.* **2022**, *269*, 112844. [CrossRef]
51. Wissler, C. The Spearman Correlation Formula. *Science* **1905**, *22*, 309–311. [CrossRef]
52. Breiman, L. Random Forests. *Mach. Learn.* **2001**, *45*, 5–32. [CrossRef]
53. Friedman, J.H. Greedy Function Approximation: A Gradient Boosting Machine. *Ann. Stat.* **2001**, *29*, 1189–1232. [CrossRef]
54. Guo, G.; Wang, H.; Bell, D.; Bi, Y.; Greer, K. KNN Model-Based Approach in Classification. In *On The Move to Meaningful Internet Systems 2003: CoopIS, DOA, and ODBASE.*; Meersman, R., Tari, Z., Schmidt, D.C., Eds.; Springer: Berlin/Heidelberg, Germany, 2003; Volume 2888, pp. 986–996. ISBN 978-3-540-20498-5.
55. Dimitriadou, E.; Hornik, K.; Leisch, F.; Meyer, D.; Weingessel, A. Misc Functions of the Department of Statistics (E1071), TU Wien. *R Package* **2008**, *1*, 5–24.
56. Kuhn, M. Building Predictive Models in *R* Using the Caret Package. *J. Stat. Softw.* **2008**, *28*, 1–26. [CrossRef]
57. Yang, Y.; Watanabe, M.; Li, F.; Zhang, J.; Zhang, W.; Zhai, J. Factors Affecting Forest Growth and Possible Effects of Climate Change in the Taihang Mountains, Northern China. *For. Int. J. For. Res.* **2006**, *79*, 135–147. [CrossRef]
58. Tao, S.; Guo, Q.; Li, C.; Wang, Z.; Fang, J. Global Patterns and Determinants of Forest Canopy Height. *Ecology* **2016**, *97*, 3265–3270. [CrossRef]
59. Lang, N.; Jetz, W.; Schindler, K.; Wegner, J.D. A High-Resolution Canopy Height Model of the Earth. *arXiv* **2022**, arXiv:2204.08322.

Disclaimer/Publisher's Note: The statements, opinions and data contained in all publications are solely those of the individual author(s) and contributor(s) and not of MDPI and/or the editor(s). MDPI and/or the editor(s) disclaim responsibility for any injury to people or property resulting from any ideas, methods, instructions or products referred to in the content.

Article

Estimating Stem Diameter Distributions with Airborne Laser Scanning Metrics and Derived Canopy Surface Texture Metrics

Xavier Gallagher-Duval [1], Olivier R. van Lier [2,*] and Richard A. Fournier [1]

[1] Department of Applied Geomatics, Centre d'Applications et de Recherche en Télédétection (CARTEL), Université de Sherbrooke, Sherbrooke, QC J1K 2R1, Canada
[2] Canadian Forest Service—Canadian Wood Fibre Centre, Natural Resources Canada, Corner Brook, NL A2H 5G4, Canada
* Correspondence: olivier.vanlier@nrcan-rncan.gc.ca

Abstract: This study aimed to determine the optimal approach for estimating stem diameter distributions (SDD) from airborne laser scanning (ALS) data using point cloud metrics (M_{als}), a canopy height model (CHM) texture metrics (M_{tex}), and a combination thereof (M_{comb}). We developed area-based models (i) to classify SDD modality and (ii) predict SDD function parameters, which we tested for 5 modelling techniques. Our results demonstrated little variability in the performance of SDD modality classification models (mean overall accuracy: 72%; SD: 2%). Our best SDD function parameter models were generally fitted with M_{comb}, with R^2 improvements up to 0.25. We found the variable Correlation, originating from M_{tex}, to be the most important predictor within M_{comb}. Trends in the performance of the predictor groups were mostly consistent across the modelling techniques within each parameter. Using an Error Index (EI), we determined that differentiating modality prior to estimating SDD improved the accuracy of estimates for bimodal plots (~12% decrease in EI), which was trivially not the case for unimodal plots (<1% increase in EI). We concluded that (i) CHM texture metrics can be used to improve the estimate of SDD parameters and that (ii) differentiating for modality prior to estimating SSD is especially beneficial in stands with bimodal SDD.

Keywords: airborne laser scanning; texture; stem diameter distributions; forest inventory; boreal forest

Citation: Gallagher-Duval, X.; van Lier, O.R.; Fournier, R.A. Estimating Stem Diameter Distributions with Airborne Laser Scanning Metrics and Derived Canopy Surface Texture Metrics. *Forests* **2023**, *14*, 287. https://doi.org/10.3390/f14020287

Academic Editors: Jianping Wu, Zhongbing Chang and Xin Xiong

Received: 10 January 2023
Revised: 27 January 2023
Accepted: 31 January 2023
Published: 2 February 2023

Copyright: © 2023 by the authors. Licensee MDPI, Basel, Switzerland. This article is an open access article distributed under the terms and conditions of the Creative Commons Attribution (CC BY) license (https://creativecommons.org/licenses/by/4.0/).

1. Introduction

In the last decade, much effort has been devoted to modelling and mapping forest inventory attributes from airborne laser scanning data (ALS) to the point where these data are being used operationally over large, continuous areas internationally (e.g., [1–3]). ALS can provide precise and reliable predictions of many stand-mean values of biophysical attributes (e.g., biomass, volume, height, and DBH [4–6]), as well as distributions thereof (e.g., stem diameter, height, and volume distributions [7–9]). Stem diameter is the most frequently modelled distribution found in the literature (e.g., [10–17]) as it provides insights on stand structure, the basis for understanding the stand's ecological and economic value. Stem diameter distributions can be used to describe forest dynamics [18], carbon stock, biomass, and wood volumes [19,20], and are known to be correlated with species diversity [21,22]. This information is an important aid for forest managers, who are planning silvicultural strategies [23] and assessing the commercial value of given stands.

Numerous functions have been described in the literature to fit stem diameter distributions (SDD). The early works of Bailey and Dell (1973) [24] proposed the Weibull probability density function (PDF) as a diameter distribution model. Since then, many studies have evaluated the effectiveness of other statistical functions. Hafley and Schreuder (1977) [25] found Johnson's S_B function to outperform the Weibull in terms of quality of fit of the distributions. Similarly, and more recently, Gorgoso-Varela et al. (2021) [26] compared the Weibull (2P and 3P), Johnson's S_B, beta, generalized beta, and gamma-2P functions, and although the Weibull (2P and 3P) and Johnson's S_B yielded the poorest fits to the data, they concluded that all six assessed PDF produced reasonable results. Similarly, Consenza et al. (2019) [27] demonstrated that Johnson's S_B presented comparable performances to the Weibull for two forest types: slightly better for a *Eucalyptus globulus* plantation and slightly worse for a *Pinus radiata* plantation. Although these studies have demonstrated that the Weibull is not always the best-fitting PDF, it has been most widely used in forestry (e.g., [7,14,28,29]), namely due to the function's flexibility in shape and relative simplicity of mathematical implementation [24]. The Weibull, however, is better suited to represent homogenous stands, given that it contains only one mode. Recent studies have demonstrated improvements in SDD predictions by fitting the bimodal SDD of heterogeneous stands to two PDF in structurally diverse forests [28,30]. The accuracy in representing SDD is therefore inevitably dependent on the forest structure being assessed.

Parametric and non-parametric approaches have been used to model SDD regardless of the distribution's modality (e.g., [8,19,20,31]). As PDFs are multivariate, it is often necessary to use multiple models developed with methods that can handle high-dimensional space [32]. Although many approaches to predict SDD have been proposed in recent decades, the current trends have been based on the PDF parameter prediction [28,33–35] and recovery methods [12,36–38]. In the 1990s, studies found *k*-nearest neighbor (*k*-NN) regression to be more accurate and flexible than methods based on parametric distributions in predicting stand-level diameter distributions [39,40]. With the advent of ALS, *k*-NN approaches were implemented using the area-based approach [41] to produce sub-stand-level diameter predictions with similar results [42–44]. Although *k*-NN estimation has long been used to predict SDD, the large amounts of training data required can limit its application. Many other approaches have also been proposed. For example, Kangas and Maltamo (2000) [45] suggested a model that first predicted diameters at 12 percentiles, then the basal area diameter distribution was interpolated using a rational spline. Liu et al. (2009) [46] later assessed the percentile-based approach [47] against five other methods in predicting parameters for SDD represented by a Weibull function for white spruce plantations and found the percentile-based parameter recovery method performed best. In another study, Bollandsås and Naesset (2007) [19] proposed to use partial least squares regression to effectively predict diameters at percentiles of basal area in uneven-sized Norway spruce stands. In a most recent study, Strunk and McGaughey (2023) [36] compared post-stratification, ordinary least squares regression, *k*-NN, and random forest to predict diameter class-specific volumes and found that random forest produced overall better results for a managed southern white pine forest. The complex distributions associated with more heterogeneous forest structures are, however, often better represented within a Finite Mixture Model (FMM) by combining two or more PDFs [28,33,34,48]. For example, Mulverhill et al. (2018) [34] developed maximum likelihood estimation models for both unimodal and bimodal SDD to appropriately characterize the simple and irregular distributions found in stands of boreal mixedwood forests (Canada). Though the estimation approaches continue to evolve, no consensus on a singularly favoured modelling method has yet been established.

Spatially explicit and exhaustive characterizations of SDD are made possible with remote sensing. Tarp-Johansen (2002) [49] used a 3D model and digital aerial photographs to estimate stem diameters for monospecific English oak (*Quercus robur* L.) stands in Denmark. With the development of ALS, Gobakken and Næsset (2004) [50] used various ALS height metrics to estimate Weibull parameters accurately (R^2 ranging between 0.6–0.9 with an RMSE of 0.15) to predict SDD for the boreal forest in southeast Norway. Multi-source remote sensing data can also be combined to improve prediction accuracy. Peuhkurinen et al. (2018) [30] combined ALS data and SPOT5 imagery to make accurate predictions (Reynold's Error Index for all plots ranged from 17.99 to 122.94) of SDD for coniferous boreal forests of Russia's Perm Region with the non-parametric *k*-Most Similar Neighbour method. In addition to height metrics, intensity metrics can be derived from ALS data, thereby providing indications of the strength of backscattered energy. Shang et al. (2017) [51] used ALS height and intensity metrics to predict SDD for a hardwood forest in

Ontario, Canada. They found that combining intensity and height metrics improved the model's performance beyond employing either height-only or intensity-only metrics.

Texture metrics that are derived from remote sensing can provide additional information regarding canopy structure that is independent of spectral features regarding spatial variations [52]. Haralick's Grey-Level Co-occurrence Matrix (GLCM) [53] is one common approach to calculating texture features from a given raster surface. GLCM uses second-order statistics, which are defined as the probability of observing a certain pair of pixel values within a predefined angle and observation window size [54]. Studies have demonstrated that texture metrics derived from optical data can be used successfully to predict forest attributes for a range of forest types (e.g., for boreal and Great Lakes—St. Lawrence forests of Canada [55]; temperate forests of Ontario, Canada [8]; boreal forests of Finland [56]). Dube and Mutanga (2015) [57] compared aboveground biomass models for three medium-density plantation forest species in South Africa that were derived from Landsat-8 spectral bands, spectral band ratios, vegetation indices, texture bands, and texture band ratios. The study demonstrated that models developed from multiple texture band ratios yielded the highest R^2. Several studies have incorporated canopy height model (CHM)-derived texture metrics in predicting forest attributes. Ozdemir and Donoghue (2013) [58] used CHM-derived texture metrics to explain tree diversity for a broad range of stand types (pure conifer, mixed conifer, pure deciduous, mixed deciduous, and conifer, different age classes) and found that the combination of ALS metrics with texture metrics explained up to 85% of the measured tree height diversity. Niemi and Vauhkonen (2016) [59] demonstrated that using texture metrics improved prediction of total stem volume and basal area over models that were developed solely from ALS metrics for boreal forests in southern Finland. Similarly, van Ewijk et al. (2019) [55] found that combining ALS, CHM texture, and intensity metrics improved R^2 by 0.19 for the prediction of stem density when compared with models that were developed solely with ALS metrics.

The studies provide meaningful insight into potential improvements for predicting forest attributes using a variety of modelling approaches and predictor variables that are derived from remote sensing data. To date, no studies have specifically examined whether the inclusion of canopy surface texture metrics can improve the characterization of SDD from ALS data. In this study, we compared the accuracy of SDD predictions that were modelled independently from commonly used ALS metrics, CHM-derived texture metrics, and a combination of the two using multiple statistical modelling techniques. We first hypothesized that models using texture-derived metrics would more accurately predict SDD parameters than ones using ALS metrics alone. Second, based upon past research, we hypothesized that developing differentiated modality-specific models (unimodal or bimodal) would improve SDD predictions. We tested these hypotheses by developing two modelling approaches: the first considers a priori knowledge regarding the modality of the SDD, while the second considers all SDD to be unimodal. We then evaluated the contribution of texture metrics in both approaches and determined which approach is best suited for estimating SDD in the eastern boreal forests of Quebec and western Newfoundland.

2. Materials and Methods

2.1. Study Area

Two study areas were selected based on their similarity in forest composition: both are conifer-dominated and lie within the eastern extent of the North American boreal forest [60] (Figure 1). The forests are comprised of balsam fir (*Abies balsamea* (L.) Miller), black spruce (*Picea mariana* [Miller] Britton), white spruce (*Picea glauca* [Moench] Voss), paper or white birch (*Betula papyrifera* Marshall), yellow birch (*Betula alleghaniensis* Britton) and, to a lesser extent, tamarack, or eastern larch (*Larix laricina* [Du Roi] K. Koch). Balsam fir and white spruce-dominated mixed stands are found south of the 50th parallel in our first study area (123,140 km^2), located in the province of Quebec. As we move north, the presence of black spruce increases until it completely dominates the landscape above the 52nd parallel. The second study area (977 km^2) is in the most eastern extent of the Boreal Shield Ecozone, in the province of Newfoundland and Labrador, and is dominated by balsam fir. The climate at both sites is favorable for forest growth due to abundant precipitation and warm summers. The primary silvicultural treatments practiced in these areas are pre-commercial thinning and clear-cut harvesting, which generally yield even-aged, homogeneous forest stands.

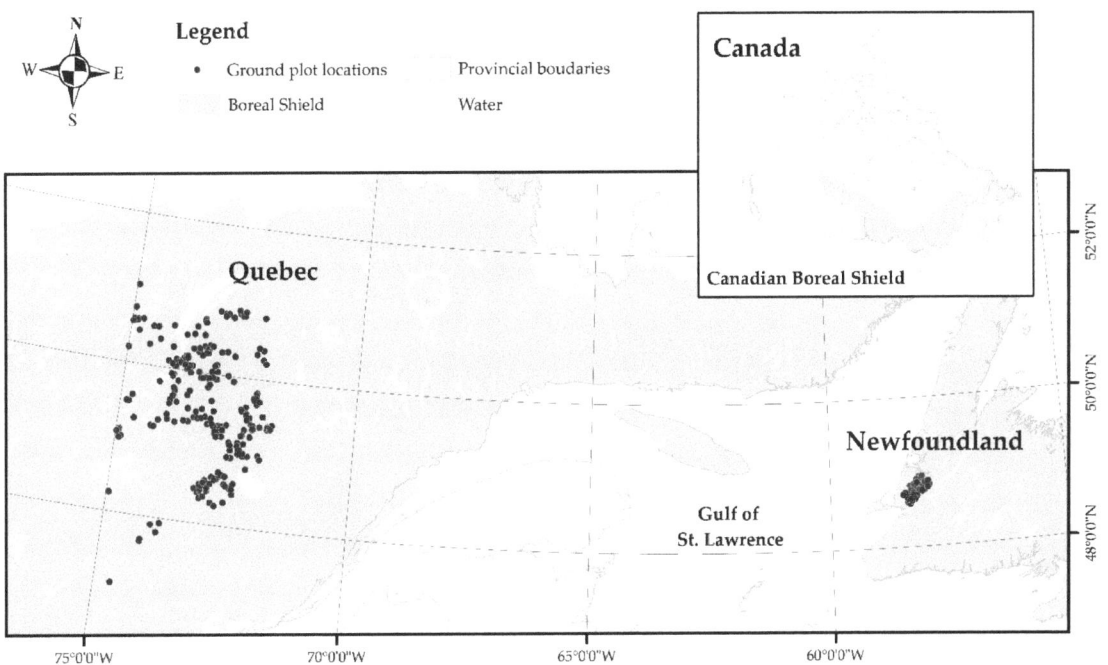

Figure 1. Plot distribution across two sites within the eastern Boreal Shield, Canada.

2.2. Ground Plots

Fixed-area circular plots were established with radii of 11.28 m where species, diameter at breast height (DBH), height, and status (live or dead) were recorded for all merchantable trees (trees $\geq$ 9 cm DBH). We retained plots having a total basal area $\geq$ 75% associated with balsam fir or black spruce with a presence of $\leq$10% hardwoods. We then identified and removed outlier plots by performing a multivariate local outlier factor analysis with the R package *DMwR* [61]. The analysis was based upon mean DBH and gross merchantable volume, together with the shape and scale parameters of a fitted Weibull function. We differentiated the SDD of each retained plot as unimodal or bimodal using the Bimodality Coefficient (BC) [62], given that its validity has been demonstrated in boreal forest environments [34] (Figure 2). The BC is proportional to the ratio between squared skewness and uncorrected kurtosis [63]. We associated plots having BC values $\leq 5/9$ with unimodal distributions, while bimodal distributions were associated with BC values $> 5/9$ [64]. In total, we retained 307 plots differentiated as unimodal and 120 as bimodal for the analysis of our hypotheses.

2.3. ALS Data and Metrics

All ALS data were acquired within 2 years of ground-plot measurements between 2012 and 2016. We calculated the mean point densities from plot locations to be 5.8 points m^{-2} and 4.9 points m^{-2} for the Quebec and Newfoundland sites, respectively. We created a CHM at a 1 m $\times$ 1 m resolution from first returns that were classified as vegetation using a natural neighbor interpolation. Binning cell assignment was set to the maximum value, and zeros replaced negative values. We calculated ALS metrics that are commonly used to describe the height, structure, and density of the canopy using the *lidR* package [65] in the R programming environment [66], using only returns ≥ 2 m that were classified as vegetation. We calculated the GLCM edge (contrast and dissimilarity) and patch interior texture metrics from the CHM, i.e., correlation, homogeneity, mean, and angular second moment [67]. We considered three window sizes, 3 × 3, 5 × 5 and 7 × 7, for the GLCM texture feature calculations and determined that the 3 × 3 window produced metrics that explained the most variation in our response variables (i.e., Weibull parameters). We computed the GLCM features in all directions and limited the number of grey levels to 32. We then averaged the 1 m × 1 m resolution texture feature values for each ground plot location to produce associated metrics of texture. To evaluate our hypotheses, we grouped the predictor variables into three sets of ALS metrics based upon: (i) point cloud metrics (M_{als}); (ii) CHM texture metrics (M_{tex}); and (iii) a combination thereof (M_{comb}) (Table 1).

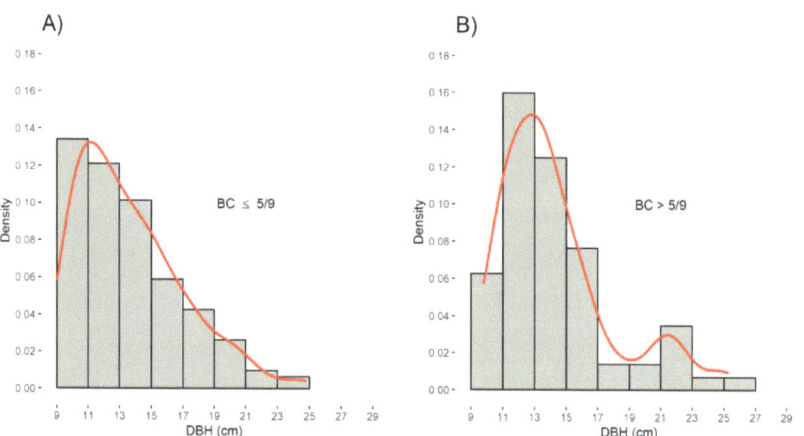

Figure 2. Example of Stem Diameter Distribution (SDD) differentiated according to the Bimodality Coefficient (BC) as (**A**) unimodal and (**B**) bimodal, and respectively fitted with a Weibull distribution and a Finite Mixture Model (red lines).

Table 1. Description of metrics and associated groupings used as predictor variables: ALS metrics (M_{als}), texture metrics (M_{tex}), and combined ALS and texture metrics (M_{comb}).

Group	Metric	Units	Description
M_{als}	MAX	m	Maximum height
	MEAN	m	Mean height [68]
	P25, P75, P90	m	Height percentiles. E.g., P25 is the height of the 25th percentile. [69]
	SKEW		Skewness
	VAR		Variance [68]
	COVAR	%	Coefficient of variation: standard deviation/mean [70]
	VDR		Vertical Distribution Ration: (MAX-MEAN)/MAX [71]
	VCI		Vertical Complexity Index [72]
	ENT		Entropy: normalized Shannon diversity index [73]
	RI		Rumple Index of roughness [74]
	D2, D5, D8	%	Proportion of all vegetation returns found in sections divided within the range of heights of all returns for each plot. [75]
	COVER		Ratio of the number of vegetated returns above 2 m to the total number of ground and vegetated returns [76]
	LPI		Light Penetration Index, Ground returns/(Ground returns + Canopy returns). [69]
	LPI1st		Light Penetration Index (first returns): Ground first returns/(Ground returns + Canopy returns) [77]
	FR		First return ratio: number of first return heights below a specified height threshold/total number of first return heights [68]

Table 1. *Cont.*

Group	Metric	Units	Description
	RR		All return ratio: all returns < 2 m/all returns [78]
	LAI		Sum of Leaf Area Density [68]
	cvLAI		Coefficient of variation of Leaf Area Density [68]
M_{tex}	CON		Contrast (edge texture) [67]
	COR		Correlation (interior textures) [67]
	DIS		Dissimilarity (edge textures) [67]
	HOM		Homogeneity (interior textures) [67]
	MEAN		Mean (interior textures) [67]
M_{comb}			Combination of all metrics (M_{als} and M_{tex})

2.4. Overview of the Methods

Figure 3 provides an overview of the methodological approach of the study. We used the ground-plot data to develop area-based models (i) to classify SDD modality and (ii) to predict SDD function parameters. We first defined three sets of ALS metrics from the ground plot locations (M_{als}, M_{tex}, and M_{comb}). We then created three ground plot datasets: the first two, unimodal and bimodal, were differentiated based on the modality of the SDD, while the third group was undifferentiated and assumed all plots were unimodal. Within each of the differentiated modality groups, we randomly selected 70% of plots for model development and used the remaining 30% as test cases. We developed models using 70% of the model development data for training and the remaining 30% for evaluating model performances. We generated three sets of models for each of the ground-plot groups using the ALS metrics sets. We used the modality and associated Weibull parameters as response variables for the SDD modality classification models and the SDD parameter prediction models, respectively. We implemented our best-performing models on our reserved test case data and analyzed the contribution of the CHM texture metrics to both groups of models (classification and prediction). Finally, we compared the predicted SDD that was obtained from the differentiated and undifferentiated modality models to assess whether modality differentiation improved the prediction of SDD in our data. All calculations were performed in R [66].

2.5. Development of SDD Modality Classification Models

We developed classification models to classify the modality of SDD using the differentiated SDD modality plot datasets (unimodal and bimodal). We constructed models independently using the three metrics groups (M_{als}, M_{tex}, and M_{comb}) as predictor variables. Herein, we evaluated four statistical techniques: random forest (RF); generalized linear model (Logit); support vector machine (SVM); and generalized linear model through penalized maximum likelihood (GLMNET), which uses the elastic net penalty that mixes the lasso and ridge penalties [79]. These contained internal feature selection mechanisms for selecting the best predictors and models with the *caret* package [80]. We developed the RF models with the *randomForest* package [81] and optimized the parameter *mtry*, which controls the number of predictors that were randomly picked at each split, by testing five values, viz., 1, 2, 3, 4, and 5. Logit models were developed with the *MASS* package [82] and used stepwise model selection based upon the Akaike Information Criterion (AIC). We defined the family parameter as a binomial and conducted no grid search for parameter optimization. SVM models were developed with the *kernlab* package [83] and used a radial basis function. We tuned two parameters for SVM, **sigma**, which controls the rigidity of the decision boundaries, and C, which controls the influence of misclassification. The values for **sigma** were 2^{-25}, 2^{-20}, 2^{-15}, 2^{-10}, 2^{-5}, and 2^0, while those for C were 2^0, 2^1, 2^2, 2^3, 2^4, and 2^5. Finally, GLMNET models were developed with the *glmnet* package [84]. GLMNET corresponds to a ratio between model regularization levels L1 and L2, affecting the penalty coefficient, and allows the selection of relevant predictors [85]. The two parameters that were tuned were **lambda**, which controls the overall strength of the penalty, and **alpha**, which controls the gap between the L1 and L2 regularization. We tested **alpha** values ranging from 0 to 1 with 0.1 increments and the following **lambda** values: 0.0001, 0.1112, 0.2223, 0.3334, 0.4445, 0.5556, 0.6667, 0.7778, 0.8889, and 1. We repeated cross-validation five times, using 70% of the model development data for training and 30% for validation. Finally, we averaged the overall accuracies within each technique and ALS metric group and applied the best performing models to our test case dataset and assessed the contribution of CHM texture metrics.

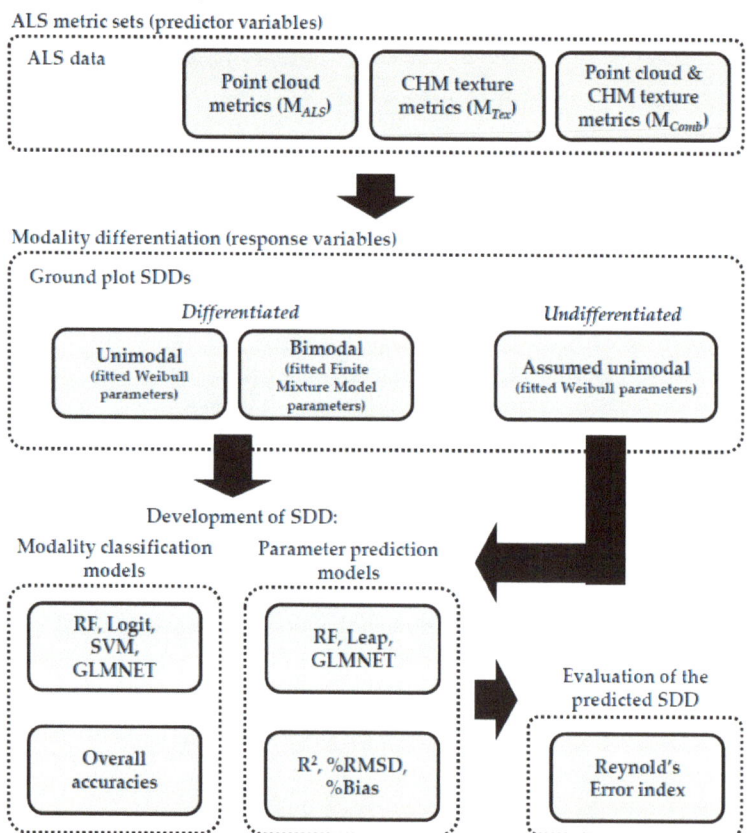

Figure 3. Overview of the methodological approach for assessing the contribution of CHM texture metrics and modality differentiation in predicting stem diameter distribution (SDD) parameters. CHM = Canopy Height Model; RF = Random Forest; Logit = generalized linear model with stepwise feature selection; SVM = Support Vector Machine; GLMNET = Generalized linear model through penalized maximum likelihood; Leap = Best subset regression with branch-and-bound algorithm; R^2 = Coefficient of determination; %RMSD = relative root-mean-squared deviation expressed as a percentage of the mean; %Bias = relative Bias expressed as a percentage of the mean.

2.6. Development of SDD Prediction Models

We developed three sets of models to predict SDD function parameters using (i) differentiated unimodal, (ii) differentiated bimodal, and (iii) undifferentiated SDD modality plot datasets. Using the differentiated unimodal plot data, we fitted a truncated Weibull function over the measured SDD and estimated the two function parameters (i.e., shape and scale) using the *fitdistrplus* package [86]. We implemented the same analysis for the undifferentiated plot data, for which all plots were treated as having a unimodal SDD distribution. From the differentiated bimodal plot data, we fitted a FMM composed of two Weibull functions over the SDD. The first Weibull related to smaller stem diameters relative to the second Weibull, which described the probability distribution of larger stems. The FMM can be represented by either the scale and shape, or the mean and standard deviation, of each of the two Weibull components and their associated proportions. We estimated the parameters of each function using the *mixR* package [87]. We assessed three modelling techniques within each model set, which included feature selection that was based on optimizing the root-mean-square deviation (RMSD) using the *caret* package. Again, the three metric groups (M_{als}, M_{tex}, and M_{comb}) were used independently as predictor variables. The maximization option for RMSD was set to FALSE to ensure that the best combination of parameters produced the lowest RMSD. The first technique that was used was RF from the *randomForest* package. Again, the only optimized parameter with grid search was *mtry*, with values 1, 2, 3, 4, and 5. The second technique was GLMNET, with two parameters to optimize, i.e., **alpha** and **lambda**. The **alpha** that was tested ranged from 0 to 1 in 0.1 increments; **lambda** values were 0.0001, 0.1112, 0.2223, 0.3334, 0.4445, 0.5556, 0.6667, 0.7778, 0.8889, or 1. We implemented the third and final technique, i.e., best subset regression with branch-and-bound algorithm (LEAP) [88], with the R package *leaps* [89]. This best subset regression used the branch-and-bound algorithm [90], which solves and optimizes combinatorial problems to select the best subset of predictors. In this study, we defined the number of predictors allowed in each subset to range between 2 and 6 predictors.

We evaluated the best-tuned models from the repeated 5-time cross-validation with the reserved test case dataset not used for model development. We compared the coefficient of determination (R^2), the absolute and relative RMSD (Equations (1) and (2)), and the absolute and relative bias (Equations (3) and (4)) for both the model development and test case datasets to assess our two hypotheses:

$$RMSD = \sqrt{\frac{\sum_{i=1}^{n}(y_i - \hat{\hat{y}}_i)^2}{n-1}} \quad (1)$$

$$RMSD\% = \frac{RMSD}{\bar{y}} \times 100 \quad (2)$$

$$Bias = \frac{\sum_{i=1}^{n}(y_i - \hat{y}_i)}{n} \quad (3)$$

$$Bias\% = \frac{Bias}{\bar{y}} \times 100 \quad (4)$$

where y_i is the observed value, $\hat{y}_i$ is the predicted value for case i, n is the number of observations, and $\bar{y}$ is the mean.

To evaluate the composition of metrics used in the best-performing models developed with M_{comb}, we calculated the associated variable importance. Since methods to characterize variable importance are dependent on the modelling technique implemented, we first scaled values between 0 and 100 to finally derive an average for each parameter modelled. For random forest models, we calculated the variable importance as the percent increase in mean square error (noted %IncMSE) [91]. For GLMNET models, we scaled variable coefficients as a representation of variable importance since they are proportionally indicative of the variables' importance [85] due to the penalization that reduces the coefficients of less-important variables [84]. Finally, we calculated variable importance for LEAP models as the absolute value of the t-statistic for each parameter in the final model [80].

2.7. Evaluation of the Predicted SDD

The quality of the predicted SDD was estimated with the Reynolds Error Index (EI) [92]. To do so, we predicted the SDD's parameters with the models demonstrating the highest R^2 and lowest RMSD% for the unimodal, bimodal, and undifferentiated plots from both model development and test case datasets. We then grouped the predicted tree DBH into

2-centimetre-wide bins to limit variability at larger intervals [93]. Finally, we evaluated the goodness-of-fit between the predicted SDD and the observed SDD of each plot with EI as follows:

$$\text{EI} = \sum_{i=1}^{m} 100 \left| \frac{f_{refi} - f_{alsi}}{N_{ref}} \right| \quad (5)$$

where m is the total number of bins, f_{refi} is the reference stem count for DBH bin i, f_{alsi} is the predicted stem count for DBH bin i, and N_{ref} is the true stem count of all DBH bins. EI values ranged between 0 and 200, where an EI of 0 indicated a perfect fit between predicted and observed SDD and an EI of 200 indicated a completely different SDD. To assess the effects of modality differentiation, we averaged the EI from all plots that had been derived independently for both the differentiated (unimodal and bimodal) and undifferentiated modelling approaches.

3. Results

3.1. SDD Modality Classification Models

Table 2 denotes the overall accuracies of the modality classification models using the three ALS metric sets as predictor variables and four modelling techniques for both model development and test case datasets. During model development, we observed M_{als} and M_{comb} to perform best using RF and GLMNET (overall accuracy of 74%). Surprisingly, the M_{tex} predictor set was used in both the best (using Logit) and worst (using RF) performing models in our test case. Overall, we observed little variability in the overall model accuracies regardless of the ALS predictor variable set or modelling technique used during model development or in our test case (mean: 72%; SD: 2% in both scenarios).

Table 2. Overall accuracies (%) of the SDD modality differentiation models using predictor variables that were derived from the three ALS metrics sets (M_{als}, M_{tex}, and M_{comb}) for both model development and test case datasets.

ALS Metric Set	RF	SVM	Logit	GLMNET
Model development				
M_{als}	74	72	71	74
M_{tex}	73	72	68	68
M_{comb}	74	71	70	74
Test case				
M_{als}	72	73	72	71
M_{tex}	66	72	74	73
M_{comb}	71	71	72	71

3.2. SDD Prediction Models

We developed model sets to estimate probability distribution function parameters from the differentiated unimodal, differentiated bimodal, and undifferentiated SDD modality plot datasets. We developed models within each model set using the three ALS metrics sets (M_{als}, M_{tex}, and M_{comb}) and three modelling techniques (RF, GLMNET, and LEAP). The model performance measures (R^2, RMSD%) that were derived from cross-validation are presented as Supplementary Material (Figure S1), as we observed for the most part the same trends in results with our case study illustrated in Figure 4. The results of our test case show that the proportion of the variance in the parameters describing the differentiated unimodal SDD were variable (R^2: 0–0.62). We observed associated errors ranging between 9.9% and 13.4% and 16.4% and 23.8% for models predicting scale and shape, respectively. For both parameters, the results indicate, with one exception (Shape $\sim f(M_{als})$ using RF), that models developed with M_{comb} consistently outperformed models that were developed with either M_{als} or M_{tex}. Both parameters were best predicted with RF; scale was best predicted using M_{comb} (R^2: 0.62; RMSD%: 9.9%), while shape, using M_{als} (R^2: 0.39; RMSD%: 16.4%).

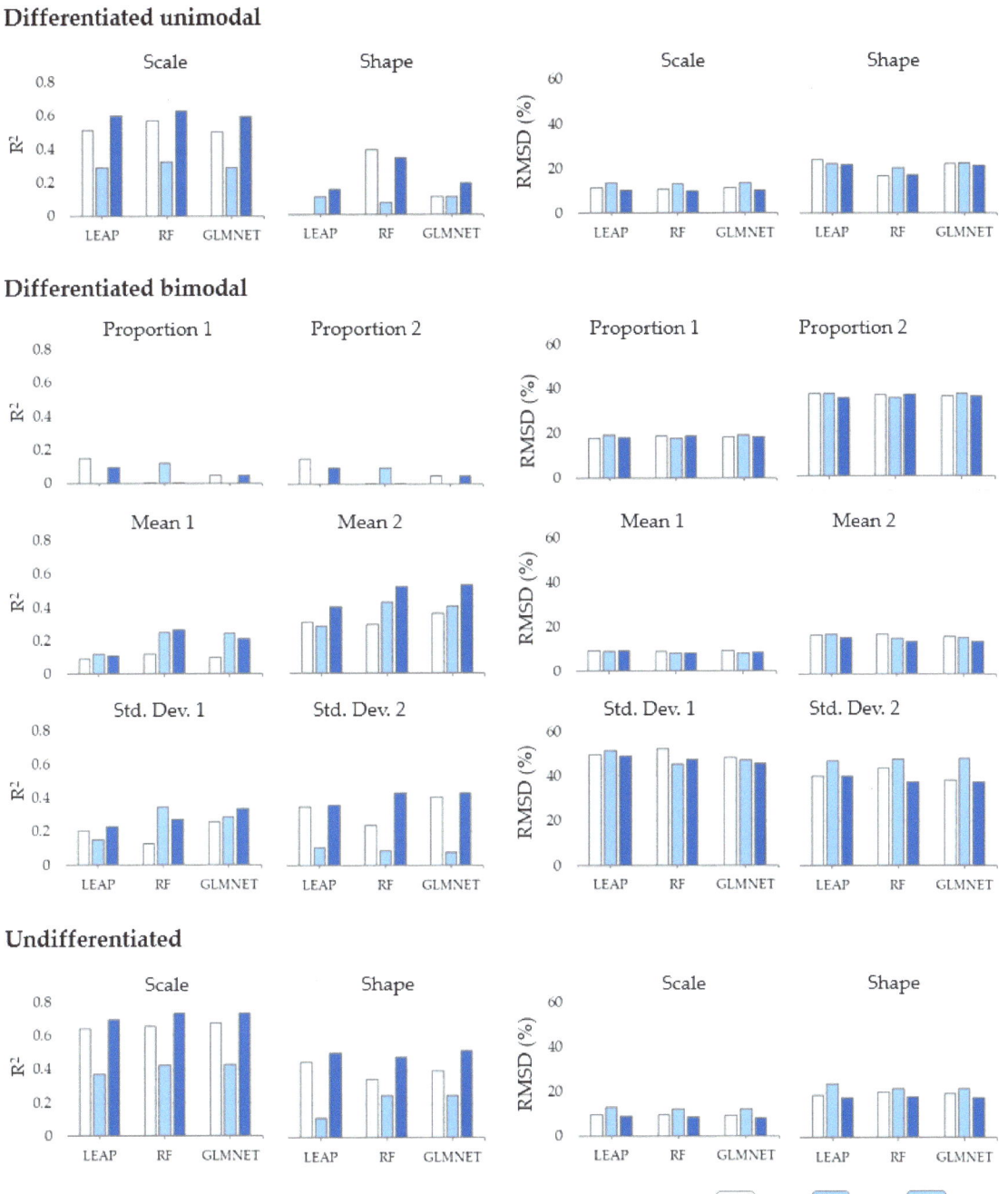

Figure 4. Coefficient of determination (R^2) and relative root-mean-squared deviation (RMSD%) that was derived from the application of the SDD prediction models to the test case data using the differentiated unimodal, differentiated bimodal, and undifferentiated SDD modality plot groupings; three ALS metrics sets (M_{als}, M_{tex}, and M_{comb}) and three modelling techniques (RF, GLMNET, LEAP) were used.

The performance of models that were developed using the differentiated bimodal SDD modality plot data were again variable (R^2: 0–0.53; RMSD%: 8.2%–52.1%). The results indicated that the FMM could not be represented by the parameters' scale and shape; the parameter shape of the first Weibull component could not be predicted given that the resulting models could never explain any of the variation in the parameter around its mean (R^2: 0), regardless of the ALS metric set or modelling approach. We therefore used the parameters mean and standard deviation to describe each component of the FMM. As expected, variation in the two proportion parameters was very poorly explained, if at all, by the predictor sets (R^2: 0–0.15), with associated errors ranging from 17.5% to 36.9%. As expected, the two Weibull component proportions of the FMMs were poorly predicted, with the best predictions modeled with RF using M_{als} (R^2: 0.15, 0.15; RMSD%: 17.5% and 33.8% for the proportions of the first and second components, respectively). The parameter mean was best predicted using M_{comb} for both components (R^2: 0.27, 0.53; RMSD%: 8.2%, 14.4%; using LEAP and GLMNET for means 1 and 2, respectively). Of note, GLMNET only marginally outperformed LEAP for the mean of the second FMM component (increase in R^2 < 0.01, decrease in RMSD% < 0.13%), both using M_{comb}. Standard deviation was best predicted with LEAP using M_{tex} for the first Weibull component (R^2: 0.34; RMSD%: 45.13%) and M_{comb} for the second (R^2: 0.43; RMSD%: 37.6%) with either LEAP or GLMNET.

The development of models using the undifferentiated modality SDD plot data involved applying the unimodal fitting analysis to all plots, regardless of modality. Herein, models performed better for the scale parameter (R^2: 0.37–0.73; RMSD%: 8.4%–12.9%) than for shape (R^2: 0.12–0.52; RMSD%: 17.7%–23.9%). We consistently observed improvements in model performance associated with models that have been developed with M_{comb}. Scale was best predicted with LEAP (R^2: 0.73; RMSD%: 8.4%), while shape was best predicted with GLMNET (R^2: 0.52; RMSD%: 17.7%). For these models, we observed a mean increase in R^2 of 0.08 (SD: 0.03) and a mean decrease in RMSD% of 1.3% (SD: 0.6%) with models that were developed using M_{comb} over those developed using M_{als}.

Analysis of the variable importance indicated that the correlation metric from M_{tex} is holistically the most important predictor within M_{comb} (Figure 5). The most important predictors thereafter are, for the majority, from M_{als}. In summary, we generally observed higher R^2 and lower RMSD% to be associated with models that were developed with M_{comb} compared with those using M_{als} or M_{tex}, regardless of the parameter being modelled or modelling technique being used. We found the variable correlation, originating from M_{tex}, to be the most important predictor within M_{comb}. Relative biases remained very low regardless of the parameter being modelled, the ALS metric set that was used, or the modelling approach that was employed (min.: −8.8; max.: 9.2; mean: 1.0; SD: 2.7 in absolute values of bias; data not shown). We observed no trend in the performance of the modelling techniques across all parameters.

3.3. Goodness-of-Fit of the Predicted SDD

We applied the best model within each model set independently to each plot and calculated mean Error Indices (EIs) from the predicted SDD parameters for both the model development and test case datasets (Table 3). We observed the same trends in both datasets. Surprisingly, we observed an increase in EI by applying differentiated unimodal models to unimodal plots, although the increase is negligible (<1%). Differentiating modalities prior to estimating SDD most improved the accuracy of estimates for bimodal plots (~12% decrease in EI). Of the 120 plots that were used to test our models, 50 (41.7%) had a better EI when derived from differentiated modality model predictions (31 and 19 plots within the differentiated unimodal and bimodal plots, respectively). Overall, we observed a marginally better fit (~4% decrease in EI) for SDD that were estimated from the differentiated modality model set in comparison with those estimated from the undifferentiated modality model set. The results therefore indicate improvements in SDD predictions by using differentiated modality-specific models, namely for heterogeneous (bimodal) stands.

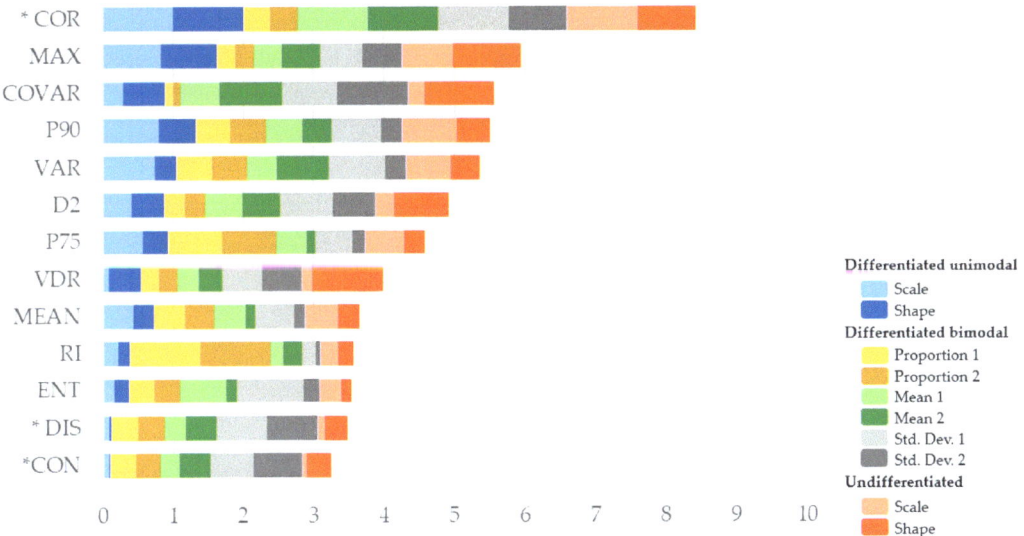

Figure 5. Cumulative variable importance values for metrics used in the best SDD parameter models which used M_{comb} during model development. Individual values represent the average variable importance across the three modelling techniques within each parameter and was scaled between 0 and 1. Only metrics with a cumulative value > 3 are shown. Asterix (*) denotes metrics originating from M_{tex}.

Table 3. Plot-level Reynold's Error Index means for each ground plot dataset and model set. EI values ranged between 0 and 200, where an EI of 0 indicated a perfect fit between predicted and observed SDD, which an EI of 200 indicated a completely different SDD.

		Model Set	
Plot Dataset	n	Differentiated	Undifferentiated
Model development			
Differentiated as unimodal	215	50.4	50.3
Differentiated as bimodal	92	65	74
Undifferentiated modality	307	54.8	57.4
Test case			
Differentiated as unimodal	88	50.8	50.5
Differentiated as bimodal	32	59.1	67
Undifferentiated modality	120	53	54.9

4. Discussion

From our first hypothesis, we expected models that were developed with CHM texture metrics to outperform SDD prediction models developed solely with ALS metrics. This expectation was based upon previous studies that related CHM texture metrics (M_{tex}) to properties of the growing stock, such as the spatial pattern of trees [94], and furthermore, demonstrated that their inclusion as predictors in modelling forest attributes improved predictions over using ALS metrics alone [55,58,59]. For example, van Ewijk et al. (2019) [55] tested multiple predictor sets using height metrics with combinations of CHM texture and intensity metrics and found that the addition of texture metrics improved prediction accuracy for basal area, quadratic mean DBH, and stem density. To our knowledge, no published studies have directly assessed the contribution of CHM texture metrics in estimating SDD using ALS data. Hence, the innovative aspects of our study make direct comparisons with past research challenging, especially regarding the attributes that we assessed (i.e., SDD modality and parameters), together with the CHM texture metrics that were included in our analyses. Nevertheless, our study demonstrated comparable results in classifying SDD modality with Zhang et al. (2019) [33] and Mulverhill et al. (2018) [34] using M_{als} (range in overall accuracies: 71%–73% vs. 49%–76% and 47%–78%, respectively). Our results for estimating SDD were generally comparable with those presented in Mulverhill et al. (2018) [34] for the differentiated unimodal distributions' modelled parameters, albeit with consistently lower error. Consistent with Thomas et al. (2008) [28] and Zhang et al. (2019) [33], the second component of the FMM that was associated with differentiated bimodal distributions was better predicted than the first. As highlighted by Thomas et al. (2008) [28], the main drawback of FMM is the increase in parameters that are needed to describe it. With the increase in modelled parameters, it becomes unlikely that each can be predicted accurately with M_{als}. Apart from the proportions associated with the FMM's components, the parameters of the differentiated bimodal distributions were best predicted with M_{comb}. Unlike Zhang et al. (2019) [33] and Mulverhill et al. (2018) [34], who developed models solely from M_{als}, our best SDD prediction models were generally developed with M_{comb}. Therefore, we could confirm our first hypothesis given that our study demonstrated that SDD prediction models developed with M_{comb} usually outperformed those developed with M_{als} (Figure 4). Inevitably, the contribution of CHM texture metrics will be dependent on the complexity of the forest environment assessed. Further research is warranted to determine the consistency of these results across varied forest types.

Our second hypothesis stated that developing differentiated modality-specific models (i.e., unimodal or bimodal) would improve SDD predictions for heterogeneous stands in our study site. The literature demonstrates improvements in estimating SDD with approaches that differentiate stand modality over approaches that do not (e.g., [33,34]). Our results indicated a similar trend. Yet, when interpreted globally, the improvements were marginal (~4 decrease in EI). Surprisingly, within our differentiated plot datasets, we observed that SDD was marginally better predicted by the undifferentiated modality model set that was intended for unimodal plots. Notably, and in support of our hypothesis, we observed SDD to be better predicted by the differentiated bimodal model set for bimodal plots (mean EI of 59.1 vs. 67.0). Our results therefore support the idea that developing model sets based on the modality of stands can improve SDD predictions for bimodal stands. Given this, we can confirm our hypothesis that differentiating for modality prior to estimating SSD improved the accuracy of estimates for the bimodal SDD conifer stands of our study site.

The accurate differentiation of the SDD modalities was assumed in our analyses, and therefore, potential errors in differentiation would directly impact model performances. Of the multiple available approaches to differentiate SDD modalities, we implemented BC as it has been successfully implemented in similar studies (e.g., [34]). Yet, it should be noted that BC is directly influenced by the kurtosis and, more so, by the skewness of a given distribution [64]. A distribution with high skewness and low kurtosis can inflate BC and subsequently differentiate the distribution as bimodal. Left-skewed distributions are observed when larger diameter trees dominate, while right-skewed distributions are associated with stands that are dominated by smaller diameter trees. Both situations will yield, however, a skewness value greater than zero. The closer that observed skewness is to zero, the more homogeneous the distribution will be and the stand can be described as having an even-aged distribution [48]. Freeman and Dale (2013) [63] evaluated the effect of the skewness, the proportion, and the distance between the modes on the BC value. In their study, BC produced 21% of false positives where simulated unimodal distributions had BC values greater than the bimodality threshold of 5/9 and were subsequently classified as bimodal. The BC relies upon the basic assumption that bimodality involves an increase in distribution asymmetry; therefore, an increase in skewness within a unimodal context can increase the BC and produce misclassification. Furthermore, the BC is not calibrated to proportion size; a small proportion in either component of a bimodal distribution can also produce false positives when the former is combined with a small distance between associated means. Of the 124 (92 in model development and 32 in test case) plots that were differentiated as bimodal in our study, 81 had skewness estimates > 1 and, thus, can be considered substantially skewed. Furthermore, the proportions that were associated with the second component of the bimodal distributions of our bimodal plots were low, as were the distances between the observed means (mean 5.8 cm). Given these results, it is possible that the combination of these factors could have inflated the BC and, therefore, mis-differentiated plots as bimodal. We can advance this as a plausible explanation, given the observed better fit for SDD that was estimated from the differentiated modalities model set was minimal (decrease in RI ~4%). These effects on the BC suggest that relying solely on this differentiation method may not be advisable for all forest types. Zhang et al. (2019) [33] used a combination of the Gini Coefficient and the asymmetry of the Lorenz curve to differentiate SDD modality, given that both measures are related to stand heterogeneity and the skewness of the diameter distribution [28,68]. Additional research is required to determine the optimal approach for differentiating the modality of SDD for a given forest type.

Nevertheless, the research presented here is important for several reasons. First, the methodology is used to differentiate the SDD modality and to develop the modality classification model, which can be used by foresters to improve the differentiation of stand structure types and to select the most appropriate models for accurately estimating diameter distributions across large ALS coverages. Second, we demonstrated that models fitted with M_{comb} yielded higher R^2 and lower RMSD% in comparison with those using solely M_{als}, thereby indicating that textural metrics contain additional information useful for the estimation of SDD.

5. Conclusions

In this study, we demonstrated that SDD probability function parameters were generally best estimated using a combination of ALS and texture metrics, thereby emphasizing the additional information contained in CHM texture metrics. As expected, we confirmed that developing modality-specific models improved SDD predictions for bimodal distributions, which, surprisingly, was not the case for unimodal distributions. For forest managers who rely on timely and detailed information, more accurate assessments of the distribution of diameters across a land base can therefore be made by differentiating modalities and adding texture metrics to modelling and mapping efforts. These results may provide for operational efficiencies in modelling and mapping SDD in these balsam fir or spruce-dominated forest environments.

Supplementary Materials: The following supporting information can be downloaded at: https://www.mdpi.com/article/10.3390/f14020287/s1, Figure S1: Average of 5 repeated cross-validation performance measures (R^2, RMSD%) derived during model development using the various SDD modality plot groupings, ALS metric sets and modelling techniques.

Author Contributions: Conceptualization, X.G.-D., O.R.v.L. and R.A.F.; methodology, X.G.-D., O.R.v.L. and R.A.F.; validation, X.G.-D.; formal analysis, X.G.-D.; investigation, X.G.-D. and O.R.v.L.; resources, O.R.v.L. and R.A.F.; data curation, X.G.-D., O.R.v.L. and R.A.F.; writing—original draft preparation, X.G.-D. and O.R.v.L.; writing—review and editing, X.G.-D., O.R.v.L. and R.A.F.; visualization, X.G.-D. and O.R.v.L.; supervision, R.A.F. and O.R.v.L.; project administration, R.A.F.; funding acquisition, R.A.F. All authors have read and agreed to the published version of the manuscript.

Funding: This work was supported by Natural Resources Canada's Canadian Forest Service—Canadian Wood Fibre Centre; and the Assessment of Wood Attributes using Remote Sensing Project (National Sciences and Engineering Research Council of Canada Collaborative Research and Development Grant PJ-462973-14, grantee N.C. Coops, UBC); in collaboration with Corner Brook Pulp and Paper Limited; and the Newfoundland and Labrador Department of Fisheries and Land Resources.

Data Availability Statement: The data underlying this article will be shared on reasonable request to the corresponding author.

Acknowledgments: This research was mainly developed in the Centre d'Applications et de Recherche en TÉLédétection of the Université de Sherbrooke, Canada. We thank Faron Knott and Kim Childs of Corner Brook Paper Limited for their input and assistance with the project. We thank the journal's associate editor and anonymous reviewers for their constructive feedback and suggestions for improving the manuscript.

Conflicts of Interest: The authors declare no conflict of interest. The funders had no role in the design of the study; in the collection, analyses, or interpretation of data; in the writing of the manuscript, or in the decision to publish the results.

References

1. Hyyppä, J.; Hyyppä, H.; Leckie, D.; Gougeon, F.; Yu, X.; Maltamo, M. Review of Methods of Small-Footprint Airborne Laser Scanning for Extracting Forest Inventory Data in Boreal Forests. *Int. J. Remote Sens.* **2008**, *29*, 1339–1366. [CrossRef]
2. Næsset, E. Area-Based Inventory in Norway—From Innovation to an Operational Reality. In *Forestry Applications of Airborne Laser Scanning*; Springer: Dordrecht, The Netherlands, 2014; Volume 27, pp. 215–240.
3. White, J.C.; Coops, N.C.; Wulder, M.A.; Vastaranta, M.; Hilker, T.; Tompalski, P. Remote Sensing Technologies for Enhancing Forest Inventories: A Review. *Can. J. Remote Sens.* **2016**, *42*, 619–641. [CrossRef]
4. Badreldin, N.; Sanchez-Azofeifa, A. Estimating Forest Biomass Dynamics by Integrating Multi-Temporal Landsat Satellite Images with Ground and Airborne LiDAR Data in the Coal Valley Mine, Alberta, Canada. *Remote Sens.* **2015**, *7*, 2832–2849. [CrossRef]
5. Luther, J.E.; Fournier, R.A.; van Lier, O.R.; Bujold, M. Extending ALS-Based Mapping of Forest Attributes with Medium Resolution Satellite and Environmental Data. *Remote Sens.* **2019**, *11*, 1092. [CrossRef]
6. Mielcarek, M.; Kamińska, A.; Stereńczak, K. Digital Aerial Photogrammetry (DAP) and Airborne Laser Scanning (ALS) as Sources of Information about Tree Height: Comparisons of the Accuracy of Remote Sensing Methods for Tree Height Estimation. *Remote Sens.* **2020**, *12*, 1808. [CrossRef]
7. Cao, L.; Zhang, Z.; Yun, T.; Wang, G.; Ruan, H.; She, G. Estimating Tree Volume Distributions in Subtropical Forests Using Airborne LiDAR Data. *Remote Sens.* **2019**, *11*, 97. [CrossRef]
8. Spriggs, R.A.; Coomes, D.A.; Jones, T.A.; Caspersen, J.P.; Vanderwel, M.C. An Alternative Approach to Using LiDAR Remote Sensing Data to Predict Stem Diameter Distributions across a Temperate Forest Landscape. *Remote Sens.* **2017**, *9*, 944. [CrossRef]
9. Tompalski, P.; Coops, N.C.; White, J.C.; Wulder, M.A. Enriching ALS-Derived Area-Based Estimates of Volume through Tree-Level Downscaling. *Forests* **2015**, *6*, 2608–2630. [CrossRef]
10. Zhang, L.; Gove, J.H.; Liu, C.; Leak, W.B. A Finite Mixture of Two Weibull Distributions for Modeling the Diameter Distributions of Rotated-Sigmoid, Uneven-Aged Stands. *Can. J. For. Res.* **2001**, *31*, 1654–1659. [CrossRef]
11. Cao, Q.V. Predicting Parameters of a Weibull Function for Modeling Diameter Distribution. *For. Sci.* **2004**, *50*, 682–685. [CrossRef]
12. Siipilehto, J.; Mehtätalo, L. Parameter Recovery vs. Parameter Prediction for the Weibull Distribution Validated for Scots Pine Stands in Finland. *Silva Fenn.* **2013**, *47*, 22. [CrossRef]
13. Mcgarrigle, E.; Kershaw Jr, J.A.; Lavigne, M.B.; Weiskittel, A.R.; Ducey, M. Predicting the Number of Trees in Small Diameter Classes Using Predictions from a Two-Parameter Weibull Distribution. *Forestry* **2011**, *84*, 431–439. [CrossRef]
14. Poudel, K.P.; Cao, Q.V. Evaluation of Methods to Predict Weibull Parameters for Characterizing Diameter Distributions. *For. Sci.* **2013**, *59*, 243–252. [CrossRef]
15. Palahí, M.; Pukkala, T.; Trasobares, A. Modelling the Diameter Distribution of Pinus Sylvestris, Pinus Nigra and Pinus Halepensis Forest Stands in Catalonia Using the Truncated Weibull Function. *For. Int. J. For. Res.* **2006**, *79*, 553–562. [CrossRef]
16. Duan, A.G.; Zhang, J.G.; Zhang, X.Q.; He, C.Y. Stand Diameter Distribution Modelling and Prediction Based on Richards Function. *PLoS ONE* **2013**, *8*, e62605. [CrossRef]
17. Guo, H.; Lei, X.; You, L.; Zeng, W.; Lang, P.; Lei, Y. Climate-Sensitive Diameter Distribution Models of Larch Plantations in North and Northeast China. *For. Ecol. Manag.* **2022**, *506*, 119547. [CrossRef]
18. West, G.B.; Enquist, B.J.; Brown, J.H. A General Quantitative Theory of Forest Structure and Dynamics. *Proc. Natl. Acad. Sci. USA* **2009**, *106*, 7040–7045. [CrossRef]
19. Martin Bollandsås, O.; Næsset, E. Estimating Percentile-Based Diameter Distributions in Uneven-Sized Norway Spruce Stands Using Airborne Laser Scanner Data. *Scand. J. For. Res.* **2007**, *22*, 33–47. [CrossRef]
20. Rana, P.; Vauhkonen, J.; Junttila, V.; Hou, Z.; Gautam, B.; Cawkwell, F.; Tokola, T. Large Tree Diameter Distribution Modelling Using Sparse Airborne Laser Scanning Data in a Subtropical Forest in Nepal. *ISPRS J. Photogramm. Remote Sens.* **2017**, *134*, 86–95. [CrossRef]
21. Xu, Q.; Hou, Z.; Maltamo, M.; Tokola, T. Calibration of Area Based Diameter Distribution with Individual Tree Based Diameter Estimates Using Airborne Laser Scanning. *ISPRS J. Photogramm. Remote Sens.* **2014**, *93*, 65–75. [CrossRef]
22. Fries, C.; Johansson, O.; Pettersson, B.; Simonsson, P. Silvicultural Models to Maintain and Restore Natural Stand Structures in Swedish Boreal Forests. *For. Ecol. Manag.* **1997**, *94*, 89–103. [CrossRef]
23. Packalén, P.; Maltamo, M. Estimation of Species-Specific Diameter Distributions Using Airborne Laser Scanning and Aerial Photographs. *Can. J. For. Res.* **2008**, *38*, 1750–1760. [CrossRef]
24. Bailey, R.L.; Dell, T.R. Quantifying Diameter Distributions with the Weibull Function. *For. Sci.* **1973**, *19*, 97–104. [CrossRef]

25. Hafley, W.L.; Schreuder, H.T. Statistical Distributions for Fitting Diameter and Height Data in Even-Aged Stands. *Can. J. For. Res.* **1977**, *7*, 481–487. [CrossRef]
26. Gorgoso-Varela, J.J.; Ponce, R.A.; Rodríguez-Puerta, F. Modeling Diameter Distributions with Six Probability Density Functions in Pinus Halepensis Mill. Plantations Using Low-Density Airborne Laser Scanning Data in Aragón (Northeast Spain). *Remote Sens.* **2021**, *13*, 2307. [CrossRef]
27. Nepomuceno Cosenza, D.; Soares, P.; Guerra-Hernández, J.; Pereira, L.; González-Ferreiro, E.; Castedo-Dorado, F.; Tomé, M. Comparing Johnson's S B and Weibull Functions to Model the Diameter Distribution of Forest Plantations through ALS Data. *Remote Sens.* **2019**, *11*, 2792. [CrossRef]
28. Thomas, V.; Oliver, R.D.; Lim, K.; Woods, M. LiDAR and Weibull Modeling of Diameter and Basal Area. *For. Chron.* **2008**, *84*, 866–875. [CrossRef]
29. Hao, Y.; Widagdo, F.R.A.; Liu, X.; Quan, Y.; Liu, Z.; Dong, L.; Li, F. Estimation and Calibration of Stem Diameter Distribution Using UAV Laser Scanning Data: A Case Study for Larch (Larix Olgensis) Forests in Northeast China. *Remote Sens. Environ.* **2022**, *268*, 112769. [CrossRef]
30. Peuhkurinen, J.; Tokola, T.; Plevak, K.; Sirparanta, S.; Kedrov, A.; Pyankov, S. Predicting Tree Diameter Distributions from Airborne Laser Scanning, SPOT 5 Satellite, and Field Sample Data in the Perm Region, Russia. *Forests* **2018**, *9*, 639. [CrossRef]
31. Maltamo, M.; Malinen, J.; Kangas, A.; Härkönen, S.; Pasanen, A.M. Most Similar Neighbour-Based Stand Variable Estimation for Use in Inventory by Compartments in Finland. *Forestry* **2003**, *76*, 449–463. [CrossRef]
32. Mauro, F.; Frank, B.; Monleon, V.J.; Temesgen, H.; Ford, K.R. Prediction of Diameter Distributions and Tree-Lists in Southwestern Oregon Using LiDAR and Stand-Level Auxiliary Information. *Can. J. For. Res.* **2019**, *49*, 775–787. [CrossRef]
33. Zhang, Z.; Cao, L.; Mulverhill, C.; Liu, H.; Pang, Y.; Li, Z. Prediction of Diameter Distributions with Multimodal Models Using LiDAR Data in Subtropical Planted Forests. *Forests* **2019**, *10*, 125. [CrossRef]
34. Mulverhill, C.; Coops, N.C.; White, J.C.; Tompalski, P.; Marshall, P.L.; Bailey, T. Enhancing the Estimation of Stem-Size Distributions for Unimodal and Bimodal Stands in a Boreal Mixedwood Forest with Airborne Laser Scanning Data. *Forests* **2018**, *9*, 95. [CrossRef]
35. Saad, R.; Wallerman, J.; Lämås, T. Estimating Stem Diameter Distributions from Airborne Laser Scanning Data and Their Effects on Long Term Forest Management Planning. *Scand. J. For. Res.* **2015**, *30*, 186–196. [CrossRef]
36. Strunk, J.L.; McGaughey, R.J. Stand Validation of Lidar Forest Inventory Modeling for a Managed Southern Pine Forest. *Can. J. For. Res.* **2023**, *53*, 1–19. [CrossRef]
37. Peuhkurinen, J.; Mehtätalo, L.; Maltamo, M. Comparing Individual Tree Detection and the Areabased Statistical Approach for the Retrieval of Forest Stand Characteristics Using Airborne Laser Scanning in Scots Pine Stands. *Can. J. For. Res.* **2011**, *41*, 583–598. [CrossRef]
38. Arias-Rodil, M.; Diéguez-Aranda, U.; Álvarez-González, J.G.; Pérez-Cruzado, C.; Castedo-Dorado, F.; González-Ferreiro, E. Modeling Diameter Distributions in Radiata Pine Plantations in Spain with Existing Countrywide LiDAR Data. *Ann. For. Sci.* **2018**, *75*, 36. [CrossRef]
39. Tomppo, E.; Katila, M. Satellite Image-Based National Forest Inventory of Finland for Publication in the Igarss'91 Digest. In Proceedings of the IGARSS'91 Remote Sensing: Global Monitoring for Earth Management, Espoo, Finland, 3–6 June 1991; Volume 3, pp. 1141–1144.
40. Maltamo, M.; Kangas, A. Methods Based on K-Nearest Neighbor Regression in the Prediction of Basal Area Diameter Distribution. *Can. J. For. Res.* **1998**, *28*, 1107–1115. [CrossRef]
41. Næsset, E. Predicting Forest Stand Characteristics with Airborne Scanning Laser Using a Practical Two-Stage Procedure and Field Data. *Remote Sens. Environ.* **2002**, *80*, 88–99. [CrossRef]
42. Bollandsås, O.M.; Maltamo, M.; Gobakken, T.; Næsset, E. Comparing Parametric and Non-Parametric Modelling of Diameter Distributions on Independent Data Using Airborne Laser Scanning in a Boreal Conifer Forest. *Forestry* **2013**, *86*, 493–501. [CrossRef]
43. Peuhkurinen, J.; Maltamo, M.; Malinen, J. Estimating Species-Specific Diameter Distributions and Saw Log Recoveries of Boreal Forests from Airborne Laser Scanning Data and Aerial Photographs: A Distribution-Based Approach. *Silva Fenn.* **2008**, *42*, 625–641. [CrossRef]
44. Strunk, J.L.; Gould, P.J.; Packalen, P.; Poudel, K.P.; Andersen, H.-E.E.; Temesgen, H. An Examination of Diameter Density Prediction with K-NN and Airborne Lidar. *Forests* **2017**, *8*, 444. [CrossRef]
45. Kangas, A.; Maltamo, M. Percentile Based Basal Area Diameter Distribution Models for Scots Pine, Norway Spruce and Birch Species. *Silva Fenn.* **2000**, *34*, 371–380. [CrossRef]
46. Liu, C.; Beaulieu, J.; Prégent, G.; Zhang, S.Y. Applications and Comparison of Six Methods for Predicting Parameters of the Weibull Function in Unthinned Picea Glauca Plantations. *Scand. J. For. Res.* **2009**, *24*, 67–75. [CrossRef]
47. Borders, B.E.; Souter, R.A.; Bailey, R.L.; Ware, K.D. Percentile-Based Distributions Characterize Forest Stand Tables. *For. Sci.* **1987**, *33*, 570–576. [CrossRef]
48. Zhang, L.; Liu, C. Fitting Irregular Diameter Distributions of Forest Stands by Weibull, Modified Weibull, and Mixture Weibull Models. *J. For. Res.* **2006**, *11*, 369–372. [CrossRef]
49. Tarp-Johansen, M.J. Stem Diameter Estimation from Aerial Photographs. *Scand. J. For. Res.* **2002**, *17*, 369–376. [CrossRef]

50. Gobakken, T.; Næsset, E. Estimation of Diameter and Basal Area Distributions in Coniferous Forest by Means of Airborne Laser Scanner Data. *Scand. J. For. Res.* **2004**, *19*, 529–542. [CrossRef]
51. Shang, C.; Treitz, P.; Caspersen, J.; Jones, T. Estimating Stem Diameter Distributions in a Management Context for a Tolerant Hardwood Forest Using ALS Height and Intensity Data. *Can. J. Remote Sens.* **2017**, *43*, 79–94. [CrossRef]
52. Hou, Z.; Xu, Q.; Tokola, T. Use of ALS, Airborne CIR and ALOS AVNIR-2 Data for Estimating Tropical Forest Attributes in Lao PDR. *ISPRS J. Photogramm. Remote Sens.* **2011**, *66*, 776–786. [CrossRef]
53. Haralick, R.M.; Shanmugam, K.; Its'Hak, D. Textural Features for Image Classification. *IEEE Trans. Syst. Man. Cybern.* **1973**, *SMC-3*, 610–621. [CrossRef]
54. Tuceryan, M.; Jain, A.K. Texture Analysis. In *Handbook of Pattern Recognition and Computer Vision*; Chen, C.H., Pau, L.F., Wang, P.S.P., Eds.; World Scientific: Singapore, 1999; pp. 207–248.
55. van Ewijk, K.; Treitz, P.; Woods, M.; Jones, T.; Caspersen, J. Forest Site and Type Variability in ALS-Based Forest Resource Inventory Attribute Predictions over Three Ontario Forest Sites. *Forests* **2019**, *10*, 226. [CrossRef]
56. Tuominen, S.; Pekkarinen, A. Performance of Different Spectral and Textural Aerial Photograph Features in Multi-Source Forest Inventory. *Remote Sens. Environ.* **2005**, *94*, 256–268. [CrossRef]
57. Dube, T.; Mutanga, O. Investigating the Robustness of the New Landsat-8 Operational Land Imager Derived Texture Metrics in Estimating Plantation Forest Aboveground Biomass in Resource Constrained Areas. *ISPRS J. Photogramm. Remote Sens.* **2015**, *108*, 12–32. [CrossRef]
58. Ozdemir, I.; Donoghue, D.N.M. Modelling Tree Size Diversity from Airborne Laser Scanning Using Canopy Height Models with Image Texture Measures. *For. Ecol. Manag.* **2013**, *295*, 28–37. [CrossRef]
59. Niemi, M.T.; Vauhkonen, J. Extracting Canopy Surface Texture from Airborne Laser Scanning Data for the Supervised and Unsupervised Prediction of Area-Based Forest Characteristics. *Remote Sens.* **2016**, *8*, 582. [CrossRef]
60. Rowe, J.S. *Forest Regions of Canada. Based on W. E. D. Halliday's "A Forest Classification for Canada" 1937*; Publication No 1300; Department of the Environment, Canadian Forestry Service: Ottawa, ON, Canada, 1972.
61. Torgo, L. *Data Mining with R: Learning with Case Studies*, 2nd ed.; Chapman and Hall/CRC: Boca Raton, FL, USA, 2017; ISBN 978-1482234893.
62. Ellison, A.M. Effect of Seed Dimorphism on the Density-Dependent Dynamics of Experimental Populations of Atriplex Triangularis (Chenopodiaceae). *Am. J. Bot.* **1987**, *74*, 1280–1288. [CrossRef]
63. Freeman, J.B.; Dale, R. Assessing Bimodality to Detect the Presence of a Dual Cognitive Process. *Behav. Res. Methods* **2013**, *45*, 83–97. [CrossRef]
64. Pfister, R.; Schwarz, K.A.; Janczyk, M.; Dale, R.; Freeman, J. Good Things Peak in Pairs: A Note on the Bimodality Coefficient. *Front. Psychol.* **2013**, *4*, 700. [CrossRef]
65. Roussel, J.-R.; Auty, D. LidR: Airborne LiDAR Data Manipulation and Visualization for Forestry Applications. Available online: https://cran.r-project.org/web/packages/lidR/index.html (accessed on 15 May 2020).
66. R Core Team R: A Language and Environment for Statistical Computing. Available online: https://www.r-project.org/ (accessed on 22 May 2020).
67. Hall-Beyer, M. Practical Guidelines for Choosing GLCM Textures to Use in Landscape Classification Tasks over a Range of Moderate Spatial Scales. *Int. J. Remote Sens.* **2017**, *38*, 1312–1338. [CrossRef]
68. Bouvier, M.; Durrieu, S.; Fournier, R.A.; Renaud, J.-P.P. Generalizing Predictive Models of Forest Inventory Attributes Using an Area-Based Approach with Airborne LiDAR Data. *Remote Sens. Environ.* **2015**, *156*, 322–334. [CrossRef]
69. Peduzzi, A.; Wynne, R.H.; Fox, T.R.; Nelson, R.F.; Thomas, V.A. Estimating Leaf Area Index in Intensively Managed Pine Plantations Using Airborne Laser Scanner Data. *For. Ecol. Manag.* **2012**, *270*, 54–65. [CrossRef]
70. Pope, G.; Treitz, P. Leaf Area Index (LAI) Estimation in Boreal Mixedwood Forest of Ontario, Canada Using Light Detection and Ranging (LiDAR) and Worldview-2 Imagery. *Remote Sens.* **2013**, *5*, 5040–5063. [CrossRef]
71. Goetz, S.; Steinberg, D.; Dubayah, R.; Blair, B. Laser Remote Sensing of Canopy Habitat Heterogeneity as a Predictor of Bird Species Richness in an Eastern Temperate Forest, USA. *Remote Sens. Environ.* **2007**, *108*, 254–263. [CrossRef]
72. van Ewijk, K.Y.; Treitz, P.M.; Scott, N.A. Characterizing Forest Succession in Central Ontario Using Lidar-Derived Indices. *Photogramm. Eng. Remote Sens.* **2011**, *77*, 261–269. [CrossRef]
73. Pretzsch, H. Description and Analysis of Stand Structures. In *Forest Dynamics, Growth and Yield*; Springer: Berlin/Heidelberg, Germany, 2010; pp. 223–289. ISBN 9783540883067.
74. Jenness, J.S. Calculating Landscape Surface Area from Digital Elevation Models. *Wildl. Soc. Bull.* **2004**, *32*, 829–839. [CrossRef]
75. Woods, M.; Pitt, D.; Penner, M.; Lim, K.; Nesbitt, D.; Etheridge, D.; Treitz, P. Operational Implementation of a LiDAR Inventory in Boreal Ontario. *For. Chron.* **2011**, *87*, 512–528. [CrossRef]
76. Beets, P.N.; Reutebuch, S.; Kimberley, M.O.; Oliver, G.R.; Pearce, S.H.; McGaughey, R.J. Leaf Area Index, Biomass Carbon and Growth Rate of Radiata Pine Genetic Types and Relationships with LiDAR. *Forests* **2011**, *2*, 637–659. [CrossRef]
77. Solberg, S.; Brunner, A.; Hanssen, K.H.; Lange, H.; Næsset, E.; Rautiainen, M.; Stenberg, P. Mapping LAI in a Norway Spruce Forest Using Airborne Laser Scanning. *Remote Sens. Environ.* **2009**, *113*, 2317–2327. [CrossRef]
78. Hopkinson, C.; Chasmer, L. Testing LiDAR Models of Fractional Cover across Multiple Forest Ecozones. *Remote Sens. Environ.* **2009**, *113*, 275–288. [CrossRef]

79. Zou, H.; Hastie, T. Regularization and Variable Selection via the Elastic Net. *J. R. Stat. Soc. Ser. B Stat. Methodol.* **2005**, *67*, 301–320. [CrossRef]
80. Kuhn, M. Building Predictive Models in R Using the Caret Package. *J. Stat. Softw.* **2008**, *28*, 1–26. [CrossRef]
81. Liaw, A.; Wiener, M. Classification and Regression by RandomForest. *R News* **2002**, *2*, 18–22.
82. Venables, W.N.; Ripley, B.D. *Modern Applied Statistics with S*, 4th ed.; Springer: New York, NY, USA, 2002.
83. Karatzoglou, A.; Smola, A.; Hornik, K.; Zeileis, A. Kernlab—An S4 Package for Kernel Methods in R. *J. Stat. Softw.* **2004**, *11*, 1–20. [CrossRef]
84. Friedman, J.; Hastie, T.; Tibshirani, R. Regularization Paths for Generalized Linear Models via Coordinate Descent. *J. Stat. Softw.* **2010**, *33*, 1–22. [CrossRef]
85. Friedman, J.; Hastie, T.; Tibshirani, R. *The Elements of Statistical Learning: Data Mining Inference and Prediction*, 2nd ed.; Springer Series in Statistics New York; Springer: New York, NY, USA, 2001; Volume 1, ISBN 978-0-387-21606-5.
86. Delignette-Muller, M.L.; Dutang, C. Fitdistrplus: An R Package for Fitting Distributions. *J. Stat. Softw.* **2015**, *64*, 1–34. [CrossRef]
87. Yu, Y. MixR: An R Package for Finite Mixture Modeling for Both Raw and Binned Data. *J. Open Source Softw.* **2022**, *7*, 4031. [CrossRef]
88. Furnival, G.M.; Wilson, R.W. Regressions by Leaps and Bounds. *Technometrics* **1974**, *16*, 499–511. [CrossRef]
89. Lumley, T. Leaps: Regression Subset Selection, Based on Fortran Code by Alan Miller, R Package Version 3.1. Available online: https://cran.r-project.org/web/packages/leaps/leaps.pdf (accessed on 18 May 2020).
90. Land, A.H.; Doig, A.G. An Automatic Method of Solving Discrete Programming Problems. *Econometrica* **1960**, *28*, 497–520. [CrossRef]
91. Oliveira, S.; Oehler, F.; San-Miguel-Ayanz, J.; Camia, A.; Pereira, J.M.C. Modeling Spatial Patterns of Fire Occurrence in Mediterranean Europe Using Multiple Regression and Random Forest. *For. Ecol. Manag.* **2012**, *275*, 117–129. [CrossRef]
92. Reynolds, M.R.; Burk, T.E.; Huang, W.-C. Goodness-of-Fit Tests and Model Selection Procedures for Diameter Distribution Models. *For. Sci.* **1988**, *34*, 373–399. [CrossRef]
93. Coomes, D.A.; Duncan, R.P.; Allen, R.B.; Truscott, J. Disturbances Prevent Stem Size-Density Distributions in Natural Forests from Following Scaling Relationships. *Ecol. Lett.* **2003**, *6*, 980–989. [CrossRef]
94. Pippuri, I.; Kallio, E.; Maltamo, M.; Peltola, H.; Packalén, P. Exploring Horizontal Area-Based Metrics to Discriminate the Spatial Pattern of Trees and Need for First Thinning Using Airborne Laser Scanning. *Forestry* **2012**, *85*, 305–314. [CrossRef]

Disclaimer/Publisher's Note: The statements, opinions and data contained in all publications are solely those of the individual author(s) and contributor(s) and not of MDPI and/or the editor(s). MDPI and/or the editor(s) disclaim responsibility for any injury to people or property resulting from any ideas, methods, instructions or products referred to in the content.

Article

A Novel Approach for Simultaneous Localization and Dense Mapping Based on Binocular Vision in Forest Ecological Environment

Lina Liu [1,2], Yaqiu Liu [1,2,*], Yunlei Lv [1,3] and Xiang Li [1,3]

[1] College of Computer and Control Engineering, Northeast Forestry University, Harbin 150040, China; lln@nefu.edu.cn (L.L.); yunleilv@nefu.edu.cn (Y.L.); nefu_lx@nefu.edu.cn (X.L.)
[2] Key Laboratory of Sustainable Management of Forest Ecosystems, Ministry of Education, Northeast Forestry University, Harbin 150040, China
[3] National and Local Joint Engineering Laboratory for Ecological Utilization of Biological Resources, Northeast Forestry University, Harbin 150040, China
* Correspondence: yaqiuLiu@nefu.edu.cn

Abstract: The three-dimensional reconstruction of forest ecological environment by low-altitude remote sensing photography from Unmanned Aerial Vehicles (UAVs) provides a powerful basis for the fine surveying of forest resources and forest management. A stereo vision system, D-SLAM, is proposed to realize simultaneous localization and dense mapping for UAVs in complex forest ecological environments. The system takes binocular images as input and 3D dense maps as target outputs, while the 3D sparse maps and the camera poses can be obtained. The tracking thread utilizes temporal clue to match sparse map points for zero-drift localization. The relative motion amount and data association between frames are used as constraints for new keyframes selection, and the binocular image spatial clue compensation strategy is proposed to increase the robustness of the algorithm tracking. The dense mapping thread uses Linear Attention Network (LANet) to predict reliable disparity maps in ill-posed regions, which are transformed to depth maps for constructing dense point cloud maps. Evaluations of three datasets, EuRoC, KITTI and Forest, show that the proposed system can run at 30 ordinary frames and 3 keyframes per second with Forest, with a high localization accuracy of several centimeters for Root Mean Squared Absolute Trajectory Error (RMS ATE) on EuRoC and a Relative Root Mean Squared Error (RMSE) with two average values of 0.64 and 0.2 for t_{rel} and R_{rel} with KITTI, outperforming most mainstream models in terms of tracking accuracy and robustness. Moreover, the advantage of dense mapping compensates for the shortcomings of sparse mapping in most Simultaneous Localization and Mapping (SLAM) systems and the proposed system meets the requirements of real-time localization and dense mapping in the complex ecological environment of forests.

Keywords: binocular vision SLAM; pose estimation; dense mapping; keyframe selection; spatial clue compensation; forest 3D reconstruction

1. Introduction

With the advantage of quickly and accurately obtaining three-dimensional spatial information, Light Detection and Ranging (LiDAR) is widely used in forest resource surveys. Solares-Canal et al. [1] proposed a methodology based on Machine Learning (ML) techniques to automatically detect the positions of and dasometric information about individual Eucalyptus trees from a point cloud acquired with a portable LiDAR system. Gharineiat et al. [2] summarized the methods of feature extraction and classification using ML techniques for laser point cloud data, which have good applications in scene segmentation, vegetation detection, and tree species classification.

Citation: Liu, L.; Liu, Y.; Lv, Y.; Li, X. A Novel Approach for Simultaneous Localization and Dense Mapping Based on Binocular Vision in Forest Ecological Environment. *Forests* **2024**, *15*, 147. https://doi.org/10.3390/f15010147

Academic Editor: Giorgos Mallinis

Received: 22 November 2023
Revised: 2 January 2024
Accepted: 6 January 2024
Published: 10 January 2024

Copyright: © 2024 by the authors. Licensee MDPI, Basel, Switzerland. This article is an open access article distributed under the terms and conditions of the Creative Commons Attribution (CC BY) license (https://creativecommons.org/licenses/by/4.0/).

In recent years, with the rapid development of UAVs technology, UAVs carrying LiDAR or visual sensors have greatly assisted in forest ecological exploration and forest management [3], and the forest information they captured can provide an essential basis for the three-dimensional reconstruction of forest ecological models. UAVs do not have a priori information about the relevant environment and their own position before executing the task, and they know nothing about the environment they are in, so there is no way to talk about UAV path planning and autonomous navigation, and how to solve the navigation problem of UAVs in the unknown environment of the forest is a difficult problem for forestry ecological exploration. Simultaneous Localization and Mapping (SLAM) [4] is one of the key algorithms for realizing fully autonomous navigation and real intelligence of mobile robots by means of sensor-equipped motion carriers that can move in unknown environments, senses and build environment maps, and estimate their own positions in the constructed maps at the same time, which empowers the robots to autonomously localize themselves and build real-time maps in unknown environments.

2. Related Work

Visual SLAM [5–7] has the advantages of being low cost, easy to use, and rich in information compared with LiDAR SLAM [8–10], so there is a huge potential for the development of visual SLAM. The mainstream methods for visual SLAM are the direct method and the feature point method. Dense Tracking and Mapping (DTAM) [11] based on the direct method uses the image pixels to construct a cost function and describes the depth using the inverse depth, constructing a 3D dense map in a global optimization. Large-scale direct monocular SLAM (LSD-SLAM) [12] directly matches image luminosity, uses a probabilistic model to represent semi-dense depth maps, and generates maps with global consistency. Direct sparse odometry (DSO) [13] is an improved version of LSD-SLAM that combines the direct method with sparse synchronization optimization, which can be applied in the case where RAM and CPU resources are lacking. Large-scale direct sparse visual odometry with stereo cameras (Stereo DSO) [14] integrates the constraints of the fixed binoculars into the Bundle Adjustment (BA) of the multi-view binoculars, which solves the scale drift problem while mitigating the optical flow sensitivity and the roll-up shutter effect of the conventional direct method.

The direct methods track directly on the image grayscale information, which have the advantages of fast speed and good real-time performance, however, they are based on the assumption of grayscale invariance and are limited to the narrow baseline motion. Feature-based methods use an indirect representation of the image, usually in the form of point features tracked along consecutive frames, recover poses of the camera by minimizing the projection error, which are more robust, and currently dominate the field of vision SLAM. Real-time single camera SLAM (MonoSLAM) [15] is the first monocular SLAM system, which achieves real-time drift-free performance from the structure to the motion model, but the feature stability is greatly affected by motion. Parallel Tracking and Mapping for small AR workspaces (PTAM) [16] is the first visual SLAM to be solved by an optimization method, which pioneers a keyframe mechanism and a dual-threaded parallel processing task to simultaneously handle tracking and mapping. A series of Oriented FAST and Rotated BRIEF (ORB) SLAM algorithms built by the SLAM group at the University of Zaragoza, Spain, is currently the most popular feature point method solution. A versatile and Accurate Monocular SLAM System (ORB-SLAM) [17] was first proposed in 2015, which is based on PTAM and uses ORB descriptors, and it only supports monocular cameras, thus it suffers from the scale uncertainty problem. An Open-Source SLAM System for Monocular, Stereo, and RGB-D Cameras (ORB-SLAM2) [18] improves the efficiency and robustness of ORB feature extraction and descriptor matching, and it adds functions such as closed-loop detection and pose map optimization to enable it to cope with more complex environments and faster pose change. However, the keyframe selection conditions are more lenient leading to the high redundancy between frames in uniform linear motion, which will bring a higher cost of maintenance and deletion of keyframes in the later stage, thus affecting the performance of the system. Cameras with excessive cornering, fast speed changes, severe scene occlusion, and large changes in lighting can cause tracking to be lost if keyframes are not inserted in time. In addition, ORB-SLAM2 can only construct sparse maps, which does not enable tasks such as autonomous navigation and obstacle avoidance for robots.

Monocular SLAM is based on spatial geometric relationships and suffer from the disadvantage of scale drift. SLAM based on RGB Depth Camera (RGBD SLAM) [19,20] is susceptible to interference from varying light intensities, making it unsuitable for outdoor scenes. Additionally, the high cost of these cameras hinders their widespread adoption in the industry. Conventional vision SLAM based on geometric transformations has poor robustness when lighting changes, fast carrier motion, and low texture grayscales are not obvious, and are poorly applied in the scenes, as well as having drawbacks such as large amounts of calculations and large cumulative error. Therefore, it is generally used in indoor small target scenes and its application in outdoor complex scenes is limited. Deep learning methods [21–24] are not constrained by the above environmental conditions, and are able to quickly estimate the more accurate disparity in outdoor complex environments, enhancing the robustness of pose estimation and 3D scenes reconstruction. In this work, combine with deep learning, a stereo vision system, D-SLAM, is proposed to realize the simultaneous localization and dense mapping for UAVs in complex forest ecological environment. The main work is as follows.

(1) Using the temporal clue of binocular images as the main clue, the six Degrees of Freedom (6-DoF) rigid body pose of the UAV is estimated by utilizing the minimized visual feature reprojection error.

(2) Using the spatial clue of binocular images as an auxiliary clue, a binocular spatial compensation strategy is proposed to increase the robustness of the algorithm's tracking when the camera corner is too large.

(3) Taking the relative motion amount and data association between frames as the important conditions for filtering keyframes, the keyframe filtering strategy is improved to enhance the system's localization and mapping accuracy as well as running speed.

(4) By increasing the 3D dense map construction thread, using the LANet network to predict the disparity map of the keyframes, and combining the poses of the keyframes to generate a dense point cloud, a dense map of the complex ecological environment of the forest is constructed by utilizing techniques such as point clouds registration, point clouds fusion and point clouds filtering.

3. Study Area and Data

The experimental forest farm of Northeast Forestry University was selected as the study area, using a ZED2 binary camera (Stereolabs, San Francisco, CA, USA) to collect image data and video bag data to create a Forest dataset for the training and testing of generating disparity maps and dense mapping, respectively.

3.1. Study Area

The study area is located at 126°37′ E, 45°43′ N in the Harbin Experimental Forestry Farm of Northeast Forestry Universit, at an altitude of 136–140 m, with a mesothermal continental monsoon climate. The forest is covered with 18 species of plantation forests, including *Larix gmelinii* Rupr., *Quercus mongolica* Fisch. ex Ledeb., *Betula platyphylla* Suk., and *Fraxinus mandshurica* Rupr. The structural types in the sample plots include the tree layer and herb layer, and the canopy density exceeds 0.70. The experimental environment is characterized by the easy lock–lose of Global Navigation Satellite System (GNSS) satellites), large sample plot coverage, and obvious topographic relief, which is representative of the field forest exploration tasks.

3.2. Forest Dataset

Five types of forest vegetation, *Larix gmelinii*, *Pinus sylvestris* var. mongolica Litv., *Pinus tabulaeformis* Carr., *Fraxinus mandschurica*, and *Betula platyphylla*, were selected for the experiments, and a ZED2 binocular camera was used with a baseline length of 20 cm, while RGB binocular image pairs and their corresponding disparity maps were collected with an original pixel resolution of 1280 × 720, which was cropped to a resolution of 1240 × 426, totaling 5000 pairs to produce the Forest dataset used for the Neural network model training and testing. Out of this, 80% are used as training data, 10% as the validation set and 10% as the test set. The details are shown in Table 1.

Table 1. Forest dataset details.

Variety	Training Set	Validation Set	Test Set	Total	Resolution	Sparse /Dense	Synthetic /Real
Larix gmelinii	960	120	120	1200	1240 × 426	dense	real
Pinus sylvestris	1200	150	150	1500	1240 × 426	dense	real
Pinus tabulaeformis	800	50	50	500	1240 × 426	dense	real
Fraxinus mandschurica	640	80	80	800	1240 × 426	dense	real
Betula platyphylla	800	100	100	1000	1240 × 426	dense	real

The Forest dataset includes not only the disparity map dataset (Figure 1) used for training and testing LANet networks, but also the bag video dataset (Figure 2) and binocular image dataset (Figures 3 and 4) used for testing the D-SLAM system, including three resolutions: High Definition (HD) 1080:1920 × 1080, HD720:1280 × 720, and Video Graphic Array (VGA): 672 × 376, as shown below.

Figure 1. Disparity maps in Forest dataset.

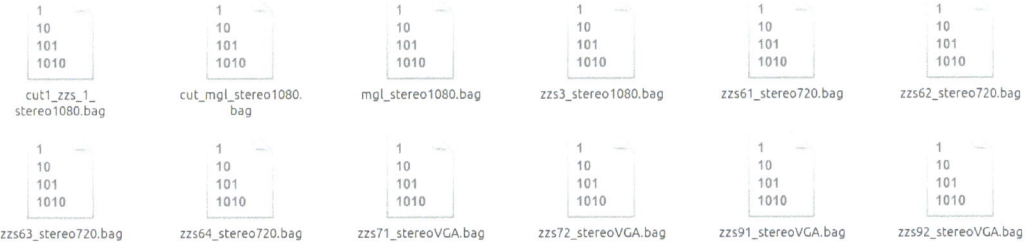

Figure 2. Bag in Forest dataset.

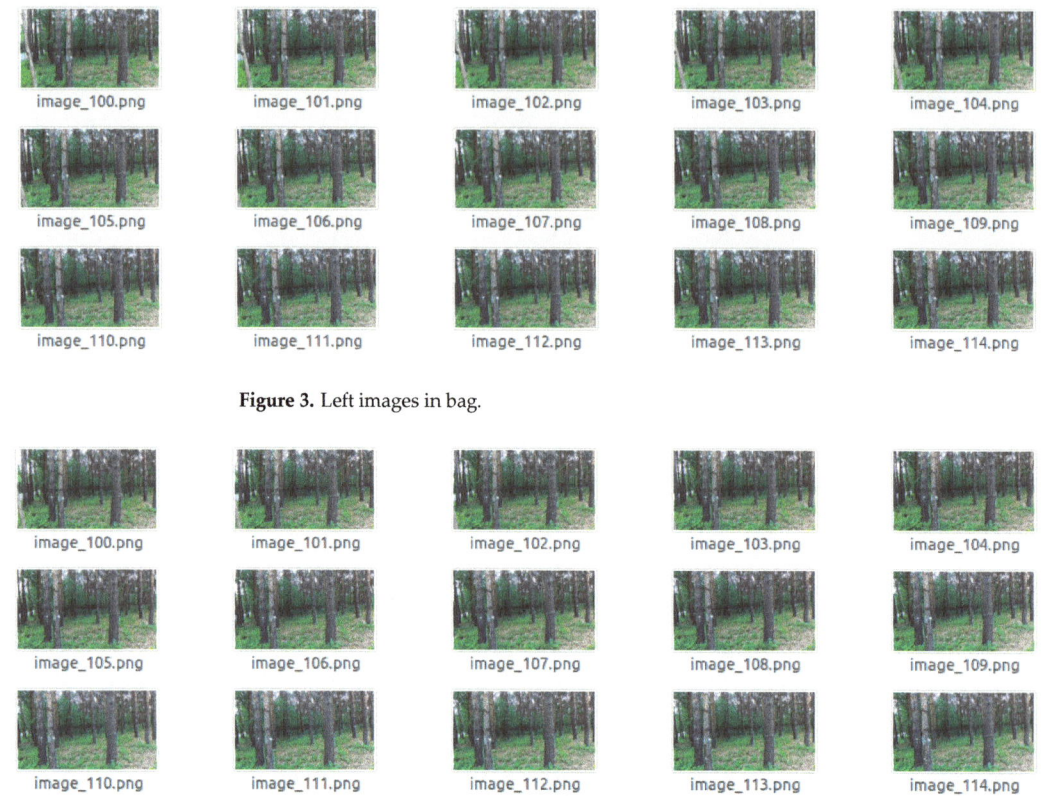

Figure 3. Left images in bag.

Figure 4. Right images in bag.

4. Methods

The structure of D-SLAM system is shown in Figure 5. The tracking thread searches for feature points to match with the local map in each frame and uses motion-only BA to minimize the reprojection error to optimize the pose of the current frame to realize the camera's location and tracking in each frame, and at the same time determines whether the current frame is a keyframe or not according to the conditions. The local mapping thread receives the keyframes from the tracking thread, eliminates redundant map points, generates new map points, optimizes the local map points and the poses of keyframes, and deletes redundant keyframes. The dense mapping thread receives the disparity map generated by the LANet, combines it with the pose of the keyframe to obtain a 3D point cloud, and then generates the dense point cloud map through point clouds registration, point clouds fusion and point clouds filtering. The loop closing thread corrects the cumulative drift through pose-graph optimization and starts the full BA thread for the BA optimization of all map points and keyframes.

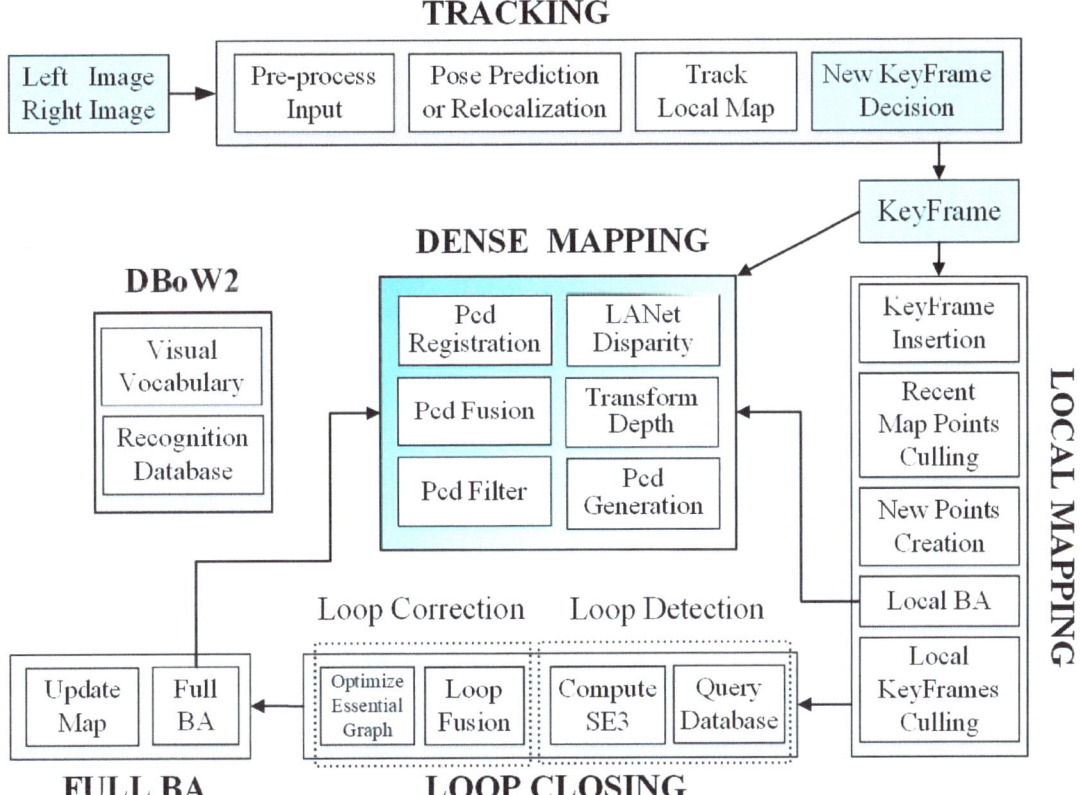

Figure 5. The D-SLAM system consists of four main parallel threads, tracking, local mapping, dense mapping, and loop closing, where the acronyms are defined as follows: Preprocessing (Pre-process), Llocal Bundle Adjustment (Local BA), Full Bundle Adjustment (Full BA), Special Euclidean group (SE3), Point cloud (Pcd), Linear Attention Network (LANet), and bags of binary Words for fast place recognition in image sequences (DboW2).

4.1. Tracking

(1) Binocular initialization

Monocular initialization requires two or more image frames with both rotation and translation necessary for successful initialization. Binocular initialization is conducted in the first frame and absolute scale information is obtained. The binocular camera performs stereo matching through the left and right images of the first frame, using the principle of triangulation to obtain the depth information of the feature points, according to the current frame of the pose to obtain the world coordinates, so the binocular camera can generate 3D map points and create an initial map in the first frame, and then tracking is conducted directly in the next frame.

(2) Pose estimation

In binocular mode, two consecutive frames of the left image in the temporal dimension perform feature matching to find the corresponding data-associated feature points, whereby the left image frames are consecutive in the temporal clue which serves as the main clue for the pose estimation. The continuity of the right image frames in the spatial dimension is used as an auxiliary clue to search for feature matching points on the right spatial clue frame corresponding to the left image frame, and if a match can be made and the effective depth value of the feature point can be obtained, the 3D-3D Iterative Closest Point (ICP) method can be used to estimate the pose of the current frame.

The two consecutive frames of the left image in the temporal dimension find the corresponding data-associated feature points through the feature matching, assuming that the 3D map points corresponding to these two sets of feature points are $p = \{p_1, p_2, \ldots p_n\}$ and $p' = \{p'_1, p'_2, \ldots p'_n\}$, where $p, p' \in \mathbb{R}^3$. Because each point in p and p' has already been associated with the data through subscripts one by one, $p'_i = R \cdot p_i + t$ is satisfied between p and p' in the ideal case, where $R \in SE(3)$ and $t \in \mathbb{R}^3$, but in fact $p'_i \neq R \cdot p_i + t$ due to the presence of noise. At this time, (R, t) can be solved by constructing the least squares problem [25] expressed as follows.

$$\operatorname*{argmin}_{R,t} \sum_{i=1}^{n} \|p'_i - (R \cdot p_i + t)\|^2 = \operatorname*{argmin}_{R,t} \sum_{i=1}^{n} \left\{ \|(p'_i - \overline{p'}) - R \cdot (p_i - \overline{p})\|^2 + \|\overline{p'} - R \cdot \overline{p} - t\|^2 \right\} \\ = \operatorname*{argmin}_{R,t} \sum_{i=1}^{n} \left\{ \|q'_i - R \cdot q_i\|^2 + \|\overline{p'} - R \cdot \overline{p} - t\|^2 \right\} \quad (1)$$

where the point clouds p and p' are moved towards the center, let $q_i = p_i - \overline{p}$, $q'_i = p'_i - \overline{p'}$, $\overline{p} = \frac{1}{n}\sum_{i=1}^{n} p_i$, and $\overline{p'} = \frac{1}{n}\sum_{i=1}^{n} p'_i$.

In Equation (1), the first additive term is expanded and simplified as follows.

$$\operatorname*{argmin}_{R} \sum_{i=1}^{n} \|q'_i - R \cdot q_i\|^2 = \operatorname*{argmin}_{R} \sum_{i=1}^{n} (q'^T_i q'_i - 2q'^T_i R q_i + q_i^T R^T R q_i) \\ \Updownarrow \\ \operatorname*{argmin}_{R} \sum_{i=1}^{n} -q'^T_i R q_i = \operatorname*{argmin}_{R} \left[-tr(R \sum_{i=1}^{n} q'^T_i q_i) \right] \quad (2)$$

Let $M = \sum_{i=1}^{n} q'^T_i q_i$, Singular Value Decomposition (SVD) decomposition is utilized to obtain $SVD(M) = USV^T$ and then $R = UV^T$, and by substituting the resulting R into the second additive term $\operatorname*{argmin}_{t} \sum_{i=1}^{n} \|\overline{p'} - R \cdot \overline{p} - t\|^2$, it will be easy to obtain $t = \overline{p'} - R \cdot \overline{p}$.

If there is no right feature point on the spatial cue that can match on the left feature point, or the effective depth value of the left feature point can not be obtained, at this time, it can only be triangulated by the multi-frame view. Based on the known 3D positions of the feature points in the local sliding window, and their 2D observations in the image, the pose of the current frame can be solved by using the 3D-2D Perspective-n-Point (PnP) method.

The coordinates of n 3D spatial points and their 2D point observations are known, n known map points p_i^w (i = 1, 2, ..., n) are selected as reference points from the world coordinate system, and 4 known map points c_j^w (j = 1, 2, 3, 4) are selected as control points, which are associated with the reference points by means of a weighted sum expressed as follows.

$$p_i^w = \sum_{j=1}^{4} \alpha_{ij} c_j^w \quad (3)$$

where $\sum_{j=1}^{4} \alpha_{ij} = 1$.

p_i^c and c_j^c are map points and control points under the camera coordinate system, and because only the values of the coordinates are taken differently, the relative spatial positions between the points have not changed, so the relationship of the weighted sum also holds, then there is

$$p_i^c = \sum_{j=1}^{4} \alpha_{ij} c_j^c \quad (4)$$

In the camera coordinate system, the reference point p_i^c with its corresponding pixel point $p_i^u(u_i, v_i)$ can be described by the projection equation [26] as follows

$$w_i \begin{bmatrix} p_i^u \\ 1 \end{bmatrix} = K \cdot p_i^c = K \cdot \sum_{j=1}^{4} \alpha_{ij} c_j^c \quad (5)$$

there is

$$w_i \begin{bmatrix} u_i \\ v_i \\ 1 \end{bmatrix} = \begin{bmatrix} f_u & 0 & u_c \\ 0 & f_v & v_c \\ 0 & 0 & 1 \end{bmatrix} \sum_{j=1}^{4} \alpha_{ij} \begin{bmatrix} x_j^c \\ y_j^c \\ z_j^c \end{bmatrix} \Leftrightarrow \begin{cases} \sum_{j=1}^{4} \alpha_{ij} f_u x_j^c + \alpha_{ij}(u_c - u_i) z_j^c = 0 \\ \sum_{j=1}^{4} \alpha_{ij} f_v y_j^c + \alpha_{ij}(v_c - v_i) z_j^c = 0 \end{cases} \Leftrightarrow A_{2 \times 12} \cdot h_{12 \times 1} = 0 \quad (6)$$

The coefficient matrix $A_{2\times 12}$ is constructed from the weighting coefficients of the reference points, the pixel coordinates and the camera internal parameters, and the vector $h_{12\times 1}$ is constructed from the 12 coordinate values x_1^c、y_1^c、z_1^c, x_2^c、y_2^c、z_2^c, x_3^c、y_3^c、z_3^c, and x_3^c、y_3^c、z_3^c of the 4 control points in the camera coordinate system. From this, a linear equation $A_{2\times 12} \cdot h_{12\times 1} = 0$ can be constructed by projecting a reference point p_i^c to a camera pixel point u_i. Then n reference points p_i^c projected to pixel points u_i can construct the linear equations $A_{2n\times 12} \cdot h_{12\times 1} = 0$, followed by solving the equation for the vector h. The solution process is based on least squares and SVD methods, solving the equations to obtain the coordinate values of the four control points in the camera coordinate system, combined with the known coordinate values of the four control points in the world coordinate system, and then utilizing the ICP algorithm to find the transformation relationship (R, t) between the four control points in the two coordinate systems.

The camera poses solved by the above methods are subject to errors due to noise, computation and other factors, and BA optimization can be used to further optimize the poses and improve the accuracy. Construct a nonlinear least squares problem on reprojection error: 3D points are projected to 2D points which are combined with the observed 2D points to construct the reprojection error equation, and the optimal solution is obtained by iterations using the Gaussian Newton method. In the world coordinate system, the map point p_i^w is converted to camera coordinates $p_i^c = T \cdot p_i^w$ by the transfer matrix $T = \begin{bmatrix} R & t \\ 0 & 1 \end{bmatrix}$, and the pixel coordinates are obtained by projecting the camera coordinates using the camera model

$$w_i \begin{bmatrix} \hat{p}_i^u \\ 1 \end{bmatrix} = K \cdot p_i^c \quad (7)$$

Establish the relationship from world coordinates to pixel coordinates: $w_i \hat{p}_i^u = K \cdot p_i^c = K \cdot T \cdot p_i^w$, i.e., $\hat{p}_i^u = \frac{1}{w_i} K \cdot T \cdot p_i^w$, obtain the error term by subtracting the value $p_i^u(u_i, v_i)$ from the observed coordinates of the 2D point, construct the least squares problem by using the sum of all the error terms, minimize the reprojection error and the camera pose is solved by using Gauss-Newton optimization algorithm [25] as follows.

$$T = \underset{T}{\arg\min} \frac{1}{2} \sum_{i=1}^{n} \left\| p_i^u - \frac{1}{w_i} K \cdot T \cdot p_i^w \right\|_2^2 \quad (8)$$

(3) Keyframes selection

ORB-SLAM2 obtains keyframes under more relaxed conditions to ensure that it can "keep up" with the tracking thread in the early stage, however, the quality of the keyframes is not taken into consideration. For instance, the high redundancy between frames when the camera in uniform linear motion will lead to the high cost of maintaining and deleting the keyframes in the later stage, which would affect the performance of the system. The untimely insertion of keyframes can easily lead to loss of tracking when turning or changing speeds quickly. Furthermore, tracking loss also occurs easily when there is serious occlusion and large changes in lighting.

According to the actual complex ecological environment of the forest, the keyframe selection strategy is designed to avoid the introduction of too much information redundancy due to the excessive image overlap. At the same time, the image overlap should not be too small to ensure that there are some covisibility feature points to avoid tracking loss. Under the constraints of the covisibility graph, the quality of keyframe tracking is guaranteed while achieving the goal of having both constraints and less information redundancy between keyframes and other keyframes in the local map. Based on the above requirements, the amount of relative motion between frames (rotation angles and translation changes) and data association (number of matched feature points) are considered as the important basis for the selection of keyframes.

The amount of relative motion $Tran\ (R, t)$ between the current frame and the previous keyframe is a function of the pose (R, t), and it is defined as follows:

$$Tran\ (R, t) = (1 - \alpha)\ ||t|| + \alpha min(2\pi - ||R||, ||R||) \tag{9}$$

where $\alpha = (\frac{\tan \omega}{\tan \omega + 1})^{\frac{1}{4}}$, and $\omega = min(2\pi - ||R||, ||R||)$.

Because $R \in SE(3)$, $t \in \mathbb{R}^3$, $Tran$ represent the relative motion between frames, their Euclidean spatial distances are taken as the amount of translation and rotation changes between frames, respectively. $\alpha \in [0, 1]$ is a motion transformation factor between frames, the size of which increases exponentially with the increase in the angle of camera rotation, and it has a corner of the amplification function and at the same time has a translation suppression function. When α is large, the suppression $1 - \alpha$ of the translation can be brought close to 0, where the amount of relative motion between the frames depends mainly on α. The value of α and the specific expression of $Tran$ are determined by the range of $\omega = min(2\pi - ||R||, ||R||)$. $Tran\ (R, t)$ can be expressed as follows,

$$Tran\ (R,t) = \begin{cases} (1-\alpha)||t|| + \alpha min(2\pi - ||R||, ||R||), & \omega \in [0, \frac{\pi}{2}), \ \alpha = (\frac{\tan \omega}{\tan \omega + 1})^{\frac{1}{4}} \\ \alpha min(2\pi - ||R||, ||R||) & , \omega \in [\frac{\pi}{2}, \pi], \ \alpha = (\frac{|\tan \omega|}{|\tan \omega| + 1})^{\frac{1}{4}} \end{cases} \tag{10}$$

When the angle of camera rotation $\omega \in [\frac{\pi}{2}, \pi]$, the camera almost loses the perspective, at which time the feature points cannot be matched on the temporal clue causing the camera tracking to fail. In order to increase the robustness of the system in tracking, the spatial clue compensation strategy is adopted: the previous frame of the right image of the spatial clue is used as the clue connection, and it is inserted to compensate for the lost field of view on the temporal clue, continuing the tracking, as shown in Figure 6.

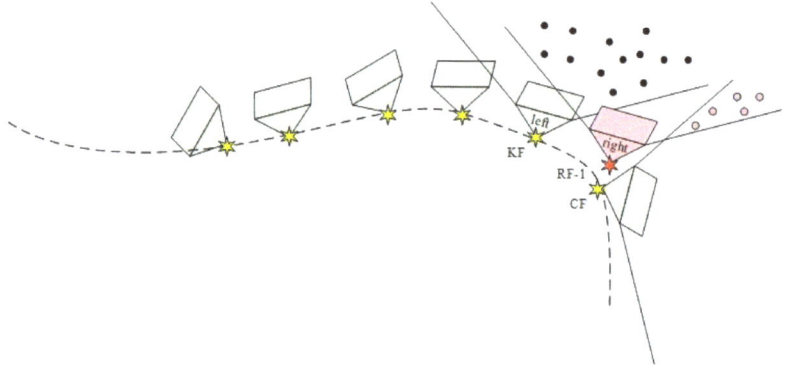

Figure 6. Compensation of spatial clue in the right image frame, where KF represents the key frame, CF represents the current frame, and RF-1 represents the previous frame of the right image.

The system is preset with a maximal threshold of η and a minimal threshold of ξ, comparing the $Tran$ with the threshold are follows:

when $Tran < \xi$, $Frame_{ckey} \neq Frame_{cur}$;
when $\xi \leq Tran \leq \eta$, $Frame_{ckey} = Frame_{cur}$;
when $Tran > \eta$, if $\omega < \frac{\pi}{2}$, $Frame_{key} = Frame_{cur}$, else $\omega \geq \frac{\pi}{2}$, $Frame_{key} = Frame_{rcur} - 1$.

$Frame_{cur}$ is the current frame of the left image, $Frame_{rcur} - 1$ is the previous frame of the right image, $Frame_{ckey}$ is the candidate keyframe, and $Frame_{key}$ is the keyframe.

In the above analysis, when the amount of relative motion between frames $Tran > \eta$, it indicates a significant change in the camera's view, and keyframes should be inserted in time, otherwise the tracking will be lost. When $\xi \leq Tran \leq \eta$, it is the normal motion range of the camera, in which redundant keyframes should be avoided. Taking the amount of relative motion between frames and data association as constraints, combining the candidate keyframes obtained from the above calculation, the keyframes selection for the system needs to satisfy the following two conditions:

(1) The number of tracked covisibility feature points between $Frame_{ckey}$ and the previous keyframe satisfies the following condition: $Track(Frame_{key} - 1, Frame_{ckey}) > \tau_f$.
(2) The number of near points tracked by $Frame_{ckey}$ is less than the threshold τ_t and more than τ_c new near points can be created.

In outdoor environments, such as forest scenes, where most areas are far away from the sensors, the introduction of near and far points for binocular vision as conditions for filtering keyframes is particularly important for improving the localization accuracy of the system. The near point is the feature point in binocular mode where the depth value is less than 40 times the binocular baseline distance, otherwise it is called the far point.

The 3D coordinates obtained by triangulation for the near point are more accurate which can provide information about orientation, translation, and scale. In contrast the far point carries less information which provides only relatively accurate information about orientation. It is very challenging in a large forest scene and at a distance from the camera, the system needs enough near points to accurately estimate the camera's translation, so the system has certain requirements for the number of tracked near points and the number of generated new near points. It works better by setting $\tau_t = 90$, $\tau_c = 50$ in the experiments.

The steps of keyframes selection for the system are as follows.

Step 1: Determine whether the prerequisites for inserting keyframes are met: the system is not currently in localization mode and the local mapping is free, while it is far from the last relocation, and the number of internal points must be greater than the minimum threshold of 15, i.e., mnMatchesInliers > 15.

Step 2: Calculate the relative motion $Tran(R,t)$ between frames and determine the candidate keyframe $Frame_{ckey}$ or keyframe $Frame_{ckey}$ based on the comparison between $Tran(R,t)$ and the thresholds.

Step 3: If it is a candidate keyframe $Frame_{ckey}$, calculate the number of tracked feature points between $Frame_{ckey}$ and the previous keyframe $Frame_{key} - 1$ and perform the judgment of condition 1: $Track(Frame_{key} - 1, Frame_{ckey}) > \tau_f$.

Step 4: Calculate the number of near points tracked by $Frame_{ckey}$ and perform the judgment of condition 2: The number of near points tracked by $Frame_{ckey}$ is less than the threshold τ_t and more than τ_c new near points can be created.

Step 5: If both conditions 1 and 2 are satisfied, which indicates high matching and correlation between frames and the high quality of feature points, the candidate keyframe $Frame_{ckey}$ is set as the keyframe $Frame_{key}$, i.e., $Frame_{key} = Frame_{ckey}$.

Correspondingly, the algorithm of keyframes selection for the system is as follows (Algorithm 1).

Algorithm 1 Keyframe Selection

Input: the binocular image frames $Frame_{cur}$ and $Frame_{rcur}$
Parameter: threshold $\tau_t, \tau_c, \tau_f, \xi, \eta$
Output: Keyframe $Frame_{key}$

1: **for** each vailable new $Frame_{cur}$ **do**
2:     calculate $\omega = min\,(2\pi - ||R||, ||R||)$
3:     **if** $\omega \in [0, \frac{\pi}{2})$ **then**
4:         $\alpha = (\frac{\tan \omega}{\tan \omega + 1})^{\frac{1}{4}}$
5:         $Tran\,(R, t) = (1 - \alpha)\,||\,t|| + \alpha min\,(2\pi - ||R||, ||R||)$
6:     **else**
7:         $\alpha = (\frac{|\tan \omega|}{|\tan \omega| + 1})^{\frac{1}{4}}$
8:         $Tran\,(R, t) = \alpha min(2\pi - ||R||, ||R||)$
9:     **end if**
10:    **if** $Tran < \xi$ **then**
11:        $Frame_{ckey} \neq Frame_{cur}$
12:    **else**
13:        **if** $\xi \leq Tran \leq \eta$ **then**
14:            $Frame_{ckey} = Frame_{cur}$
15:        **else**
16:            **if** $\omega < \frac{\pi}{2}$ **then**
17:                $Frame_{key} = Frame_{cur}$
18:            **else**
19:                $Frame_{key} = Frame_{rcur} - 1$
20:            **end if**
21:        **end if**
22:    **end if**
23:    calculate the number of covisibility feature points $Track$ between $Frame_{ckey}$ and the last keyframe
24:    **if** $Track(Frame_{key} - 1, Frame_{ckey}) > \tau_f$ **then**
25:        calculate the number of near points tracked and the number of new near points created in $Frame_{ckey}$
26:        **if** the number of near points tracked in $Frame_{ckey}$ is less than the threshold τ_t and more than τ_c new near points are created **then**
27:            $Frame_{key} = Frame_{ckey}$
28:        **end if**
29:    **end if**
30: **end for**

4.2. Local Mapping

The local mapping thread implements mid-term data association, it receives keyframes imported from the tracking thread, eliminates substandard map points, generates new map points, performs local map optimization, removes redundant keyframes, and sends optimized keyframes to the loop closing thread. Only the information of adjacent common frames or keyframes is used in the tracking thread; moreover, only the pose of the current frame is optimized, and there is no joint optimization of multiple poses and no optimization of the map points. The Local BA optimizes both multiple keyframes that satisfy a certain covisibility relationship and the corresponding map points, so as to make the keyframes more accurate in terms of poses and map points. More new map points are obtained by re-matching between covisibility keyframes, increasing the number of map points while improving the tracking stability. Removing redundant keyframes helps to reduce the scale and number of Local BAs and improve the real-time performance of the system.

4.3. Loop Closing

Loop closing is divided into two steps: loop closing detection and loop closing correction. Loop closing detection uses Bag of Words (BoW) to accelerate matching, queries the dataset to detect whether the loop is closed or not, and then computes the Special Euclidean Group (SE3) poses between the current keyframe and the loop closing candidate keyframe. Monocular vision suffers from scale drift while binocular vision easily obtains depth information making the scale observable, so there is no need to deal with scale drift in geometric validation and pose-graph optimization. The loop closing correction focuses on loop closing fusion and essential graph optimization to correct cumulative drift, and starts the Full BA thread for the BA optimization of all map points and keyframes, which is more costly and therefore a separate thread is needed.

4.4. Dense Mapping

The tracking thread calculates the pose for each frame, if the dense map is constructed using each frame in the tracking thread, it not only increases running time and storage space overhead for the system, but it also affects the localization accuracy due to the heavy computation which slows down the system's running speed; hence, the keyframes are used to construct the dense map. Firstly, the disparity map is calculated for each keyframe; secondly, the point cloud is generated by combining the more accurate keyframe poses optimized by Local BA, and then the initial dense map is formed through point clouds registration, point clouds fusion, and point clouds filtering; and thirdly, the dense map is updated by global BA optimization.

Obtaining disparity maps is a key step in dense mapping, while the traditional stereo matching methods have a poor matching effect in the regions of weak texture, occlusion and other features that are not obvious, and the generated disparity map is insufficiently robust. In this work, LANet [27], a linear attention stereo matching network is embedded into D-SLAM as one of the modules to generate dense disparity maps, which are transformed to generate depth maps and point clouds to realize the construction of dense maps of the forest ecological environment. LANet networks are capable of optimizing depth estimation by efficiently utilizing environmental global and local information to improve stereo matching accuracy in ill-posed regions, such as those with weak texture, poor lighting, and occlusion and achieve efficient disparity inference prediction. Because LANet is one of the research findings of the authors of this paper, it has been published in a public paper "LANet: Stereo matching network based on linear-attention mechanism for depth estimation optimization in 3D reconstruction of inter-forest scene" https://www.frontiersin.org/articles/10.3389/fpls.2022.978564/full (accessed on 2 September 2022), and the overview of the LANet as shown in Figure 7.

(1) Feature extraction

ResNet [28] is adopted as the backbone network for feature extraction; all layers use a 3×3 convolutional kernel, and the first stage uses three convolutional layers conv0_1, conv0_2, and conv0_3, to extract the primary features of the image. The second stage uses four sets of basic residual blocks, conv1_x, conv2_x, conv3_x, and conv4_x, to extract the deep semantic features of the image. Downsampling with stride 2 was used in conv0_1 and conv2_1, and the input image size is reduced to 1/4 of the original size after two downsamplings. The dilated convolution is applied to enlarge the receptive field in conv3_x and conv4_x, and the dilation rates of these two layers are 2 and 4, respectively.

(2) Attention Module (AM)

AM can better integrate local and global information to obtain richer feature representations at the pixel level, and the AM consists of two parts: the Spatial Attention Module (SAM) and Channel Attention Module (CAM). SAM captures long-range dependencies between global contexts, seeks correlations between pixels at different locations, and models semantic correlations in the spatial dimension.

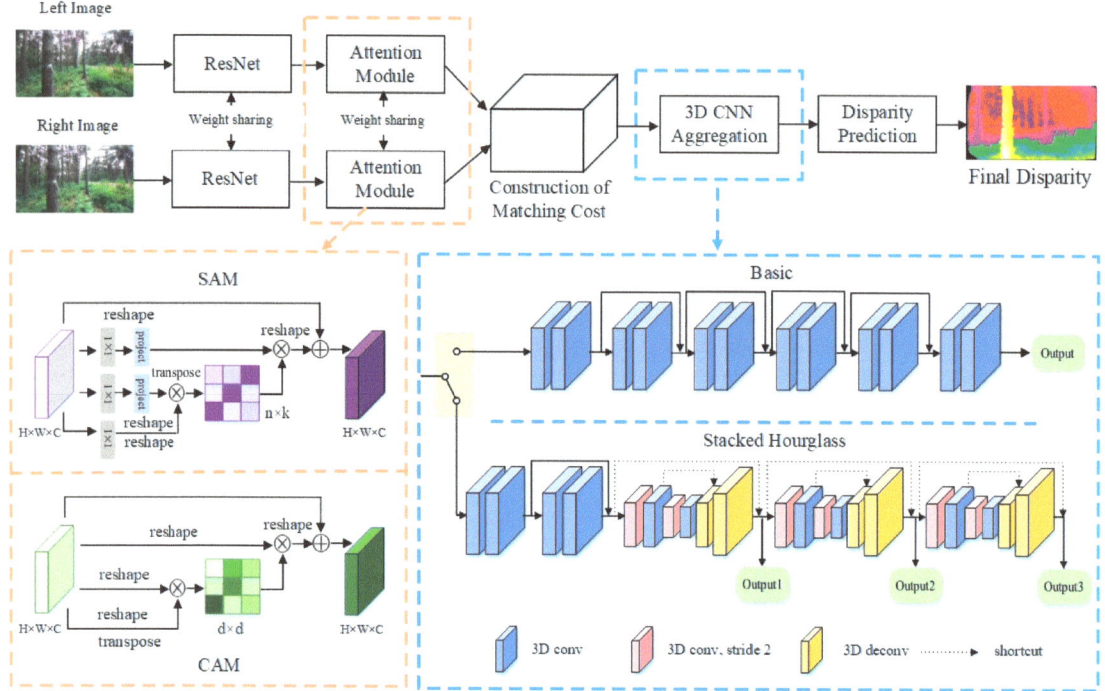

Figure 7. Network structure of LANet. LANet consists of five main parts: feature extraction ResNet, Attention Module (AM), Construction of Matching cost, Three Dimensional Convolutional Neural Network aggregation (3D CNN aggregation), and disparity prediction. The AM consists of two parts: Spatial Attention Module (SAM) and Channel Attention Module (CAM); the 3D CNN aggregation consists of two structures: the basic structure is used for ablation experiments to test the performance of various parts of the network and the stacked hourglass structure is used to optimize the network.

Because the time and space complexity of self-attention [29] is $O(n^2)$, the cost of training and deploying the model is very high when it is used on large-size images. Linear-attention is proposed to be able to reduce the overall complexity of self-attention from $O(n^2)$ to $O(n)$ while retaining high accuracy. The correlation matrix $P \in \mathbb{R}^{n \times n}$ in self-attention is low rank, where most of the information is concentrated in a small number of maximum singular values; hence, a low rank matrix $\overline{P}$ is used to approximate P to reduce the complexity of self-attention by changing its structure. The details are as follows:

Let $X \in \mathbb{R}^{n \times d_m}$ be the input sequence, $W^Q, W^K \in \mathbb{R}^{d_m \times d_k}, W^V \in \mathbb{R}^{d_m \times d_v}$ are three learnable matrices, and $Q = XW^Q, K = XW^K, V = XW^V$, the query matrix, the key matrix and the value matrix $Q, K, V \in \mathbb{R}^{n \times d_m}$ embedded in the input sequence are obtained respectively. where n is the length of the sequence and d_m, d_k, d_v are the dimensions of the hidden layers of the projection space. Two low dimensional linear projection matrices $E \in \mathbb{R}^{n \times k}$ and $F \in \mathbb{R}^{n \times k}$ are constructed, which are fused with K and V to reduce their dimensionality. E or F performs matrix multiplication with K or V to reduce K and V from their original $n \times d$-dimension to the $k \times d$-dimension, as shown in Figure 8.

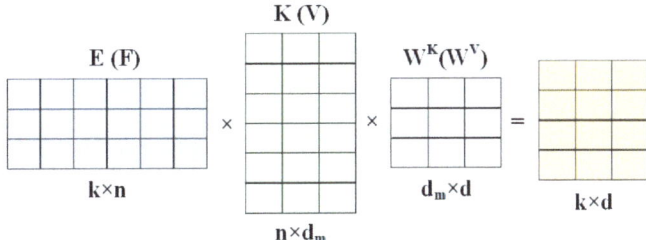

Figure 8. Linear mapping layers.

The correlation matrix $\overline{P} \in \mathbb{R}^{n \times k}$ is computed by the scaled dot product method, and the value of linear-attention is

$$\overline{P} \cdot (FVW^V) \tag{11}$$

where

$$\overline{P} = softmax\left[\frac{QW^Q(EKW^K)^T}{\sqrt{d}}\right] \tag{12}$$

and the complete form of linear-attention is

$$Linear-Attention\left(QW^Q, KW^K, VW^V\right) = softmax\left[\frac{QW^Q(EKW^K)^T}{\sqrt{d}}\right] \cdot (FVW^V) \tag{13}$$

The complexity of linear-attention is mainly determined by $\overline{P}$, $O(\overline{P}) = O(nk)$, and if a very small mapping dimension k is choosen and set to $k << n$, the overall complexity of $\overline{P}$ will decrease to linear $O(n)$. It can be proven that when $k = O(nd/\varepsilon^2)$, the value of $\overline{P} \cdot (FVW^V)$ approaches $P \cdot (VW^V)$, and the value of linear-attention can be approximately equivalent to that of self-attention, with an error of no more than ε.

The feature values obtained through linear-attention are multiplied by the scale factor α and then summed bit-wise with the original features $X \in \mathbb{R}^{H \times W \times C}$ to obtain the spatial attention feature map $Y \in \mathbb{R}^{H \times W \times C}$ as follows.

$$Y_j = \alpha \sum_{i=1}^{n} (\overline{P}_{ij} F_i V_i W_i^V) + X_j \tag{14}$$

where feature Y_j is the weighted sum of the features at all locations and the original location feature X_j. Therefore, it has global contextual information, and spatial attention can fuse similar features in the global spatial range, which is conducive to the consistent expression of feature semantics, and likewise it enhances the robustness of feature extraction in ill-posed regions.

Each channel corresponds to a feature map of a specific category of semantics. CAM models semantic relevance in the channel dimension, capturing long-range semantic dependencies between channel features, enabling global correlations between each channel, which is beneficial for obtaining stronger semantic feature responses and improving feature recognition. CAM is calculated on the original feature map based on the self-attention mechanism, without involving the complexity of $O(n^2)$.

The input feature $X \in \mathbb{R}^{H \times W \times C}$ is reshaped into $Q', K', V' \in \mathbb{R}^{n \times d}$ and there exists $Q' = K' = V'$, where $n = \frac{1}{4}H \times \frac{1}{4}W$, $d = C$, and the channel correlation matrix $P' \in \mathbb{R}^{d \times d}$ is obtained by multiplying the matrices between Q'^T and K' as follows.

$$P'_{ji} = softmax\left[\frac{Q'^T K'}{\sqrt{d}}\right] = \frac{\exp\left[\frac{Q'^T_i K'_j}{\sqrt{d}}\right]}{\sum_{i=1}^{C} \exp\left[\frac{Q'^T_i K'_j}{\sqrt{d}}\right]} \tag{15}$$

In the Equation (15), P'_{ji} denotes the correlation between the i th channel and the j th channel, and the higher the correlation between the two channel features, the greater the value of P'_{ji}. The final feature for each channel is a weighted sum of the features of all channels and the original feature as follows.

$$Z_j = \beta \sum_{i=1}^{C} (V'_i P'_{ji}) + X_j \qquad (16)$$

where Z is the final feature. The self-attention feature map of $\mathbb{R}^{n \times d}$ is obtained by multiplying the matrices between V' and p', which is reshaped into the form of $\mathbb{R}^{H \times W \times C}$ multiplied by a scale factor β, and it is then summed bit-wise with the original feature map $X \in \mathbb{R}^{H \times W \times C}$ to finally obtain the channel attention feature map $Z \in \mathbb{R}^{H \times W \times C}$.

(3) Construction of Matching cost

The feature information from the four parts of conv2_16, conv4_3, SAM and CAM is cascaded to form a 2D 1/4H × 1/4w × 320 feature map which is fused by two convolutional layers of 3 × 3 and 1 × 1 while the channels are compressed to 32, connecting the left 2D feature map with the right feature map corresponding to each disparity to construct a 4D matching cost-volume of 1/4D × 1/4H × 1/4W × 64.

(4) The 3D CNN aggregation

The 3D CNN aggregation module is used for cost-volume regularization, aggregating semantic and structural feature information in disparity and spatial dimensions to predict accurate cost-volume. It consists of two structures: the basic structure is used for ablation experiments to test the performance of various parts of the network, and it consists of twelve convolutional layers with a convolution kernel size of 3 × 3 × 3 performing BN and ReLU. The stacked hourglass structure is used to optimize the network and increase the robustness of disparity prediction in low-texture regions and occluded regions to obtain more accurate disparity values. The first four 3D convolutional layers contain BN and ReLU, and the 3D stacked hourglass network utilizes an "encoder-decoder" structure to reduce the parameters and computation of the network. The encoder downsamples twice by using a 3D convolution with a convolution kernel of 3 × 3 × 3 and a step size of 2. Correspondingly the decoder upsamples twice to recover the size by using an inverse convolution with a step size of 2, while the number of channels is halved. To compensate for the information loss caused by the "encoder–decoder" structure, a 1 × 1 × 1 3D convolution is used inside each hourglass module to connect features of the same size directly, which uses fewer parameters than a 3 × 3 × 3 convolution, and reduces the computational power to 1/27 of the original one, with negligible runtime; thereby, the running speed of the network is improved without increasing the computational cost.

(5) Disparity prediction

Each hourglass corresponds to one output, the total loss is a weighted sum of the losses corresponding to each output, and the last output is the final disparity map. A differentiable Soft Argmin function was utilized to obtain disparity estimation $\hat{d}$ through the regression method as follows. Equations (17)–(19) are from reference [30].

$$\hat{d} = \sum_{k=0}^{D_{\max}-1} k \cdot p_k \qquad (17)$$

where $D_{\max}$ denotes the maximum disparity. With the L1 loss function, the total loss is calculated as follows.

$$L = \sum_{i=1}^{3} \lambda_i \cdot Smooth_{L_1}(\hat{d}_i - d_i) \qquad (18)$$

where λ_i denotes the coefficient of the i th disparity prediction, d_i denotes the true value of the i th disparity map, $\hat{d}_i$ denotes the i th predicted disparity map, and the $Smooth_{L_1}(x)$ function is expressed as follows.

$$Smooth_{L_1}(x) = \begin{cases} 0.5x^2 & , if |x| < 1 \\ |x| - 0.5 & , otherwise \end{cases} \quad (19)$$

5. Results

The experiment and evaluation of the whole system is split in four parts:

- LANet performs the prediction training and evaluation of disparity maps on the Scene Flow and Forest datasets, and compares them with several mainstream methods.
- D-SLAM tests the accuracy of projection trajectories on two datasets, EuRoC and KITTI, and compares them with mainstream SLAM systems.
- D-SLAM tests the partial and overall performance of the system on three datasets, EuRoC, KITTI and Forest.
- D-SLAM performs real-time dense mapping on two datasets, KITTI and Forest, as well as analyzing and discussing the mapping results.

5.1. Experiment on Disparity Map Generation by LANet

LANet is pre-trained on the clean pass dataset of Scene Flow [31], and fine-tuned training is conducted on the Forest target dataset. Network training was based on Python 3.9.7, the PyTorch 1.11.0 framework, one Nvidia TITAN Xp GPU 3090 for the server, and Adam [32] for the optimizer, with $\beta 1 = 0.9$, $\beta 2 = 0.999$, and batch size set to eight.

(1) Ablation experiments on Scene Flow

Ablation experiments are carried out on the Scene Flow dataset to test the performance of each key module and parameter in the network. In Table 2, Res is the ResNet module, SA denotes the Spatial Attention Module using a self-attention mechanism, SAM denotes the Spatial Attention Module using a linear-attention mechanism, CAM denotes the Channel Attention Module, k is the dimensionality of E and F in the model, E and F share the same parameter, i.e., E = F, Basic denotes the basic structure, and Hourglass is stacked hourglass network. Experiments were conducted to evaluate the performance of each key module with evaluation metrics which are >1, >2, and >3pixel error, End Point Error EPE, and runtime.

Table 2. Ablation experiments of attention mechanism on Scene Flow.

Module	>1 px (%)	>2 px (%)	>3 px (%)	EPE (px)	Runtime (s)
Res_Base	12.78	8.11	6.41	1.65	0.12
Res_CAM_Base	11.12	7.02	5.36	1.21	0.14
Res_SA_Base	10.24	6.48	4.91	1.03	0.24
Res_SAM_k128_Base	10.47	6.65	5.04	1.10	0.16
Res_SAM_k256_Base	10.38	6.58	4.98	1.07	0.17
Res_SAM_k512_Base	10.29	6.52	4.93	1.05	0.18
Res_CAM_SAM_k512_Base	9.26	5.56	3.95	0.95	0.19
Res_CAM_SAM_k512_Hourglass	7.22	3.71	2.31	0.82	0.25

As shown in Table 2, the EPE of Res_Base is 1.65, and with the addition of CAM and SAM, the EPE becomes 1.21 and 1.1, respectively, resulting in a significant reduction in error rates. The error rate of disparity values shows that adding AM can significantly improve the accuracy of disparity prediction, thereby achieving the goal of improving the accuracy of dense mapping in D-SLAM systems. When the value of k becomes larger, the EPE of Res_SAM_kx_Base gradually approaches that of Res_SA_Base, and when k = 512, the EPEs of both are almost equal, while the inference time of the former changes little which is significantly faster than that of Res_SA_Base. Thereby, it is proved that the inference speed of linear-attention is significantly faster than that of self-attention when their error rates are close. Compared to Res_CAM_SAM_k512_Base, Res_CAM_SAM_k512_Hourglass has a significant advantage which reduces the error rate for the whole network > 3 px from 3.95 to 2.31 and EPE from 0.95 to 0.82.

(2) Comparative experiments on Forest

Several mainstream methods are compared in the Forest dataset, and the performance of each method was evaluated by three evaluation metrics, the proportion of pixels with prediction errors in all regions of the first frame image (D1-all), EPE and time. The test server was 3090GPU, and the image resolution was 1240 × 426.

The results in Table 3 indicate that after fine-tuning on the Forest dateset, LANet exhibits better performance than on the SceneFlow dataset, with an EPE reduction from 0.82 to 0.68 and an accuracy improvement of 20.6%. The D1 all and EPE of LANet are 2.15 and 0.68, respectively, which are better than those of the comparative model. The running speed is 0.35 s, and although it is not the fastest, it is also relatively competitive.

Table 3. Comparative experiments of disparity detection with Forest. The network models for comparison are Matching Cost with a Convolutional Neural Network (MC-CNN), Geometry and Context network (GCNet), Learning deep correspondence through prior and posterior feature constancy (iResNet), Disparity Network (DispNet), a two-stage convolutional neural network (CRL), Exploiting Semantic Information for Disparity Estimation (SegStereo), Edge Stereo network (EdgeStereo), and Pyramid Stereo Matching Network (PSMNet).

Method	Runtime (s)	D1-All (%)	EPE (px)
MC-CNN [33]	67.09	4.08	3.96
GCNet [34]	1.01	3.65	2.79
iResNet [35]	0.20	3.58	2.73
DispNet [31]	0.14	3.08	1.96
CRL [36]	0.55	2.75	1.54
SegStereo [37]	0.68	3.12	2.01
EdgeStereo [38]	0.40	2.81	1.68
PSMNet [39]	0.48	2.61	1.25
LANet	0.35	2.15	0.68

Figure 9 shows the visualization of disparity maps generated by LANet, PSMNet and GCNet with Forest, with the colors representing different disparity values, the farther the distance the smaller the disparity value, and the black color indicating the distant points, whose disparity values are so small that they can be ignored.

The rectangular box regions where the matching error of each method is large are usually found in locations containing fine structures such as branches, trunks, and leaf edges, as well as weakly textured regions and occluded regions. In column A, there are significant differences in the predictions of each model at the border between the pink trees and the crimson sky, and PSMNet and GCNet can preserve the main contours of the edges while the predictions are inaccurate at the fine structures; however, the LANet can better preserve the fine features of the edges while the predictions are closer to the true value. In column B, for the prediction of the red trunk, PSMNet and GCNet show missing trunk pixels, and for the prediction of the pink car's rear glass, LANet shows few color deviations, PSMNet shows more color deviations, and GCNet shows more errors in color. In column C, for the red trunk prediction, LANet shows pixel discontinuity and few missing pixels, PSMNet and GCNet show larger missing pixels or even missing trunks, and for the prediction of the purple leaf, LANet is able to retain the edge features better, PSMNet misses some fine edge structure features, and GCNet has too many edge predictions and a mismatched pieces.

The attention mechanism integrates local and global information, seeking correlations between pixels at different locations to obtain richer feature representations at the pixel level, which are beneficial for obtaining stronger semantic feature responses and improving feature recognition. Therefore, it can predict more reliable disparity maps in ill-posed regions such as those with weak texture, poor lighting, and occlusion. After testing, LANet has shown better performance than the comparative model in terms of accuracy and visualization, and it is also more competitive in terms of runtime.

Figure 9. The visualization of disparity maps with Forest. The yellow or green boxes are the regions with significant disparity contrast generated by various methods.

LANet is embedded into the D-SLAM system, and is lightweighted in order to ensure the real-time performance of the system, and the stacked hourglass structure used to optimize the network is cut off to improve the running speed of the system. For the purpose of making a balance between accuracy and speed, the Res_CAM_SAM_k512_Base combination modules are selected, with an EPE of 0.95 and a running time of 0.19 s, which fully meets the performance needs of the dense mapping thread of D-SLAM.

5.2. Experiment on the Location Accuracy of Visual Odometry

In this section, the performance of D-SLAM will be evaluated for several sequences on two popular datasets. In order to demonstrate the robustness of the proposed system, the estimation of camera generated trajectories and maps was compared with the Ground Truth (GT). In addition, the results are also compared with some advanced SLAM systems by using the results published by the original author and standard evaluation metrics in the literature. D-SLAM experiments are all conducted on a Dell G3 3590 portable computer, with an Intel Core i7-9750H CPU, 2.6GHz, 16GB memory, and only the CPU was used.

5.2.1. EuRoC Dataset

The EuRoC dataset [40] contains 11 stereo sequences recorded from a micro aerial vehicle (MAV) flying around two different rooms and a large industrial environment. The baseline of the binocular sensor is 11 cm, providing images at 20 Hz. The sequences are classified into three levels: easy, medium, and difficult based on the speed, illumination, and scene texture of the MAV.

(1) Estimated trajectory maps

Figure 10 is the estimated trajectory for nine sequences from EuRoC; by comparison with the GT, the D-SLAM system shows better trajectory accuracy on these sequences.

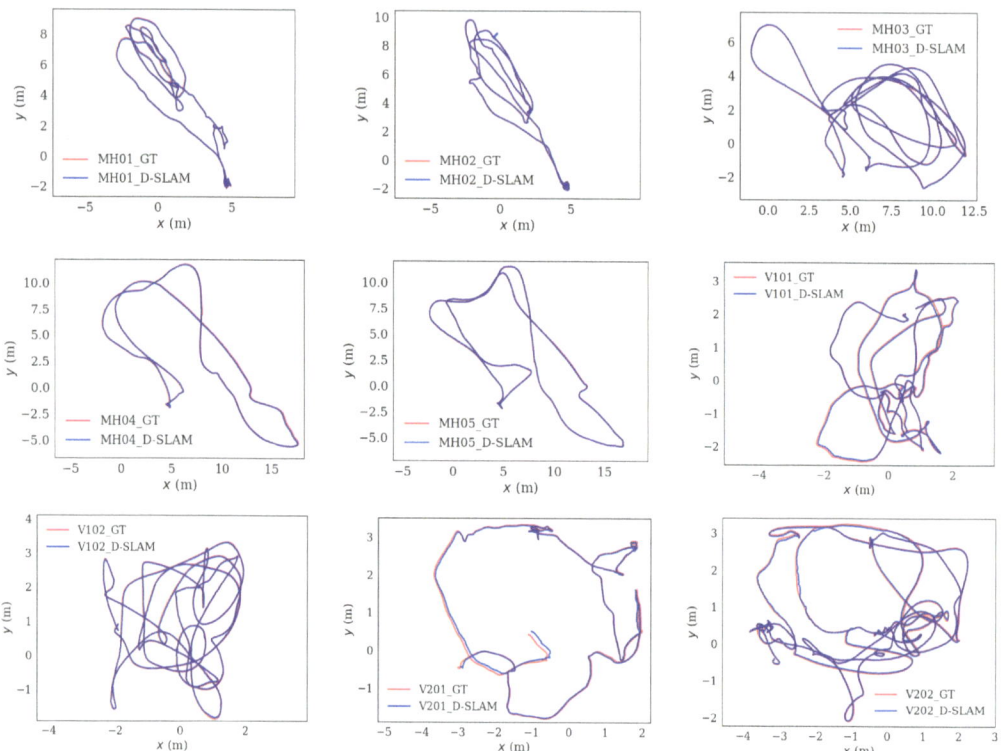

Figure 10. Estimated trajectory (dark blue) and GT (red) for 9 sequences on EuRoC.

(2) RMS ATE comparison

As is usual in the field, the accuracy is measured by RMS ATE [41]. The SE (3) transform is used to align the estimated trajectory with the GT. The result of D-SLAM is the average of five executions, while the other results are reported by the authors of each system and compared with GT for all frames in the trajectory.

As shown in Table 4 ORB-SLAM2 and VINS-Fusion all get lost in some parts of V2_03_difficult sequence due to severe motion blur. Even BASALT, a stereo vision-inertial odometry system, was not able to complete tracking on this sequence due to the loss of some frames by one of the cameras. However, due to the use of the binocular image spatial clue compensation strategy, D-SLAM could utilize the right image to compensate for the lost field of view of the camera to a certain extent when the above situations occured, and it successfully tracked and achieved an error of 0.468 on the V2_03_ difficult sequence. SVO is a semi direct visual odometry that can run in weak texture and high frequency texture environments. However, the pose estimation has significant cumulative error due to its lack of loop detection and relocation. The system is very dependent on the accuracy of pose estimation, making it difficult to relocate once tracking fails. It performs well on easy sequences, while the tracking accuracy decreases rapidly on medium and difficult sequences, with a larger RMS ATE.

Table 4. RMS ATE comparison in the EuRoC dataset (RMS ATE in m, scale error in %). The network models for comparison are ORB-SLAM2, a general optimization-based framework for local odometry estimation with multiple sensors (VINS-Fusion), Semidirect visual odometry for monocular and multicamera systems (SVO2), Visual-Inertial Mapping with Non-Linear Factor Recovery (BASALT), and D-SLAM.

Sequence	ORB-SLAM2 [18]	VINS-Fusion [42]	SVO2 [43]	BASALT [44]	D-SLAM
MH_01_easy	0.035	0.540	0.040	0.070	0.032
MH_02_easy	0.018	0.460	0.070	0.060	0.018
MH_03_medium	0.028	0.330	0.270	0.070	0.026
MH_04_difficult	0.119	0.780	0.170	0.130	0.095
MH_05_difficult	0.060	0.500	0.120	0.110	0.055
V1_01_easy	0.035	0.550	0.040	0.040	0.032
V1_02_medium	0.020	0.230	0.040	0.050	0.022
V1_03_difficult	0.048	–	0.070	0.100	0.041
V2_01_easy	0.037	0.230	0.050	0.040	0.035
V2_02_medium	0.035	0.200	0.090	0.050	0.030
V2_03_difficult	–	–	0.790	–	0.468

5.2.2. KITTI Dataset

The KITTI [45] dataset has become the standard for evaluating visual SLAM, which contains stereo images recorded from a car in urban and highway environment. The baseline of the binocular sensor is 54 cm and works at 10Hz with a rectified resolution of 1240 × 376 pixels. The D-SLAM system was tested on 11 sequences of the KITTI dataset, and the results are as follows.

(1) Error graph for the 05 sequence of KITTI

The trajectory error is calculated by Evaluation Visual Odometry (EVO) tool. Absolute Pose Error (APE) calculates the difference between the estimated value of the SLAM system and the ground truth of the camera's pose, which is suitable for evaluating the accuracy of the algorithm and the global consistency of the camera trajectory. Relative Pose Error (RPE) calculates the difference between the estimated pose change and the true pose change at the same two timestamps, which is suitable for evaluating the drift of the system and the local accuracy of the camera trajectory.

Because the KITTI dataset contains large outdoor urban and highway scenes, its overall APE error is much higher than that of the indoor dataset EuRoC, and the larger errors occur near the turns or where no loop closing occurs at the edge of the trajectory, as shown in Figure 11. In the translation direction of this sequence, the mean of APE is 1.310438 m, the median is 1.184211 m, the rmse is 1.469650 m, and the std is 0.665299. Compared with APE, RPE is smaller, with a mean of 0.015108 m, a mean of 0.013262 m, a rmse of 0.018214 m, and a std of 0.010172 m.

(2) Comparison of projection trajectories for the 08 sequence on KITTI

Figure 12 shows the comparison between the projection trajectories of D-SLAM, Large-scale direct SLAM with stereo cameras (LSD-SLAM) [46], and a stereo SLAM system through the combination of Points and Line segments (PL-SLAM) [47] on the KITTI08 sequence and the GT, from which it can be intuitively observed that the trajectory of D-SLAM is closer to the GT compared to the other two methods, while the trajectory drift of PL-SLAM is relatively large. Unlike D-SLAM, the inferior performance of PL-SLAM is mainly explained by the fact that it does not perform LBA in every frame, so the drift along the trajectory is not corrected, especially in sequences like 08 without a loop closing, resulting in a relatively large final drift of the trajectory. In addition, the translation and rotation deviations on the y-axis are relatively large for the various methods.

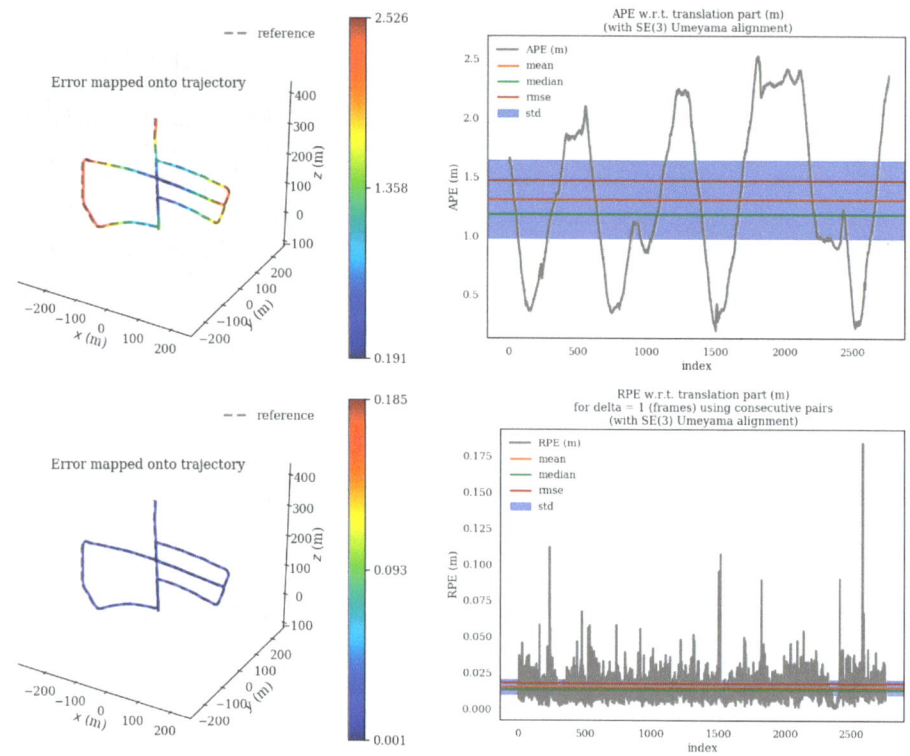

Figure 11. Error graph for the 05 sequence of KITTI.

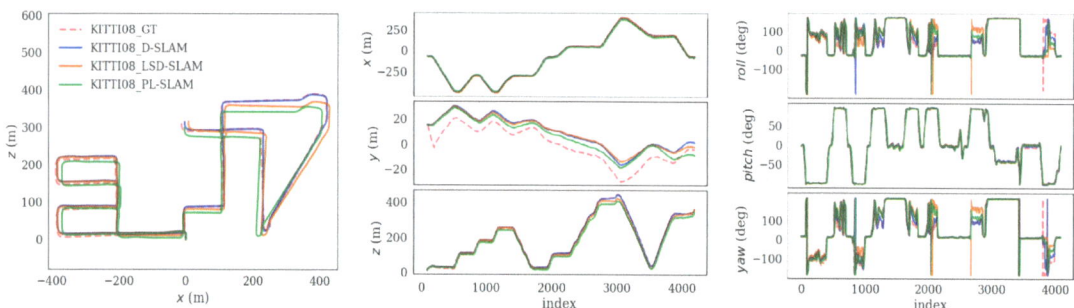

Figure 12. Comparison of projection trajectories for the 08 sequence of KITTI.

(3) Comparison of relative RMSE of KITTI

The metrics of the average relative translation error (t_{rel}) and rotation error (R_{rel}) are used to estimated the relative RMSE [41]. The translation error t_{rel} is expressed in %, the rotation error R_{rel} is also expressed deg/100 m relative to the translation, and the dash indicates that the experiment failed. The comparison of relative RMSE of KITTI is shown in Table 5.

Table 5. Relative RMSE of KITTI.

Sequence	ORB-SLAM2		LSD-SLAM		PL-SLAM		D-SLAM	
	t_{rel}	R_{rel}	t_{rel}	R_{rel}	t_{rel}	R_{rel}	t_{rel}	R_{rel}
00	0.70	0.25	0.63	0.26	2.36	0.89	0.67	0.24
01	1.39	0.21	2.36	0.36	5.80	2.32	1.05	0.19
02	0.76	0.23	0.79	0.23	2.35	0.91	0.68	0.21
03	0.71	0.18	1.01	0.28	3.74	1.54	0.65	0.16
04	0.48	0.13	0.38	0.31	2.21	0.30	0.45	0.13
05	0.40	0.16	0.64	0.18	1.74	0.88	0.38	0.15
06	0.51	0.15	0.71	0.18	3.51	2.72	0.45	0.14
07	0.50	0.28	0.56	0.29	1.83	1.03	0.44	0.25
08	1.05	0.32	1.11	0.31	2.18	1.15	0.98	0.30
09	0.87	0.27	1.14	0.25	1.68	0.92	0.75	0.24
10	0.60	0.27	0.72	0.33	1.21	0.99	0.55	0.23
Avg.	0.72	0.22	0.91	0.27	2.60	1.24	0.64	0.20

The two sequences with large errors in Table 5 are 01 and 08, neither of which show a loop closing. The 01 sequence is the only highway sequence in the KITTI dataset, in which few near points can be tracked due to the high speed and low frame rate, so it is difficult to estimate the translation, and the t_{rel} of the various methods are large. However, there are many distant points that can be tracked for long periods of time, and therefore, the rotation can be accurately estimated. ORB-SLAM2 is able to achieve a better error with an R_{rel} value of 0.21 deg/100 m, while D-SLAM is even smaller with a value of 0.19 deg/100 m. Without any loop closing in the 08 sequence, PL-SLAM is unable to correct the drift of the trajectory in time for the absence of Local BA, whereas ORB-SLAM2 and Stereo LSD-SLAM, although they perform Local BA for each frame, cannot perform Full BA due to the absence of loop closing in this sequence, which also results in the global error not being corrected, leading to a large cumulative error. The D-SLAM system is relatively accurate in the pose estimation of each previous frame, and even without loop closing correction, the drift will not be too severe. The D-SLAM system achieved an average t_{rel} of 0.64m and an average R_{rel} of 0.20, which is more accurate compared to some mainstream stereo systems and has significant advantages in most cases.

5.3. System Real-Time Evaluation

In order to evaluate the real-time performance of the proposed system, the runtimes of different resolutions of the three datasets are presented in Table 6. Because each of these sequences contains only one loop closing, the BA and Loop shown in the table are measurements where the associated task is executed only once.

Because the loop closing of the Psv_02 sequence of Forest contains more keyframes, the covisibility graph is constructed more densely, resulting in higher costs for loop fusion, as well as the higher cost of pose map optimization and Full BA tasks. In addition, the higher the density of covisibility graph, the more keyframes and points the local map contains, resulting in higher costs for local map tracking and LocalBA.

The two threads of loop closing and Full BA in Table 6 consume more time, especially Full BA, for which the D-SLAM system takes 1.42 s. However, these two operations are executed in separate threads, so they do not affect the real-time performance of the other components of the system. The real-time performance of the SLAM system is mainly determined by the speed at which the tracking thread processes each frame of the RGB image, while local mapping, dense mapping, and loop closing threads only process key frames without the need for real-time operation.

The running time of the system on the three sequences is 139.87 s, 124.9 s and 162.97 s, respectively. According to the frame rate and time of tracking threads, the D-SLAM system is able to run at 30 ordinary frames and 3 keyframes per second, which fully meets the real-time requirements of the SLAM system for forest environment location and dense map construction.

Table 6. Running time of each thread in miliseconds (ms). Where FPS represents Frames Per Second, Essential Graph Opt. represents Essential Graph Optimization, KFs represents KeyFrames, and MPs represents Map Points.

Part	Detail	EuRoC	KITTI	Forest
Settings	Sequence	V2_02	07	Psv_02
	Sensor	Stereo	Stereo	Stereo
	Resolution	752 × 480	1226 × 370	672 × 376
	Camera FPS	20Hz	10Hz	30Hz
	ORB Features	1200	2000	1200
Tracking	Stereo Rectification	2.95	–	–
	ORB Extraction	11.52	21.85	9.69
	Stereo Matching	10.54	13.64	8.81
	Pose Prediction	2.15	2.25	2.05
	Local Map Tracking	9.25	4.21	8.86
	Keframe Selection	5.65	6.12	5.29
	Total	42.06	48.07	34.70
Local Mapping	Keyframe Insertion	8.56	10.03	8.24
	Map Point Culling	0.24	0.38	0.22
	Map Point Creation	35.36	42.26	33.05
	Local BA	135.02	66.35	180.14
	Keyframe Culling	3.61	0.89	2.14
	Total	182.79	119.91	223.79
Dense Mapping	Pcd generation	123.85	138.54	95.37
	Pcd registration	18.87	22.36	15.29
	Pcd fusion	26.59	33.84	23.68
	Pcd filter	32.01	43.35	28.87
	Total	201.32	238.09	163.21
Loop Closing	Database Query	3.25	3.59	3.07
	SE3 Estimation	0.58	0.87	0.51
	Loop Fusion	20.23	79.86	298.25
	Essential Graph Opt.	71.36	175.97	268.95
	Total	95.42	260.29	570.78
Full BA	Full BA	345.71	1120.51	1420.36
	Map Update	3.09	9.65	6.58
	Total	348.8	1130.16	1426.94
Map Size	KFs	249	241	354
	MPs	14,027	26,074	17,325
	Run time	139.87s	124.9s	162.97s

5.4. Dense Mapping

This section will evaluate the dense mapping performance of D-SLAM on two challenging datasets, KITTI and Forest. Six visualized images generated during the dense mapping process are displayed on each dataset. In order to analyze the dense point cloud more comprehensively and intuitively, local detail point cloud images obtained from different perspectives are displayed also. D-SLAM supports two operating modes: online real-time dense mapping and bag video dense mapping. In order to facilitate parameter adjustment, the bag video mode was used for testing in this study.

5.4.1. Dense Mapping on KITTI Dataset

Figure 13 shows the dense map construction of the 01 sequence of the KITTI dataset. This sequence is a real-time image of a highway. Due to its high speed and low frame rate, the camera in this scene has a large translation, little rotation, and no loop closing, making it challenging. It can be observed from the Figure 13d that the projection trajectory has better accuracy in the straight section of the highway, while there is some slight drift near the turning at the end, which is due to the fact that there is no loop closing for Full BA, resulting in an increase in the cumulative error of the trajectory and ultimately an increase in drift. In Figure 13e, red points represent the covisibility observation points of the covisibility graph keyframes, i.e., reference map points, while black points represent all map points generated by keyframes. Figure 13f is the overall dense point cloud map generated after the previous steps of processing, and its local details are shown in Figure 14. It shows the local dense point cloud maps from different views of the KITTI01, in which the details of the highway can be clearly seen, including the dotted lines, crosswalks, tree shadows, and green grass along the highway. The dense point cloud maps generated from the KITTI dataset are clearer due to the fact that the KITTI dataset is a high-resolution image dataset, coupled with the long baseline of the binocular sensors, and the corrected stereo images.

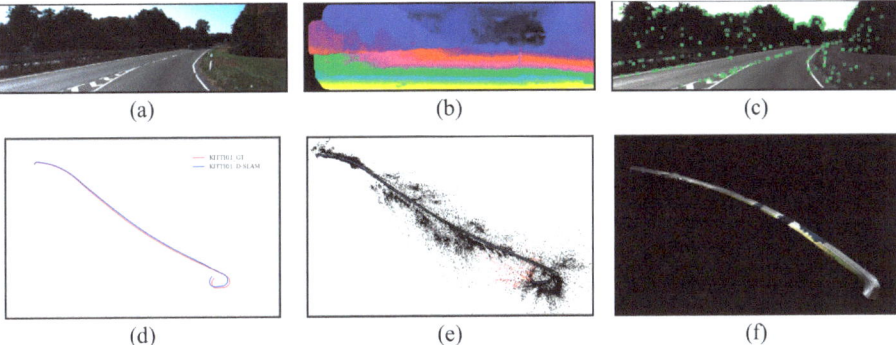

Figure 13. Dense mapping Effect for 01-sequence of KITTI. (**a**) a left RGB image, (**b**) a visual disparity map, (**c**) afeature point tracking map, (**d**) anestimated trajectory map, (**e**) asparse point cloud map, and (**f**) a dense point cloud map.

Figure 14. Local dense point cloud maps from different views for the KITTI01 sequence.

5.4.2. Dense Mapping with Forest Dataset

Forest is a large forest scene dataset with low texture images, the trunk features are very inconspicuous, and the number of feature points is not large enough, in order to have enough near point feature points to ensure the tracking accuracy and the effect of dense mapping, it is necessary to insert as many keyframes as possible, and at the same time to avoid redundancy, based on which the keyframe selection strategy has been designed previously for the characteristics of the forest scene. Figure 15 shows the dense mapping process with the Forest dataset.

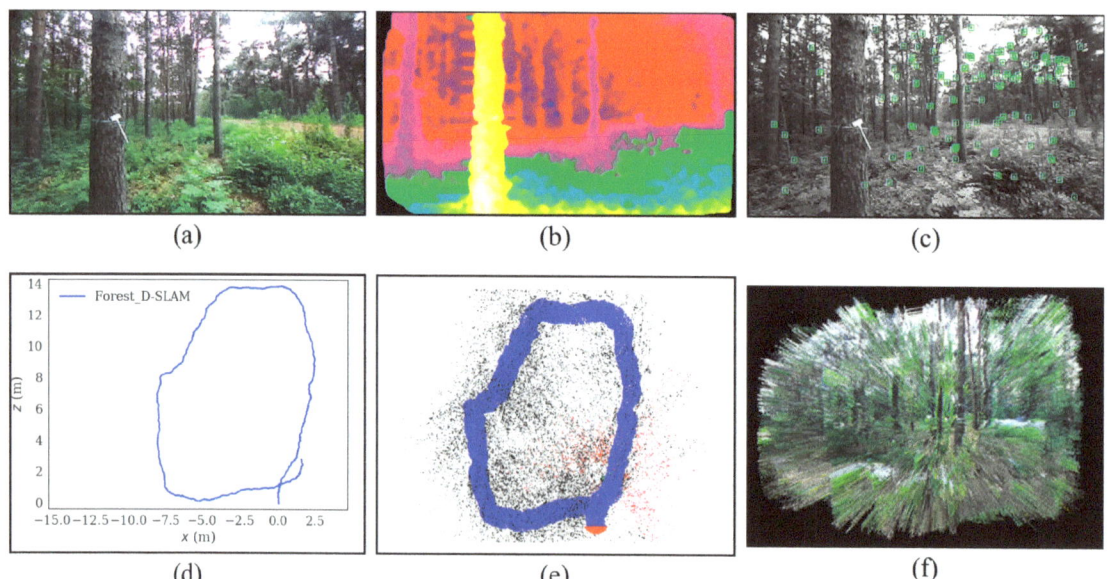

Figure 15. Dense mapping process on Forest, where (**a**) is the left RGB image, (**b**) is the visual disparity map, (**c**) is the feature point tracking map, (**d**) is the estimated trajectory map, which has a loop closing and the localization of the front end and the accuracy of the back end mapping are improved after loop closing correction, (**e**) is the sparse point cloud map, in which the blue boxes represent the keyframes, the green box represents the current frame, the red box represents the start frame, the red points represent the reference map points, and the black points represent all map points generated by keyframes. and (**f**) is the overall effect of the dense point cloud map.

Because Figure 15f shows the overall effect of the 3D point cloud of the forest scene from one perspective, the details from many perspectives are not visible, therefore, Figure 16 shows the details from different perspectives after being rotated, from which the structure of the forest scene can be clearly reproduced, including the density, poses and spatial position of the forest trees; the height, thickness, outline, color and texture of the tree trunks; the color and density of the leaves; the canopy; and the ground surface, which truly reflect the sample structure of the forest scene and provides an important basis for forestry exploration.

Figure 16. Localized dense point cloud at different angles with Forest.

6. Discussion

All tests of the D-SLAM system were run on a Dell G3 3590 portable computer, which was equipped with a ZED2 binocular camera capable of real-time localization and dense mapping in the forest scene, supportting both image dataset operation and real-time forest scene operation.

In terms of localization accuracy, D-SLAM can estimate the true scale of maps and trajectories without drift and achieve a high accuracy with an RMS ATE of 1.8 cm on EuRoC dataset (Table 4), outperforming international mainstream systems VINS Fusion, SVO2, and BASALT. Especially on the two challenging sequences V1_03_difficult and V2_03_difficult, ORB-SLAM2, VINS Fusion, and BASALT all failed to track, while the D-SLAM system adopted a binocular image spatial clue compensation strategy, which can use the right image to compensate for the lost field of view of the camera to a certain extent when the camera rotation angle is too large, so there was no tracking loss on difficult sequences. In addition, on the 11 sequences of KITTI, the D-SLAM system achieved relative RMSE with two average values of 0.64 and 0.2 for t_{rel} and R_{rel}, respectively (Table 5), which is superior to international mainstream SLAM systems ORB-SLAM2, LSD-SLAM, and PL-SLAM. Accordingly, D-SLAM is robust enough to be of great advantage in most cases. The error difference between various methods on the EuRoC and KITTI datasets is significant, which is directly related to the images in the dataset. EuRoC contains indoor small scene images with a small field of view distance, resulting in smaller errors, while KITTI is an outdoor highway large scene dataset with a larger field of view, fewer near points, and more far points, resulting in larger errors, even reaching tens of centimeters to several meters. In addition, the resolution, acquisition frequency, illumination, texture, and other factors of the image also have a significant impact on the error.

In terms of the real-time performance of the system, the dense mapping thread only processes keyframes, which does not affect the real-time performance of the system. From the frame rate and runtime of the tracking thread, it can be inferred that the D-SLAM system can run at a speed of 30 ordinary frames and 3 keyframes per second, fully meeting the real-time requirements of SLAM system forest ecological environment localization and dense mapping.

In terms of dense mapping, the construction of real-time dense maps in three-dimensional space generally uses RGBD-SLAM or LiDAR-SLAM, which obtain depth information of the scene through depth sensors or LiDAR sensors. However, they are expensive and not conducive to the popularization and application in the industry. Moreover, depth sensors cannot be used outdoors, and laser sensors can only build sparse maps. Visual sensors can overcome these shortcomings and be applied to outdoor for dense mapping. Visual SLAM is generally applied to more regular outdoor scenes such as buildings, streets, roads, and parks. However, its application in complex forest environments has been rarely reported internationally. Therefore, dense mapping poses significant challenges in forest scenes with low texture, uneven lighting, and severe occlusion. Despite many disadvantages, the experimental results (Figure 16) show that it is possible to observe the structure of the forest ecological scenes, such as the density, pose, and spatial position of the tree, and the height, thickness, outline, color, and texture of the tree trunks, which meets the general needs of forest surveys and provides an important basis for forestry ecological exploration and forestry management. This work has innovation in both technology and application, providing important reference value for related research on forest digital twins.

In terms of image texture, KITTI is a dataset of urban and highway with highly textured image sequences. Its binocular images are high-definition images, with a baseline of 54 cm for binocular sensors, and the binocular images are rectified images with a resolution of 1240 × 376 pixels, therefore the generated disparity map has high accuracy and the dense point cloud map constructed is relatively clear. While the Forest dataset is collected by the ZED2 binocular camera with a baseline length of only 20 cm, which limits the accuracy of its disparity map. There are three types of image resolutions: HD1080:1920 × 1080, HD720:1280 × 720, and VGA: 672 × 376. High resolution images have more pixels and clearer texture features, which can improve the accuracy of localization and map construction; however, at the same time, it will increase the processing time of the front-end VO and the construction time of dense point cloud maps, which will slow down the overall running speed of the system. Moreover, high-resolution images have high performance requirements for hardware platforms such as computing speed and storage space, which are difficult to meet for general consumer level platform configurations. By balancing speed and accuracy, the VGA is chosen with the smallest resolution, which is equivalent to one-half of the KITTI resolution. Forest is a sequence of forest ecological images with low texture and large scenes, due to the low image resolution with VGA, the trunk features are not obvious, the similarity of the leaves and bushes is larger, the trees are severely obstructed, the light is unstable, the forest ecological environment scene is larger, and there are less nearpoints and more farpoints, resulting in the effect of the dense point cloud maps generated from Forest being not as clear as those of the KITTI dataset. However, it still meets the general needs of forest ecological surveys and forestry management.

This research utilizes visual images from binocular cameras to construct a three-dimensional forest map. However, visual sensors are generally affected by light, and the image quality collected under conditions of high exposure or low light is poor, which affects the effect of dense mapping. The system is almost unable to work at night, rainy days, and on snowy days. In addition, the camera's movement speed should not be too fast to prevent the system's processing speed from falling behind, and the camera's angle should not exceed 180 to avoid system tracking loss. When the angle of camera rotation $\omega \geq \frac{\pi}{2}$, the camera almost loses the perspective, at which time the feature points cannot be matched on the temporal clue causing the camera tracking to fail. The spatial clue compensation strategy is adopted: the previous frame of the right image of the spatial clue is used as the clue connection, and is inserted to compensate for the lost field of view of the temporal clue, continuing the tracking. When the camera rotation angle exceeds 90 degrees, the larger the angle, the greater the challenge. Due to uneven lighting, severe tree occlusion, large field of view, and fewer features in complex forest scenes, the implementation of the system poses significant challenges. The implementation of the system in real-world scenarios should involve drones equipped with binocular cameras and software and hardware platforms, which require lightweight processing. The performance of the platform also affects the system's running speed and dense mapping accuracy. If a high-performance platform can be configured and GPU acceleration can be used, it will further improve the system's running speed and dense mapping accuracy.

Because this research mainly focuses on forest ecological scenes below the canopy, the UAV flies under the canopy of the trees and collects data mainly on the trunks, branches, leaves, bushes, and forest grasses under the canopy, without including canopy information. In future research, satellite remote sensing technology can be combined to collect canopy information to construct broader and more comprehensive 3D forest ecological models, which provide powerful basis for fine surveying of forest resources, forest management, and forest rescue through visualized digital twins of forest environments.

7. Conclusions

This study explores the use of low-cost binocular cameras for the accurate 6-DoF pose estimation of UAVs in forest ecological spatial environment in a D-SLAM system, with a lightweight localization mode that uses only Tracking threads to track unmodeled areas to achieve zero drift. A dense mapping thread is added to construct dense point cloud maps of the forest ecological spatial environment. The amount of relative motion between frames and data association are used as constraints to filter keyframes, and a binocular image spatial clue compensation strategy is adopted to improve the robustness of tracking in adverse conditions such as large rotation, fast motion, and insufficient texture. Compared with the direct methods, the proposed approach can be used for wide-baseline feature matching, which is more suitable for 3D reconstruction scenes requiring high depth accuracy. The D-SLAM system runs at a speed of 30 ordinary frames and 3 keyframes per second, achieving location accuracy of several centimeters with the EuRoC dataset and a local t_{rel} average of 0.64m and R_{rel} average of 0.20 with the KITTI dataset, which outperform some mainstream sytems in terms of location accuracy and robustness, and have significant advantages in most cases. With a consumer-level computing platform, the system is able to work in real-time on the CPU, and the dense maps constructed can clearly reproduce the structure of the forest ecological interior scenes, meeting the requirements of the UAV's localization and mapping in terms of accuracy and speed. Moreover, the system is more reliable in the case of a signal blockage and can be a powerful complement and alternative solution to the current expensive commercial GNSS/Inertial Navigation System (INS) navigation systems. However, the system is greatly affected by light, and the location and dense mapping results are poor under conditions of high exposure or low light. In addition, the system will lose tracking when the camera moves too fast and the rotation angle is too large. In the future work, various sensors such as Inertial Measurement Unit (IMU) and LiDAR can be integrated to compensate for the limitations and shortcomings of the system. Neural networks can also be used to replace some or all modules of the system, solving the problem of limited system applications to a certain extent. In addition, it is possible to combine high-altitude remote sensing to capture broader forest images and construct a more comprehensive and extensive three-dimensional map of forest ecology.

Author Contributions: Conceptualization, L.L. and Y.L. (Yaqiu Liu); methodology, L.L. and Y.L. (Yaqiu Liu); software, L.L.; validation, L.L., Y.L. (Yunlei Lv) and X.L.; formal analysis, Y.L. (Yunlei Lv); investigation, X.L.; resources, Y.L. (Yaqiu Liu); data curation, X.L.; writing—original draft preparation, L.L.; writing—review and editing, L.L. and Y.L. (Yaqiu Liu); visualization, X.L.; supervision, Y.L. (Yunlei Lv); project administration, L.L.; funding acquisition, Y.L. (Yaqiu Liu). All authors have read and agreed to the published version of the manuscript.

Funding: This research was supported by the Fundamental Research Funds for the Central Universities (Grant No. 2572023CT15-03) and the National Natural Science Foundation of China (Grant No. 32271865).

Data Availability Statement: The data presented in this study are available on request from the corresponding author.

Conflicts of Interest: The authors declare no conflict of interest.

References

1. Solares-Canal, A.; Alonso, L.; Picos, J.; Armesto, J. Automatic tree detection and attribute characterization using portable terrestrial lidar. *Trees* **2023**, *37*, 963–979. [CrossRef]
2. Gharineiat, Z.; Tarsha Kurdi, F.; Campbell, G. Review of automatic processing of topography and surface feature identification LiDAR data using machine learning techniques. *Remote Sens.* **2022**, *14*, 4685. [CrossRef]
3. Rijal, A.; Cristan, R.; Gallagher, T.; Narine, L.L.; Parajuli, M. Evaluating the feasibility and potential of unmanned aerial vehicles to monitor implementation of forestry best management practices in the coastal plain of the southeastern United States. *For. Ecol. Manag.* **2023**, *545*, 121280. [CrossRef]
4. Smith, R.C.; Cheeseman, P. On the Representation and Estimation of Spatial Uncertainty. *Int. J. Robot. Res.* **1986**, *5*, 56–68. [CrossRef]
5. Cadena, C.; Carlone, L.; Carrillo, H.; Latif, Y.; Scaramuzza, D.; Neira, J.; Reid, I.; Leonard, J.J. Past, Present, and Future of SLAM. *IEEE Trans. Robot.* **2016**, *32*, 1309–1332. [CrossRef]
6. Kazerouni, I.A.; Fitzgerald, L.; Dooly, G.; Toal, D. A survey of state-of-the-art on visual SLAM. *Expert Syst. Appl.* **2022**, *205*, 117734. [CrossRef]
7. Servières, M.; Renaudin, V.; Dupuis, A.; Antigny, N. Visual and Visual-Inertial SLAM: State of the Art, Classification, and Experimental Benchmarking. *J. Sens.* **2021**, *2021*, 2054828. [CrossRef]
8. Zhang, J.; Singh, S. Loam: Lidar odometry and mapping in real-time. In Proceedings of the Robotics: Science and Systems Conference, Berkeley, CA, USA, 14–16 July 2014. [CrossRef]
9. Khan, M.U.; Zaidi, S.A.A.; Ishtiaq, A.; Bukhari, S.U.R.; Farman, A. A Comparative Survey of LiDAR-SLAM and LiDAR based Sensor Technologies. In Proceedings of the Mohammad Ali Jinnah University Conference on Informatics and Computing, 2021 (MAJICC21), Karachi, Pakistan, 15–17 July 2021. [CrossRef]
10. Xu, M.; Lin, S.; Wang, J.; Chen, Z. A LiDAR SLAM System with Geometry Feature Group-Based Stable Feature Selection and Three-Stage Loop Closure Optimization. *IEEE Trans. Instrum. Meas.* **2023**, *72*, 8504810. [CrossRef]
11. Newcombe, R.A.; Lovegrove, S.J.; Davison, A.J. DTAM: Dense Tracking and Mapping in Real-Time. In Proceedings of the IEEE International Conference on Computer Vision, ICCV 2011, Barcelona, Spain, 6–13 November 2011. [CrossRef]
12. Engel, J.; Sturm, J.; Cremers, D. LSD-SLAM: Large-scale direct monocular SLAM. In *European Conference on Computer Vision*; Springer: Berlin/Heidelberg, Germany, 2014; pp. 834–849. [CrossRef]
13. Engel, J.; Koltun, V.; Cremers, D. Direct sparse odometry. *IEEE Trans. Pattern Anal. Mach. Intell.* **2017**, *40*, 611–625. [CrossRef]
14. Wang, R.; Schworer, M.; Cremers, D. Stereo DSO: Large-scale direct sparse visual odometry with stereo cameras. In Proceedings of the IEEE International Conference on Computer Vision, Venice, Italy, 22–29 October 2017; pp. 3903–3911. [CrossRef]
15. Davison, A.J.; Reid, I.D.; Molton, N.D.; Stasse, O. MonoSLAM: Real-time single camera SLAM. *IEEE Trans. Pattern Anal. Mach. Intell.* **2007**, *29*, 1052–1067. [CrossRef]
16. Klein, G.; Murray, D. Parallel tracking and mapping for small AR workspaces. In Proceedings of the 7th IEEE and ACM International Symposium on Mixed and Augmented Reality, ISMAR 2008, Cambridge, UK, 15-18th September 2008; 20 September 2008. [CrossRef]
17. Mur-Artal, R.; Montiel, J.M.M.; Tardós, J.D. ORB-SLAM: A versatile and Accurate Monocular SLAM System. *IEEE Trans. Robot.* **2015**, *31*, 1147–1163. [CrossRef]
18. Mur-Artal, R.; Tardós, J.D. ORB-SLAM2: An Open-Source SLAM System for Monocular, Stereo, and RGB-D Cameras. *IEEE Trans. Robot.* **2017**, *33*, 1255–1262. [CrossRef]
19. Izadi, S.; Kim, D.; Hilliges, O.; Molyneaux, D.; Newcombe, R.; Kohli, P.; Davidson, P. Kinectfusion: Real-time 3D reconstruction and interaction using a moving depth camera. In Proceedings of the 24th Annual ACM symposium on User Interface Software and Technology, Santa Barbara, CA, USA, 16–19 October 2011; pp. 559–568. [CrossRef]
20. Dai, A.; Ritchie, D.; Bokeloh, M.; Reed, S.E.; Sturm, J.; Nießner, M. BundleFusion: Real-time globally consistent 3D reconstruction using on-the-fly surface reintegration. *ACM Trans. Graph.* **2017**, *36*, 1. [CrossRef]
21. Zhang, J.; Sui, W.; Wang, X.; Meng, W.; Zhu, H.; Zhang, Q. Deep Online Correction for Monocular Visual Odometry. In Proceedings of the 2021 IEEE International Conference on Robotics and Automation (ICRA), Xi'an, China, 30 May–5 June 2021; pp. 14396–14402. [CrossRef]
22. Li, S.; Wang, X.; Cao, Y.; Xue, F.; Yan, Z.; Zha, H. Self-supervised deep visual odometry with online adaptation. In Proceedings of the IEEE/CVF Conference on Computer Vision and Pattern Recognition ((CVPR), Seattle, WA, USA, 13–19 June; 2020; pp. 6339–6348.
23. Li, S.; Wu, X.; Cao, Y.; Zha, H. Generalizing to the Open World: Deep Visual Odometry with Online Adaptation. In Proceedings of the 2021 IEEE/CVF Conference on Computer Vision and Pattern Recognition (CVPR), Nashville, TN, USA, 20–25 June 2021; pp. 13179–13188.

24. Zhang, Y.; Wu, Y.; Tong, K.; Chen, H.; Yuan, Y. Review of Visual Simultaneous Localization and Mapping Based on Deep Learning. *Remote Sens.* **2023**, *15*, 2740. [CrossRef]
25. Gao, X.; Zhang, T. *Visual SLAM Fourteen Lectures-From Theory to Practice*; Publishing House of Electronics Industry: Beijing, China, 2019; pp. 128–129, 184–185.
26. Zhang, H. *Robot SLAM Navigation*; China Machine Press: Beijing, China, 2022; pp. 292–293.
27. Liu, L.; Liu, Y.; Lv, Y.; Xing, J. LANet: Stereo matching network based on linear-attention mechanism for depth estimation optimization in 3D reconstruction of inter-forest scene. *Front. Plant Sci.* **2022**, *13*, 978564. [CrossRef]
28. He, K.; Zhang, X.; Ren, S.; Sun, J. Deep Residual Learning for Image Recognition. In Proceedings of the IEEE Conference on Computer Vision and Pattern Recognition (CVPR), Las Vegas, NV, USA, 27–30 June 2016. [CrossRef]
29. Vaswani, A.; Shazeer, N.; Parmar, N.; Uszkoreit, J.; Jones, L.; Gomez, A.N.; Kaiser, Ł.; Polosukhin, I. Attention is all you need. In Proceedings of the Advance in Neural Information Processing Systems (NIPS), Long Beach, CA, USA, 4–9 December 2017; pp. 5998–6008. [CrossRef]
30. Goodfellow, I.; Bengio, Y.; Courville, A. *Deep Learning*; The MIT Press: Cambridge, MA, USA, 2016; pp. 223–225.
31. Mayer, N.; Ilg, E.; Hausser, P.; Fischer, P.; Cremers, D.; Dosovitskiy, A.; Brox, T. A large dataset to train convolutional networks for disparity, optical flow, and scene flow estimation. In Proceedings of the IEEE Conference on Computer Vision and Pattern Recognition, Las Vegas, NV, USA, 27–30 June 2016; pp. 4040–4048. [CrossRef]
32. Diederik, P.K.; Ba, J. Adam: A method for stochastic optimization. In Proceedings of the 3rd International Conference for Learning Representations, ICLR 2015, San Diego, CA, USA, 7–9 May 2015. [CrossRef]
33. Zbontar, J.; LeCun, Y. Computing the stereo matching cost with a convolutional neural network. In Proceedings of the IEEE Conference on Computer Vision and Pattern Recognition (CVPR), Boston, MA, USA, 7–12 June 2015. [CrossRef]
34. Kendall, A.; Martirosyan, H.; Dasgupta, S.; Henry, P.; Kennedy, R.; Bachrach, A.; Bry, A. End-to-end learning of geometry and context for deep stereo regression. In Proceedings of the IEEE International Conference on Computer Vision, Venice, Italy, 22–29 October 2017; pp. 66–75. [CrossRef]
35. Liang, Z.; Feng, Y.; Guo, Y.; Liu, H.; Qiao, L.; Chen, W.; Zhou, L.; Zhang, J. Learning deep correspondence through prior and posterior feature constancy. In Proceedings of the 2018 IEEE/CVF Conference on Computer Vision and Pattern Recognition, Salt Lake City, UT, USA, 18–23 June 2018. [CrossRef]
36. Pang, J.H.; Sun, W.X.; Ren, J.S.; Yang, C.; Yan, Q. Cascade residual learning: A two-stage convolutional neural network for stereo matching. In Proceedings of the IEEE International Conference on Computer Vision Workshops, Venice, Italy, 22–29 October 2017; pp. 887–895. [CrossRef]
37. Yang, G.; Zhao, H.; Shi, J.; Deng, Z.; Jia, J. SegStereo: Exploiting Semantic Information for Disparity Estimation. In *European Conference on Computer Vision*; Springer: Cham, Switzerland, 2018. [CrossRef]
38. Song, X.; Zhao, X.; Fang, L.; Hu, H. Edgestereo: An effective multi-task learning network for stereo matching and edge detection. *Int. J. Comput. Vis.* **2020**, *128*, 910–930. [CrossRef]
39. Chang, J.R.; Chen, Y.S. Pyramid stereo matching network. In Proceedings of the IEEE Conference on Computer Vision and Pattern Recognition, Salt Lake City, UT, USA, 18–23 June 2018.
40. Burri, M.; Nikolic, J.; Gohl, P.; Schneider, T.; Rehder, J.; Omari, S.; Achtelik, M.W.; Siegwart, R. The EuRoC micro aerial vehicle datasets. *Int. J. Robot. Res.* **2016**, *35*, 1157–1163. [CrossRef]
41. Sturm, J.; Engelhard, N.; Endres, F.; Burgard, W.; Cremers, D. A benchmark for the evaluation of RGB-D SLAM systems. In Proceedings of the IEEE/RSJ International Conference on Intelligent Robots and Systems (IROS), Vilamoura-Algarve, Portugal, 7–12 October 2012; pp. 573–580. [CrossRef]
42. Qin, T.; Pan, J.; Cao, S.; Shen, S. A general optimization-based framework for local odometry estimation with multiple sensors. *arXiv* **2019**, arXiv:1901.03638.
43. Forster, C.; Zhang, Z.; Gassner, M.; Werlberger, M.; Scaramuzza, D. SVO: Semidirect visual odometry for monocular and multi-camera systems. *IEEE Trans. Robot.* **2017**, *33*, 249–265. [CrossRef]
44. Cremers, D.; Schubert, D.; Stückler, J.; Demmel, N.; Usenko, V. Visual-Inertial Mapping with Non-Linear Factor Recovery. *IEEE Robot. Autom. Lett.* **2019**, *5*, 422–429. [CrossRef]
45. Geiger, A.; Lenz, P.; Stiller, C.; Urtasun, R. Vision meets robotics: The KITTI dataset. *Int. J. Robot. Res.* **2013**, *32*, 1231–1237. [CrossRef]
46. Engel, J.; Stueckler, J.; Cremers, D. Large-scale direct SLAM with stereo cameras. In Proceedings of the 2015 IEEE/RSJ International Conference on Intelligent Robots and Systems (IROS), Hamburg, Germany, 28 September–2 October 2015. [CrossRef]
47. Gomez-Ojeda, R.; Moreno, F.A.; Zuñiga-Noël, D.; Scaramuzza, D.; Gonzalez-Jimenez, J. A Stereo SLAM System Through the Combination of Points and Line Segments. *IEEE Trans. Robot.* **2019**, *35*, 734–746. [CrossRef]

Disclaimer/Publisher's Note: The statements, opinions and data contained in all publications are solely those of the individual author(s) and contributor(s) and not of MDPI and/or the editor(s). MDPI and/or the editor(s) disclaim responsibility for any injury to people or property resulting from any ideas, methods, instructions or products referred to in the content.

Article

Aboveground Biomass and Endogenous Hormones in Sub-Tropical Forest Fragments

Chang Liu [1], Wenzhi Du [2], Honglin Cao [3], Chunyu Shen [4,5,6,*] and Lei Ma [4,5,6]

[1] School of Physical Education and Sport, Henan University, Jinming Avenue No. 1, Kaifeng 475004, China
[2] Jigongshan National Nature Reserve, Xinyang 464039, China
[3] Key Laboratory of Vegetation Restoration and Management of Degraded Ecosystems, South China Botanical Garden, Chinese Academy of Sciences, Xingke Road 723, Guangzhou 510650, China
[4] Dabieshan National Observation and Research Field Station of Forest Ecosystem at Henan, Xinyang 464000, China; lma@vip.henu.edu.cn
[5] Key Laboratory of Geospatial Technology for the Middle and Lower Yellow River Regions, Henan University, Ministry of Education, Kaifeng 475004, China
[6] The College of Geography and Environmental Science, Henan University, Jinming Avenue No. 1, Kaifeng 475004, China
* Correspondence: shency@henu.edu.cn

Abstract: Associated endogenous hormones were affected by forest fragmentation and significantly correlated with aboveground biomass storage. Forest fragmentation threatens aboveground biomass (AGB) and affects biodiversity and ecosystem functioning in multiple ways. We ask whether and how forest fragmentation influences AGB in forest fragments. We investigated differences in AGB between forest edges and interiors, and how plant community characteristics and endogenous hormones influenced AGB. In six 40 m × 40 m plots spread across three forest fragments, AGB was significantly higher in plots in the forest interior than in those at the edge of forests. The proportion of individuals with a large diameter at breast height (DBH > 40 cm) in the forest edges is higher than that in the forest interiors. Further, trees within a 15–40 cm DBH range had the highest contribution to AGB in all plots. Trees in interior plots had higher abscisic acid (ABA) and lower indole-3-acetic acid (IAA) concentrations than those in edge plots. In addition, AGB was significantly positively and negatively correlated with ABA and IAA concentrations at the community scale. In this study, we provide an account of endogenous hormones' role as an integrator of environmental signals and, in particular, we highlight the correlation of these endogenous hormone levels with vegetation patterns. Edge effects strongly influenced AGB. In the future, more endogenous hormones and complex interactions should be better explored and understood to support consistent forest conservation and management actions.

Keywords: forest fragmentation; endogenous hormones; edge effects; high-performance liquid chromatography; aboveground biomass

1. Introduction

Forests play an important role in responding to global climate change, especially the carbon cycle. Forests contributed more than half of the organic carbon to terrestrial ecosystems according to previous studies [1,2]. Therefore, accurate estimation of forest carbon storage is crucial to our response to global climate change and other unknown factors [3,4]. Intact evergreen forests store much more living carbon per unit area than fragmented forests do, most of it in AGB and soil [4]. Further, forest fragmentation is globally pervasive and increasing in extent, with forest fragments now accounting for 46% of all remaining forested areas [5,6]. Fragmentation has been a major driver of declining forest biomass and altered carbon fluxes, contributing 6%–17% of global anthropogenic CO_2 emissions to the atmosphere [7]. However, forest fragmentation likely alters forests' potential for carbon storage in ways that are not yet completely understood.

Fragmentation can result from various types of human disturbances, such as selective logging, understory fires, fragmentation, and overhunting [8–10]. Most ecological research on carbon storage of forests has either focused on monitoring change in relatively undisturbed primary forests or on quantifying deforestation and the effects of forest fragmentation on AGB [11–13]. In addition, forest fragmentation creates isolated forest patches and degrades forest edges [14]. The discontinuities fashioned between forest patches by open, deforestation habitats induce a transition zone at the border where a suite of edge effects occur [15,16].

Forest edges are ubiquitous in many fragmented landscapes, and they strongly influence biodiversity [17]. Indeed, edge effects have been reported as one of the most significant patterns structuring both flora and environmental conditions [18], making it crucial to understand how vegetation, ecological processes, and ecosystem services are affected by edges [19,20]. Edge effects can influence species composition, community structure, AGB, and nutrient cycling [21]. Bueno and Liambí [22] reported that both facilitation and edge effects influence the effectiveness of vegetation regeneration within old-field communities in the high tropical Andes. De Paula [23] suggested that fragmented forests and the consequent establishment of forest edges drastically limit forest capacity for carbon storage across human-modified landscapes, since the loss of carbon due to the reduced abundance of large trees is not compensated for by either canopy or understory trees. Forest edges also have different microclimates than interiors do, often with more light, wind, warmer temperatures, and drier air and soil than forest interiors [19]. However, higher rates of tree mortality caused by microclimatic changes in forest edges lead soil carbon stocks to increase in central Amazon Forest fragments [20].

Plants face environmental challenges including competition with neighbors for sunlight, as well as acclimation to ambient temperature fluctuations and to prevailing moisture and nutrient conditions [24]. To complete their life cycle under abiotic and biotic environmental stresses, plants have developed sophisticated mechanisms to sense and adapt to ever-changing and often adverse environmental conditions [25]. It is well known that plant hormones, such as ABA and IAA, are involved in plant adaptation to adverse environments [26]. ABA has been widely reported for its role in adaptation to different kinds of abiotic stress responses, such as high salinity, drought, high temperature, and freezing [27]. ABA has been extensively studied for its importance in the regulation of plant growth and development [28].

Auxin is also a key integrator of environmental signals, and emerging evidence implicates auxin biosynthesis as an essential component of the overall mechanisms of plants' tolerance to stress [29]. Auxin is involved in numerous biological processes ranging from control of cell expansion and cell division to tissue specification, embryogenesis, and organ development [30]. As the main auxin in higher plants, IAA plays a central role in developmental programming and environmental responses such as gravitropism, phototropism, and plastic root development [31,32]. Dinis [33] reported that environmental signals stimulate variations in IAA levels and/or their redistribution and transport in order to regulate plant growth and development. Our standing of how endogenous hormone levels shift in response to fragmentation, and how these hormone affects translate to changes in AGB potential of tree communities remains incomplete. More research is needed to understand how endogenous hormones limit the capacity of AGB in fragmented forests. Furthermore, there is a lack of estimation of the role of endogenous hormones on AGB within forest fragments.

In the interest of filling the knowledge gap, we evaluate the relationships between AGB and hormone concentrations within fragmented forests in South China. We addressed the following three questions: (1) How do edge effects influence AGB distribution within these forest fragments? (2) How are the concentrations of hormones of seven dominant tree species affected by forest edges within these fragments? (3) How do endogenous hormones influence AGB in these forest fragments? The results of this study provide new knowledge on the relationship between endogenous plant hormone levels and vegetation distributions, and its underlying mechanism. Our study could help elucidate the underlying mechanism of fragment structure and provide a basis for the development of planning strategies for the conservation of these forest fragments.

2. Materials and Methods

This study was carried out in Guangzhou City, South China (22°26′–23°56′ N, 112°57′–114°03′ E) within the most threatened region of the fengshui forests. The region is influenced by a typical sub-tropical monsoon climate. The annual mean temperature is 21.8 °C, and the annual precipitation is 1690 mm. Typhoons and thunderstorms occasionally damage trees and the mild climate permits continuous vegetation growth throughout the year. In rural areas of South China, sub-tropical forest fragments can be found near local villages. These remnants are called fengshui forests and have been protected by local residents. As a result, these fragmented forests have retained features of the original vegetation and provide a basis for testing the various theories of fragmentation in sub-tropical forests. Although these fengshui forests occur near local villages, human disturbance has had no significant effect on most of the community characteristics [34]. Certainly, such human-modified landscapes offer an interesting opportunity to examine the potential effects of habitat loss and fragmentation on AGB.

This study was carried out from September to December 2017. Three forest patches were selected in this study. These three forest fragments share similar climatic and soil conditions due to being very close in space (they are less than 10 km away from each other), leading to relatively similar soil and plant community characteristics. Previous research has shown that species turnover among these fragments is limited [34]. In the present study, we established two 40 m × 40 m plots within each forest fragment. The two plots are located in the forest core area and near the forest edge. All trees with DBH greater than 1 cm were identified and DBH was recorded within all plots. Tree species were identified by an experienced field botanist. Plant community characteristics (species richness, abundance, number of individuals within different DBH ranges, and basal area) were estimated according to the data from field censuses.

The forest floor biomass includes woody debris and surface litter in this study. Three randomly distributed 2 m × 2 m subplots were established within every plot. The fresh weight of debris and litter was obtained by using an electronic balance. In order to calculate the ratio between fresh and dry mass, subsamples of the debris and litter were then transported back to the lab and oven dried at 80 °C until constant weight.

Tree AGB was estimated using the allometric equation developed by Wen [35] for sub-tropical mixed forests in Dinghushan Nature Reserve not far away from these forest fragments:

$$TAGB = a \times DBH^b \tag{1}$$

where a and b are statistical parameters (see Table S1 for equations and summary statistics). TAGB was the sum of the dry weight of trunks, branches, leaves, and roots. This model has been successfully applied to estimate tree biomass in sub-tropical forests located in Dinghushan Nature Reserve [36]. Finally, AGB of each plot was calculated by summing forest floor biomass and TAGB.

Wood samples were collected from randomly chosen individuals of seven tree species within each plot (two or three replicates dependent on the abundance of per tree species; Table 1). In order to obtain a 1 mm diameter core, an increment borer was applied at about 1.5 m high on the main stem of each individual. In addition, the litter of seven tree species was collected within each plot with the assistance of an experienced field botanist. All these samples were immediately placed into a liquid nitrogen tank (−80 °C) and then transported to the laboratory and stored to minimize damage to the live tissue and changes to ion concentrations. We determined endogenous hormone concentrations of IAA in wood samples and ABA in litter using high-performance liquid chromatography (HPLC). IAA and ABA content in these seven tree species within each plot represent the endogenous hormone levels of each plot.

Table 1. Seven common tree species were selected in each plot within three forest patches.

No.	Species	Shade Tolerance
1	*Castanopsis chinensis* Hance	Light-demanding
2	*Aleurites moluccanus* (L.) Willd.	Light-demanding
3	*Cryptocarya concinna* Hance	Mid-tolerant
4	*Syzygium rehderianum* Merr. et Perry	Light-demanding
5	*Schima superba* Gardn. et Champ	Mid-tolerant
6	*Carallia brachiata* (*Lour.*) Merr.	Shade-tolerant
7	*Gironniera subaequalis* Planch	Mid-tolerant

Statistical analysis was performed using SPSS 20.0. Before statistical analysis, all data were tested for normality using the Shapiro–Wilk test and for homoscedasticity using the Levene test. Results were represented as the mean ± standard error. Differences among means of IAA and ABA concentration between forest edges and interiors were analyzed with one-way ANOVA tests. The statistical significance of the difference between means was determined with Duncan's new multiple range test. Pearson correlation analysis was also conducted in this study.

3. Results

3.1. Vegetation Distribution in Fragments

Both basal area and stem density were significantly lower ($p < 0.05$) near forest edges than in interior plots, respectively (Table 2). In addition, the number of stems within two DBH ranges (DBH < 15 cm; 15–40 cm) was significantly lower in forest edge plots than in forest interior plots. Smaller trees (individuals within 1–15 cm DBH range) had the highest proportion in all forest fragments. However, both the number of individuals and the percentage of larger trees (DBH > 40 cm) were higher near forest edges (191; 8.2%) than in interior plots (153; 6.4%). Aboveground biomass (AGB) was significantly lower near forest edges (80.6 Mg ha^{-1}) than in the forest core area (143.2 Mg ha^{-1}). In addition, AGB of smaller trees and medium trees (DBH: 15–40 cm) were significantly lower in forest edges than in forest interiors ($p < 0.05$). Although smaller trees had the largest stems, AGB was highest in medium trees (38.5 Mg ha^{-1}; 77.9 Mg ha^{-1}) and larger trees (33.2 Mg ha^{-1}; 36.4 Mg ha^{-1}) both in forest edges and interior within these studied plots ($p < 0.05$). In addition, medium trees had the highest contribution to AGB storage in the studied plots due to their relatively higher mean DBH and abundance.

Table 2. Community characteristics of fragmented forests from edge to interior plots.

Classification	Edge Plots	Interior Plots
Aboveground biomass (Mg ha^{-1})	80.6 ± 12.2 [a]	143.2 ± 11.9 [b]
Stem density	1735 ± 215 [a]	2969 ± 308 [b]
Basal area (m^2 ha^{-1})	30.0 ± 2.6 [a]	52.5 ± 1.4 [b]
Stems N (DBH < 15 cm)	1421 ± 125 [a]	2292 ± 168 [b]
Stems N (DBH: 15–40 cm)	161 ± 69 [a]	486 ± 112 [b]
Stems N (DBH > 40 cm)	153 ± 21 [a]	191 ± 28 [a]
Stems AGB (DBH < 15 cm)	8.9 ± 1.7 [a]	28.9 ± 2.3 [b]
Stems AGB (DBH: 15–40 cm)	38.5 ± 5.2 [a]	77.9 ± 5.4 [b]
Stems AGB (DBH > 40 cm)	33.2 ± 5.3 [a]	36.4 ± 4.2 [a]

Lowercase letters stand for significance between each row.

3.2. Endogenous Hormones Contents of IAA and ABA

ABA concentrations in the leaf litter and IAA concentrations in the trunks of seven tree species within six plots are shown in Figure 1. ABA and IAA concentrations of three tree species differed within the studied plots. ABA concentrations in all these seven tree species (Table 1), except for *Castanopsis chinensis*, were significantly lower in edge plots than in interior plots. Mid-tolerant tree species had the highest ABA concentration in interior plots (Figure 1A). IAA concentrations of all these tree species, except for *Gironniera subaequalis*, were significantly higher in edge plots than in interior plots. In addition, both *Gironniera subaequalis* and *Carallia brachiata* had the highest IAA concentration in the studied plots (Figure 1B). Further, IAA concentration differences were even larger between the tree species (*Castanopsis chinensis*, *Aleurites moluccanus*, *Cryptocarya concinna*, and *Schima superba*) than between the residual species. The ratios between IAA and ABA of these seven tree species present a similar trend: values in edge plots were higher than those in interiors plots (Figure 1C). Further, the ratios of two tree species (*Gironniera subaequalis* and *Carallia brachiata*) were the largest both in forest edge and interior in the present study.

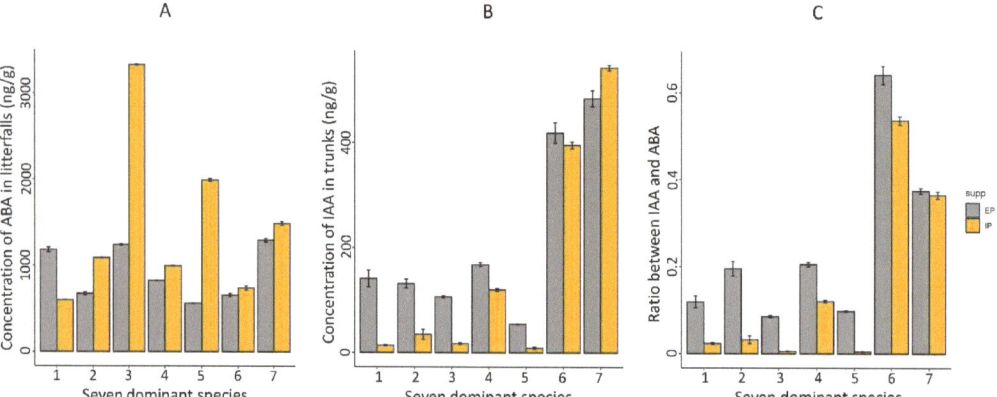

Figure 1. The concentration of endogenous hormones and their ratios in seven common species in edge (EP) and interior plots (IP). (**A**) The concentration of ABA in litter falls; (**B**) The concentration of IAA in tree trunks; (**C**) The ratios of IAA to ABA in 7 common tree species. Number 1 stands for *Castanopsis chinensis* Hance; 2 stands for *Aleurites moluccanus* (L.) Willd.; 3 stands for *Cryptocarya concinna* Hance; 4 stands for *Syzygium rehderianum* Merr. et Perry; 5 stands for *Schima superba* Gardn. et Champ; 6 stands for *Carallia brachiata* (Lour.) Merr.; 7 stands for *Gironniera subaequalis* Planch.

3.3. Relationships between AGB and Endogenous Hormones Level

The relationships between AGB and endogenous hormones are shown in Figure 2. In the present study, AGB was significantly positively correlated with mean ABA concentration among six dominant tree species (except *Castanopsis chinensis*). In addition, the gradients of four tree species (*Aleurites moluccanus*, *Syzygium rehderianum*, *Carallia brachiate*, *Gironniera subaequalis*) were even higher than the last two species, which indicated that slight changes in ABA may cause large fluctuations in AGB (Figure 2A). Mean IAA concentrations of six tree species (except *Gironniera subaequalis*) had significant negative relationships with AGB regardless of plot type (edge or interior) (Figure 2B). In addition, higher IAA:ABA ratios were significantly correlated with lower AGB in these forest fragments (Figure 2C).

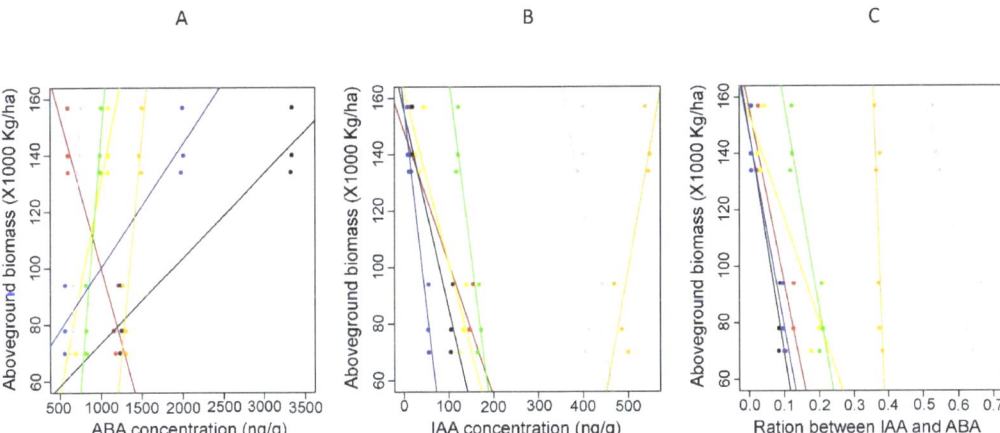

Figure 2. The relationships between aboveground biomass and the concentration of endogenous hormones. (**A**) Relationship between aboveground biomass and the concentration of ABA; (**B**) relationship between aboveground biomass and the concentration of IAA; (**C**) relationship between aboveground biomass and the IAA:ABA ratio. Red stands for *Castanopsis chinensis* Hance; yellow stands for *Aleurites moluccanus* (L.) Willd.; Black stands for *Cryptocarya concinna* Hance; Green stands for *Syzygium rehderianum* Merr. et Perry; Blue stands for *Schima superba* Gardn. et Champ; Grey stands for *Carallia brachiata* (Lour.) Merr.; Orange stands for *Gironniera subaequalis* Planch.

4. Discussion

In this study, the patterns of aboveground biomass storage and endogenous hormone concentrations of seven dominant tree species were studied. Three fragmented forests were significantly affected by edge effects, and their AGB ranges were 80.6–143.2 Mg ha^{-1} (Table 2). In addition, endogenous hormone concentrations were also significantly affected by forest edges (Figure 2). Compared with intact forests, fragmented forests showed a higher proportion of habitat edges exposed to other habitats, resulting in a higher probability of edge effects according to previous studies [21,33]. Furthermore, the forest edge can be regarded as a buffer zone, and the ecological conditions gradually change within a certain distance, which has a significant impact on AGB. Edge effects are among the primary mechanisms by which forest fragmentation can influence the link between biodiversity and ecosystem processes [21,29].

Habitat fragmentation and the consequent establishment of permanent forest edges reduce forest capacity for AGB because forest edges retain only one-third as much biomass as forest interior habitat according to former studies [37]. In the present study, AGB storage near forest edges contributes about 36.0% to the total AGB (Table 2). Our study fragments are consistent in size with previous studies, and we also found that the changes in community structure along fragment transects were consistent with knowledge of forest edge effects. The higher basal areas in the forest interior suggest that AGB in the forest interior could contain more biomass, were it not for the edge effects [38]. In this study, the AGB storage within different DBH ranges was also significantly affected by forest edge. In addition, AGB storage within different DBH ranges differed within different types of plots (edge plots and interior plots). Larger trees (DBH > 40 cm) accounted for a greater proportion of all trees at the forest edge than in the interior (Table 2), where the mean basal area of individual trees was lower than at the forest edge habitat. However, in contrast to our results, other studies have reported that tree density at forest edges generally exceeds that in the interior [39]. Our results suggested that AGB reduction in edge-affected habitats results from reduced larger tree individuals together with insufficiency of biomass make up by residual trees (Table 2). Furthermore, larger trees are likely particularly important for maintaining AGB at the forest edge, as we observed little changes in the amount of

AGB held in smaller trees at forest edges in these fragments. Therefore, any impacts on larger trees, either by global change or other disturbances that affect the abundance and persistence of these large stems, are therefore likely to have a major impact on forest AGB [40]. Our results agreed with other tropical forest studies, wherein the lower number of large and medium emergent trees near the forest edges is a major contributor to the loss of AGB [41,42].

AGB is an important ecosystem function altered by edge effects, with implications for the management of micro-environmental conditions in forest fragments [22]. The establishment of forest edges during fragmentation and the ensuing alteration in microclimate affect plant populations [4]. We observed a significant increase in the proportion of larger trees with proximity to fragment edges, likely a consequence of increased temperature and light availability. Higher biomass storage is predicted at the edge compared to the interior as a result of increased productivity resulting from increased sunlight and temperature at the forest edge. Edges expose organisms to dry, windy, high-light conditions that differ considerably from the dark, humid forest interior [43]. Elevated wind stress in fragmented forests is widely reported and has been proposed as a likely factor in reducing AGB in fragmented forests. In addition, the relaxing of competition for light as a result of lower stem density, more open canopies, and increased lateral light penetration from habitat edges might further reduce AGB in fragments [44]. Moreover, forest edges were dominated by shade-intolerant, fast-growing pioneer species, as compared to the more shade-tolerant maple, ironwood, and elm species that dominated the fragment interiors according to former studies [45]. Barros [17] reported that fragment edges had greater exposure to harsh winds compared to the forest interior, directly increasing tree mortality. As a result, there is increased biomass loss due to the mortality of large trees [27,35]. This might be an explanation for our result that plants in edge plots experience a relaxing of competition for light and have decreased biomass density.

The present study has focused specifically on the ways in which edge effects can alter the link between endogenous hormones and ecosystem functions occurring within subtropical forest fragments. As sessile organisms, plants have evolved mechanisms allowing them to control their growth and development in response to environmental changes [46]. As a primary source of energy, light is one of the most important environmental factors for plant growth [17]. The number of stems within interior plots was significantly higher than that in edge plots in the present study (Table 2). Distances between two adjacent plants are reduced, creating changes in environmental factors [47]. Moreover, the distance and size of neighboring plants determine the type of stress the plant will suffer. If a plant is exposed to intense neighboring shade, it will receive limited light input, but in open areas, it is more likely to be exposed to heat and oxidative stress caused by the high radiation load [20]. Competition for light determines the success of individual plants in dense vegetation, and the presence of neighbors is an important environmental factor inhibiting plant growth [45]. It is well known that ABA is an essential mediator in triggering the plant responses to many environmental stresses including shade [46]. Such analysis was already reported in competition among Arabidopsis plants, suggesting the involvement of plant hormones in responses to the presence of neighbors [47,48]. In the present study, higher stem density was accompanied by higher ABA concentration in leaves in the interior areas of forest patches, probably due to competition for light. Moreover, the inhibition of lettuce plant growth under increased planting density was accompanied by the accumulation of ABA in the shoots of competing plants [49]. These results confirm the important role of ABA in the growth-inhibiting effect of increased planting density.

ABA concentration is closely related to IAA concentration according to former studies [15]. Vysotskaya et al. [49] suggested that ABA is involved in the allocation of IAA in competing plants. Shkolnik-Inbar [50] reported the role of ABA accumulation in the reduction of polar auxin transport and a resulting decrease in root auxin. Our results are in accordance with theirs: concentration of ABA increased in leaves within interior plots, accompanied by a decline in the concentration of IAA in trunks. Moreover, higher stem density leads to shade avoidance syndrome, which decreased the IAA content and auxin polar transport [51,52]. The same results were observed in this study: the decline in the proportion of larger trees (DBH > 40 cm) and simultaneous increase in stem density was accompanied by a decrease in the concentration of IAA in leaves within interior plots.

It is thus not surprising that auxin has emerged as an important regulator of adaptive growth responses to environmental stresses [53]. It was discovered that local auxin biosynthesis maintains optimal plant growth in response to environmental signals, including light, temperature, and humidity [54]. Auxin is one of the most important plant hormones mediating endogenous developmental signals and exogenous environmental cues to control various plant growth and developmental responses [55]. Strong evidence for induced auxin production are indications that auxin sensitivity is also increased in response to stress [56]. Light and temperature are arguably two of the most important signals regulating the growth and development of plants [57]. Meanwhile, light and temperature patterns are often correlated under natural plant growth conditions [58]. Islam [59] reported that light quality is sensed by different photoreceptors in plants, which are involved in a wide range of developmental processes, and IAA is an important determinant of shoot elongation in poinsettia, as shown for a wide range of species. Earlier, it was generally believed that drought results in a decrease in IAA content. At present, however, it became gradually clear that the adaptation to drought is accompanied by an increase in the IAA content [60,61]. In our results, the higher concentration of IAA in plants at edge plots might result from decreased soil moisture at the forest edges compared to the forest interior. Our results are in accordance with the reports that higher auxin content in Arabidopsis might create positive regulation of drought stress resistance [62]. However, the present study focuses only on the AGB storage, endogenous hormones concentration, and its relationships within the edge and interior plots in forest fragments. In the future, research will be carried out on how endogenous hormones regulate the growth of trees to affect biomass storage under the influence of forest fragmentation.

5. Conclusions

The distribution of AGB and its associated plant endogenous hormones were analyzed in three sub-tropical forest fragments in the present study. AGB and the number of individuals were considerably reduced at forest edges, however, the proportion of larger trees (DBH > 40 cm) increased near forest edges. In addition, it is evident from our work that community characteristics change from forest edges to interiors. Plant endogenous hormone concentrations were likely affected by edge effects due to micro-environmental conditions. IAA and ABA decreased and increased from the forest edge to the interior, respectively. Higher stem density was accompanied by higher ABA concentration in leaves in the core areas of forest patches while IAA concentrations of woody species were higher at edge plots. This study shows that the fragmentation of forests and thus the spread of marginal habitats drastically reduces aboveground biomass storage, resulting from the regulation of plant growth by endogenous hormones. The present study also provided key data for the development and validation of AGB conditions in subtropical forests in southern China.

Supplementary Materials: The following supporting information can be downloaded at: https://www.mdpi.com/article/10.3390/f14040661/s1, Table S1: Allometric regression equations and summary statistics.

Author Contributions: Conceptualization, L.M.; methodology, C.S.; software, H.C.; validation, C.L.; formal analysis, W.D.; investigation, L.M; resources, L.M.; data curation, H.C.; writing—original draft preparation, C.L.; writing—review and editing, C.S. All authors have read and agreed to the published version of the manuscript.

Funding: This work was supported by the NSFC-Henan Joint Fund (grant number U1904204) and China Postdoctoral Science Foundation (2020M672206).

Data Availability Statement: If there are legitimate reasons, you can contact the corresponding author to request data.

Acknowledgments: We acknowledge the work of the principal investigators and their field assistants for collecting the field data on the experiment plots.

Conflicts of Interest: The authors declare no conflict of interest.

References

1. Fauset, S.; Johnson, M.O.; Gloor, M.; Baker, T.R.; Monteagudo, A.M.; Brienen, R.J.; Feldpausch, T.R.; Lopez-Gonzalez, G.; Malhi, Y.; ter Steege, H.; et al. Hyperdominance in Amazonian forest carbon cycling. *Nat. Commun.* **2015**, *6*, 6857. [CrossRef]
2. Haddad, N.M.; Brudvig, L.A.; Clobert, J.; Davies, K.F.; Gonzalez, A.; Holt, R.D.; Lovejoy, T.E.; Sexton, J.O.; Austin, M.P.; Collins, C.D.; et al. Habitat fragmentation and its lasting impact on Earth's ecosystems. *Sci. Adv.* **2015**, *1*, e1500052. [CrossRef] [PubMed]
3. Pütz, S.; Groeneveld, J.; Henle, K.; Knogge, C.; Martensen, A.C.; Metz, M.; Metzger, J.P.; Ribeiro, M.C.; de Paula, M.D.; Huth, A. Long-term carbon loss in fragmented Neotropical forests. *Nat. Commun.* **2014**, *5*, 5037. [CrossRef] [PubMed]
4. Schmidt, M.; Jochheim, H.; Kersebaum, K.C.; Lischeid, G.; Nendel, C. Gradients of microclimate, carbon and nitrogen in transition zones of fragmented landscapes—A review. *Agr. Forest Meteorol.* **2017**, *232*, 659–671. [CrossRef]
5. Asner, G.P.; Powell, G.V.N.; Mascaro, J.; Knapp, D.E.; Clark, J.K.; Jacobson, J.; Kennedy-Bowdoin, T.; Balaji, A.; Paez-Acosta, G.; Victoria, E.; et al. High-resolution forest carbon stocks and emissions in the Amazon. *Proc. Natl. Acad. Sci. USA* **2010**, *107*, 16738–16742. [CrossRef] [PubMed]
6. Baccini, A.; Goetz, S.J.; Walker, W.S.; Laporte, N.T.; Sun, M.; Sulla-Menashe, D.; Hackler, J.; Beck, P.S.A.; Dubayah, R.; Friedl, M.A.; et al. Estimated carbon dioxide emissions from tropical deforestation improved by carbon-density maps. *Nat. Clim. Chang.* **2012**, *2*, 182–185. [CrossRef]
7. Osuri, A.M.; Kumar, V.S.; Sankaran, M. Altered stand structure and tree allometry reduce carbon storage in evergreen forest fragments in India's Western Ghats. *Forest Ecol. Manag.* **2014**, *329*, 375–383. [CrossRef]
8. Carvalho, F.D.G.D.; Costa, K.; Romitelli, I.; Barbosa, J.M.; Vieira, S.A.; Metzger, J.P. Lack of evidence of edge age and additive edge effects on carbon stocks in a tropical forest. *Forest Ecol. Manag.* **2018**, *407*, 57–65. [CrossRef]
9. Numata, I.; Silva, S.S.; Cochrane, M.A.; d'Oliveira, M.V.N. Fire and edge effects in a fragmented tropical forest landscape in the southwestern Amazon. *Forest Ecol. Manag.* **2017**, *401*, 135–146. [CrossRef]
10. Bregman, T.P.; Lees, A.C.; Seddon, N.; MacGregor, H.E.A.; Darski, B.; Aleixo, A.; Bonsall, M.B.; Tobias, J.A. Species interactions regulate the collapse of biodiversity and ecosystem function in tropical forest fragments. *Ecology* **2015**, *96*, 2692–2704. [CrossRef]
11. Baker, T.P.; Jordan, G.J.; Baker, S.C. Microclimatic edge effects in a recently harvested forest: Do remnant forest patches create the same impact as large forest areas? *Forest Ecol. Manag.* **2016**, *365*, 128–136. [CrossRef]
12. Baker, T.P.; Jordan, G.J.; Steel, E.A.; Fountain-Jones, N.M.; Wardlaw, T.J.; Baker, S.C. Microclimate through space and time: Microclimatic variation at the edge of regeneration forests over daily, yearly and decadal time scales. *Forest Ecol. Manag.* **2014**, *334*, 174–184. [CrossRef]
13. Arroyo-Rodríguez, V.; Cavender-Bares, J.; Escobar, F.; Melo, F.P.L.; Tabarell, M.; Santos, B.A. Maintenance of tree phylogenetic diversity in a highly fragmented rain forest. *J. Ecol.* **2012**, *100*, 702–711. [CrossRef]
14. Wekesa, C.; Maranga, E.K.; Kirui, B.K.; Muturi, G.M.; Gathara, M. Interactions between native tree species and environmental variables along forest edge-interior gradient in fragmented forest patches of Taita Hills, Kenya. *Forest Ecol. Manag.* **2018**, *409*, 789–798. [CrossRef]
15. Cutler, S.R.; Rodriguez, P.L.; Finkelstein, R.R.; Abrams, S. Abscisic acid: Emergence of a core signaling network. *Annu. Rev. Plant Biol.* **2010**, *61*, 651–679. [CrossRef]
16. Goosem, M.; Paz, C.; Fensham, R.; Preece, N.; Goosem, S.; Laurance, S.G.W. Forest age and isolation affect the rate of recovery of plant species diversity and community composition in secondary rain forests in tropical Australia. *J. Veg. Sci.* **2016**, *27*, 504–514. [CrossRef]
17. Barros, H.S.; Fearnside, P.M. Soil carbon stock changes due to edge effects in central Amazon forest fragments. *Forest Ecol. Manag.* **2016**, *379*, 30–36. [CrossRef]
18. Chen, D.; Fu, Y.; Liu, G.; Liu, H. Low light intensity effects on the growth, photosynthetic characteristics, antioxidant capacity, yield and quality of wheat (*Triticum aestivum*, L.) at different growth stages in BLSS. *Adv. Space Res.* **2014**, *53*, 1557–1566.
19. Ma, L.; Shen, C.; Lou, D.; Fu, S.; Guan, D. Patterns of ecosystem carbon density in edge-affected fengshui forests. *Ecol. Eng.* **2017**, *107*, 216–223. [CrossRef]
20. Mroue, S.; Simeunovic, A.; Robert, H.S. Auxin production as an integrator of environmental cues for developmental growth regulation. *J. Exp. Bot.* **2017**, *69*, 201. [CrossRef]
21. Bueno, A.; Llambi, L.D. Facilitation and edge effects influence vegetation regeneration in old-fields at the tropical Andean forest line. *Appl. Veg. Sci.* **2015**, *18*, 613–623. [CrossRef]
22. Yang, D.L.; Yang, Y.; He, Z. Roles of Plant Hormones and Their Interplay in Rice Immunity. *Mol. Plant* **2013**, *6*, 675–685. [CrossRef]
23. de Paula, M.D.; Costa, C.P.A.; Tabarelli, M. Carbon storage in a fragmented landscape of Atlantic forest: The role played by edge-affected habitats and emergent trees. *Trop. Conserv. Sci.* **2011**, *4*, 349–358. [CrossRef]

24. Vanstraelen, M.; Benková, E. Hormonal interactions in the regulation of plant development. *Annu. Rev. Cell Dev. Bl.* **2012**, *28*, 463–487. [CrossRef]
25. Zörb, C.; Geilfus, C.M.; Mühling, K.H.; Ludwig-Müller, J. The influence of salt stress on ABA and auxin concentrations in two maize cultivars differing in salt resistance. *J. Plant Physiol.* **2013**, *170*, 220–224. [CrossRef] [PubMed]
26. ÁlvarezFlórez, F.; Lópezcristoffanini, C.; Jáuregui, O.; Melgarejo, L.M.; López-Carbonell, M. Changes in ABA, IAA and JA levels during calyx, fruit and leaves development in cape gooseberry plants (*Physalis peruviana* L.). *Plant Physiol. Biochem.* **2017**, *115*, 174–182. [CrossRef] [PubMed]
27. Li, X.; Wang, L.; Wang, S.; Yang, Q.; Zhou, Q.; Huang, X. A preliminary analysis of the effects of bisphenol A on the plant root growth via changes in endogenous plant hormones. *Ecotox. Environ. Saf.* **2018**, *150*, 152–158. [CrossRef]
28. Seki, M.; Umezawa, T.; Urano, K.; Shinozaki, K. Regulatory metabolic networks in drought stress responses. *Curr. Opin. Plant Biol.* **2007**, *10*, 296–302. [CrossRef]
29. Suh, J.H.; Han, S.B.; Wang, Y. Development of an improved sample preparation platform for acidic endogenous hormones in plant tissues using electromembrane extraction. *J. Chromatogr. A* **2007**, *1535*, 1–8. [CrossRef] [PubMed]
30. Kumar, R.; Khurana, A.; Sharma, A.K. Role of plant hormones and their interplay in development and ripening of fleshy fruits. *J. Exp. Bot.* **2014**, *65*, 4561–4575. [CrossRef]
31. Peer, W.A. From perception to attenuation: Auxin signalling and responses. *Curr. Opin. Plant Biol.* **2013**, *16*, 561–568. [CrossRef] [PubMed]
32. Zhao, Y. Auxin Biosynthesis and Its Role in Plant Development. *Annu Rev. Plant Biol.* **2010**, *61*, 49–64. [CrossRef]
33. Dinis, L.T.; Bernardo, S.; Luzio, A.; Pintó, G.; Meijon, M.; Pintó-Marijuan, M.; Cotado, A.; Correia, C.; Moutinho-Pereira, J. Kaolin modulates ABA and IAA dynamics and physiology of grapevine under Mediterranean summer stress. *J. Plant Physiol.* **2017**, *220*, 181–192. [CrossRef] [PubMed]
34. Ma, L.; Huang, M.; Shen, Y.; Cao, H.; Wu, L.; Ye, H.; Lin, G.; Wang, Z. Species diversity and community structure in forest fragments of Guangzhou, South China. *J. Trop. Forest Sci.* **2015**, *27*, 148–157.
35. Wen, D.; Wei, P.; Kong, G. Biomass study of the community of Castanopsis chinensis + Cryptocarya concinna + Schima supera in a Southern China reserve. *Acta Ecol. Sin.* **1997**, *17*, 497–504, (In Chinese, English Summary).
36. Liu, S.; Luo, Y.; Huang, Y. Studies on the community biomass and its allocations of five forest types in Dinghushan Nature Reserve. *Ecol. Sci.* **2007**, *26*, 387–393, (In Chinese, English Summary).
37. Ziter, C.; Bennett, E.M.; Gonzalez, A. Temperate forest fragments maintain aboveground carbon stocks out to the forest edge despite changes in community composition. *Oecologia* **2014**, *176*, 893–902. [CrossRef]
38. Wilson, M.C.; Chen, X.Y.; Corlett, R.T.; Didham, R.K.; Ding, P.; Holt, R.D.; Holyoak, M.; Hu, G.; Hughes, A.C.; Jiang, L.; et al. Habitat fragmentation and biodiversity conservation: Key findings and future challenges. *Landsc. Ecol.* **2016**, *31*, 219–227. [CrossRef]
39. Slik, J.W.F.; Paoli, G.; McGuire, K.; Amaral, I.; Barroso, J.; Bastian, M.; Blanc, L.; Bongers, F.; Boundja, P.; Clark, C. Large trees drive forest aboveground biomass variation in moist lowland forests across the tropics. *Glob. Ecol. Biogeogr.* **2013**, *22*, 1261–1271. [CrossRef]
40. Barlow, J.; Peres, C.A.; Lagan, B.O.; Haugaasen, T. Large tree mortality and the decline of forest biomass following Amazonian wildfires. *Ecol. Lett.* **2003**, *6*, 6–8. [CrossRef]
41. Lindenmayer, D.B.; Laurance, W.F.; Franklin, J.F. Global decline in large old trees. *Science* **2012**, *338*, 1305–1306. [CrossRef]
42. Hallinger, M.; Johansson, V.; Schmalholz, M.; Sjöberg, S.; Ranius, T. Factors driving tree mortality in retained forest fragments. *For. Ecol. Manag.* **2016**, *368*, 163–172. [CrossRef]
43. Pütz, S.; Groeneveld, J.; Alves, L.F.; Metzger, J.P.; Huth, A. Fragmentation drives tropical forest fragments to early successional states: A modelling study for Brazilian Atlantic forests. *Ecol. Model.* **2011**, *222*, 1986–1997. [CrossRef]
44. Liu, J.; Wilson, M.; Hu, G.; Liu, J.; Wu, J.; Yu, M. How does habitat fragmentation affect the biodiversity and ecosystem functioning relationship? *Landsc. Ecol.* **2018**, *33*, 341–352. [CrossRef]
45. Legris, M.; Nieto, C.; Sellaro, R.; Prat, S.; Casal, J. Perception and signalling of light and temperature cues in plants. *Plant J.* **2017**, *90*, 683. [CrossRef]
46. Reinmann, A.B.; Hutyra, L.R. Edge effects enhance carbon uptake and its vulnerability to climate change in temperate broadleaf forests. *Proc. Natl. Acad. Sci. USA* **2017**, *114*, 107. [CrossRef]
47. Malmivaara-LämsÄ, M.; Hamberg, L.; Haapamäki, E.; Liski, J.; Kotze, D.; Lehvävirta, S.; Fritze, H. Edge effects and trampling in boreal urban forest fragments—Impacts on the soil microbial community. *Soil Biol. Biochem.* **2008**, *40*, 1612–1621. [CrossRef]
48. Cagnola, J.I.; Ploschuk, E.; Benech-Arnold, T.; Finlayson, S.A.; Casal, J.J. Stem Transcriptome Reveals Mechanisms to Reduce the Energetic Cost of Shade-Avoidance Responses in Tomato. *Plant Physiol.* **2012**, *160*, 1110–1119. [CrossRef]
49. Vysotskaya, L.B.; Arkhipova, T.N.; Kudoyarova, G.R.; Veselov, S.Y. Dependence of growth inhibiting action of increased planting density on capacity of lettuce plants to synthesize ABA. *J. Plant Physiol.* **2018**, *220*, 69–73. [CrossRef]
50. Shkolnik-Inbar, D.; Bar-Zvi, D. ABI4 Mediates Abscisic Acid and Cytokinin Inhibition of Lateral Root Formation by Reducing Polar Auxin Transport in Arabidopsis. *Plant Cell.* **2010**, *22*, 3560–3573. [CrossRef]
51. Li, Y.; Zhao, H.; Duan, B.; Korpelainen, H.; Li, C. Effect of drought and ABA on growth, photosynthesis and antioxidant system of Cotinus coggygria, seedlings under two different light conditions. *Environ. Exp. Bot.* **2011**, *71*, 107–113. [CrossRef]

52. Masclaux, F.G.; Bruessow, F.; Schweizer, F.; Gouhier-Darimont, C.; Keller, L.; Reymond, P. Transcriptome analysis of intraspecific competition in Arabidopsis thaliana reveals organ-specific signatures related to nutrient acquisition and general stress response pathways. *BMC Plant Biol.* **2012**, *12*, 227. [CrossRef]
53. Wit, M.D.; Lorrain, S.; Fankhauser, C. Auxin-mediated plant architectural changes in response to shade and high temperature. *Physiol. Plantarum.* **2014**, *151*, 13–24. [CrossRef]
54. Lv, B.; Yan, Z.; Tian, H.; Zhang, X.; Ding, Z. Local Auxin Biosynthesis Mediates Plant Growth and Development. *Trends Plant Sci.* **2019**, *24*, 6–9. [CrossRef] [PubMed]
55. Franklin, K.A.; Toledo-Ortiz, G.; Pyott, D.E.; Halliday, K.J. Interaction of light and temperature signalling. *J. Exp. Bot.* **2014**, *65*, 2859–2871. [CrossRef] [PubMed]
56. Pustovoitova, T.N.; Zhdanova, N.E.; Zholkevich, V.N. Changes in the Levels of IAA and ABA in Cucumber Leaves under Progressive Soil Drought. *Russ. J. Plant Physl.* **2004**, *51*, 513–517. [CrossRef]
57. Wang, T. Exogenous abscisic acid reduces water loss and improves antioxidant defense, desiccation tolerance and transpiration efficiency in two spring wheat cultivars subjected to a soil water deficit. *Funct. Plant Biol.* **2013**, *40*, 494–506.
58. Crockatt, M.E.; Bebber, D.P. Edge effects on moisture reduce wood decomposition rate in a temperate forest. *Glob. Chang. Biol.* **2014**, *21*, 698–707. [CrossRef] [PubMed]
59. Islam, M.A.; Tarkowská, D.; Clarke, J.L.; Blystad, D.; Gislerod, H.R.; Torre, S.; Olsen, J.E. Impact of end-of-day red and far-red light on plant morphology and hormone physiology of poinsettia. *Sci. Hortic.* **2014**, *174*, 77–86. [CrossRef]
60. Karageorgou, P.; Levizou, E.F.I.; Manetas, Y. The influence of drought, shade and availability of mineral nutrients on exudate phenolics of Dittrichia viscosa. *Flora* **2002**, *197*, 285–289. [CrossRef]
61. Vanneste, S.; Friml, J. Auxin: A trigger for change in plant development. *Cell* **2009**, *136*, 1005–1016. [CrossRef] [PubMed]
62. Shi, H.; Chen, L.; Ye, T.; Liu, X.; Ding, K.; Chan, Z. Modulation of auxin content in Arabidopsis confers improved drought stress resistance. *Plant Physiol. Biochem.* **2014**, *82*, 209–217. [CrossRef] [PubMed]

Disclaimer/Publisher's Note: The statements, opinions and data contained in all publications are solely those of the individual author(s) and contributor(s) and not of MDPI and/or the editor(s). MDPI and/or the editor(s) disclaim responsibility for any injury to people or property resulting from any ideas, methods, instructions or products referred to in the content.

Article

Topographic Variation in Ecosystem Multifunctionality in an Old-Growth Subtropical Forest

Jiaming Wang [1,2], Han Xu [3], Qingsong Yang [2], Yuying Li [1], Mingfei Ji [1], Yepu Li [1], Zhongbing Chang [4], Yangyi Qin [2], Qiushi Yu [2] and Xihua Wang [2,*]

[1] School of Water Resources and Environment Engineering, Nanyang Normal University, Nanyang 473061, China; wangjmecology@163.com (J.W.); lyying200508@163.com (Y.L.); jimfdy@nynu.edu.cn (M.J.); ypli2015@lzu.edu.cn (Y.L.)

[2] School of Ecological and Environmental Sciences, East China Normal University, Shanghai 200241, China; qsyang@des.ecnu.edu.cn (Q.Y.); 51253903007@stu.ecnu.edu.cn (Y.Q.); 51253903026@stu.ecnu.edu.cn (Q.Y.)

[3] Research Institute of Tropical Forestry, Chinese Academy of Forestry, Guangzhou 510520, China; ywfj@163.com

[4] Key Laboratory of Natural Resources Monitoring in Tropical and Subtropical Area of South China, Surveying and Mapping Institute Lands and Resource Department of Guangdong Province, Guangzhou 510663, China; changzb@scbg.ac.cn

* Correspondence: xhwang@des.ecnu.edu.cn

Citation: Wang, J.; Xu, H.; Yang, Q.; Li, Y.; Ji, M.; Li, Y.; Chang, Z.; Qin, Y.; Yu, Q.; Wang, X. Topographic Variation in Ecosystem Multifunctionality in an Old-Growth Subtropical Forest. *Forests* **2024**, *15*, 1032. https://doi.org/10.3390/f15061032

Academic Editors: Nikolay S. Strigul and Adriano Mazziotta

Received: 6 May 2024
Revised: 5 June 2024
Accepted: 12 June 2024
Published: 14 June 2024

Copyright: © 2024 by the authors. Licensee MDPI, Basel, Switzerland. This article is an open access article distributed under the terms and conditions of the Creative Commons Attribution (CC BY) license (https://creativecommons.org/licenses/by/4.0/).

Abstract: Exploring the relationship between topography and forest multifunctionality enhances understanding of the mechanisms maintaining forest multifunctionality and proves beneficial for managing overall forest functions across different landscapes. Leveraging census data from a 20 ha subtropical forest plot, we investigated the topographic variations in individual functions, multifunctionality, and their interrelationships. Our results revealed that relative to lower elevations, higher elevations had higher woody productivity, sapling growth, and recruitment that drove higher average forest multifunctionality (FMA). However, forest multifunctionality at the 50% threshold level (FMt50) had no significant difference between high and low elevations. Compared with the valley and slope, higher woody productivity, higher sapling recruitment, and higher soil organic carbon stock drove higher forest multifunctionality (FMA and FMt50) in the ridge. These results indicate the ridge serves as a forest multifunctionality "hotspot" within the Tiantong 20 hm² plot. Additionally, relative to the low elevation, the degree of synergy among functions at the high elevation was significantly lower, indicating difficulties in attaining high forest multifunctionality at the high elevation. Our work underscores the importance of topography in regulating subtropical forest multifunctionality and relationships between forest functions at a local scale, suggesting that future forest management strategies (such as regulating synergistic or trade-off relationships between functions) should give particular attention to topographic conditions.

Keywords: topography; forest multifunctionality; multiple individual functions; synergistic or trade-off relationships between functions

1. Introduction

Ecosystem functions refer to properties and processes of an ecosystem, such as ecosystem matter and energy cycles, that have a specific function within the ecosystem [1,2]. Ecosystem multifunctionality can be defined as the capacity of an ecosystem to fulfill multiple functions simultaneously [3–8]. The individual functions of ecosystems, such as biomass, productivity, or soil carbon stock, exhibit spatial variability or heterogeneity due to the influence of environmental factors [9–11]. In mountainous regions, topography serves as a reliable indicator of local habitat conditions, including temperature and humidity, and significantly influences light availability, hydrology, and soil development. As a result, topography gives rise to diverse forest communities with distinct properties, thereby contributing to variations in individual ecosystem functions [12–15]. However, limited research has explored how ecosystem multifunctionality varies across different topographies. Understanding such variations is crucial for comprehending ecological processes wherein multiple functions are influenced by topography and for implementing effective ecosystem management strategies aimed at maximizing overall multifunctionality [16].

Ecosystem multifunctionality involves the consideration of numerous functions, but equally important is understanding the relationships between these functions. The trade-offs or synergies between functions play a pivotal role in achieving a high level of forest multifunctionality [5,17,18]. These relationships not only shape how ecosystem multifunctionality is evaluated but also directly impact the sustainability of multiple ecosystem functions [19]. In the short term, topography presents the most challenging environmental factor to alter, with the exception of geological disasters such as debris flows [20]. Hence, identifying topographic variations in the trade-offs or synergies between functions holds meaningful practical implications. In forest management, achieving high levels of forest performance and sustainability, especially in complex topographies, entails adjusting the relationships between multiple functions based on topographic conditions [21,22].

Subtropical forests, which often feature complex topography, encompass a quarter of China's land area [23,24], holding significant ecological importance. These forests are among the world's most productive ecosystems, playing crucial roles in regulating global and regional climates and biogeochemical cycles [25–27]. Additionally, the forests provide essential services like wood production and non-wood products, directly impacting human livelihoods. However, with the onset of global climate change and increased human activities such as land use and industrialization, subtropical evergreen broad-leaved forests face mounting pressure [28], leading to a gradual decline in their service functions. Despite recognizing the significance of subtropical evergreen broad-leaved forests, the topographic variations in their multifunctionality remain unclear, posing challenges to effectively enhancing forest multifunctionality across diverse topographic conditions during restoration or reconstruction efforts in subtropical regions. Thus, our study aims to investigate the correlation between topography and forest multifunctionality in an old-growth subtropical forest system. Our objectives include (1) quantifying topographic variations in individual forest functions; (2) analyzing topographic variations in forest multifunctionality; and (3) exploring topographic variations in the trade-offs and synergies between forest functions. This research endeavors to provide valuable insights into subtropical forest restoration and management practices.

2. Materials and Methods

2.1. Study Site and 20 hm² Permanent Forest Plot

We conducted our study in Tiantong National Forest Park (Figure 1; 29°48′ N, 121°47′ E), situated in a typical subtropical monsoon climate zone. This climate is distinguished by hot and rainy summers and cold and dry winters. The area experiences a mean annual precipitation of 1374.7 mm and a mean annual temperature of 16.2 °C [29]. The soil types vary in texture from sandy to silty clay loam, with soil pH ranging from 4.4 to 5.1 [30].

Our sampling was carried out within a 20 hm² (500 m × 400 m) permanent forest plot, referred to as the Tiantong plot, established in 2010 within the core area of Tiantong Park. This area is designated as a temple fengshui forest and features a typical subtropical evergreen broad-leaved forest vegetation type, characterized by a mature plant community with a complex structure [31]. Dominant plant families include Theaceae, Lauraceae, and Fagaceae [31]. Following a standardized field protocol [31,32], trees and shrubs with a stem diameter at breast height (DBH) of ≥ 1 cm were counted. Individuals with DBH ≥ 5 cm were measured with a diameter tape, while individuals with 1 cm $\leq$ DBH $<$ 5 cm were measured with a digital vernier caliper. In the initial census conducted in 2010, a total of 94,603 individuals with DBH ≥ 1 cm were identified, representing 152 species, 94 genera, and 51 families. The second census of the plot took place in 2015. Between the two censuses, researchers recorded 27,837 surviving trees with DBH ≥ 5 cm, 85,704 surviving individuals with 1 cm ≤ DBH < 5 cm, and 58,547 newly recruited individuals whose DBH did not reach 1 cm in the first census but reached 1 cm in the re-census.

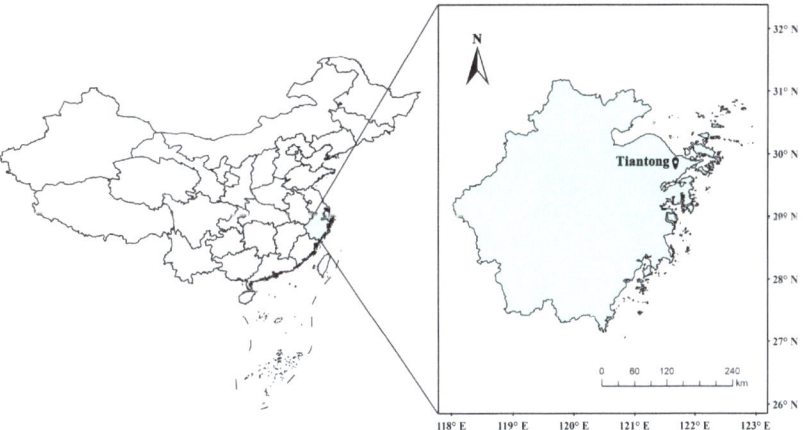

Figure 1. Location of the study site in Tiantong National Forest Park, Zhejiang Province, China.

Furthermore, the plot has a complex topography [31,33], with an elevation difference of nearly 300 m. The highest elevation is 602.89 m, and the lowest elevation is 304.26 m. The plot features two prominent ridges that span from south to north, resulting in a general elevation gradient from high in the north to low in the south, primarily oriented towards the southeast slope. According to the standardized method of Condit et al. (1998), we partitioned the 20 hm² plot into 500 subplots (20 × 20 m). The elevations of each subplot were determined by obtaining the average of the four corner elevations using a total station instrument. Subsequently, the slope and convexity of each subplot were calculated based on the elevation data [34,35].

2.2. Topography Classification

Wang et al. (2023) found that the slope did not affect forest multifunctionality in this plot due to the slope being large overall [33]. Thus, we analyzed the topographic variation in ecosystem multifunctionality in two scenarios. In the first scenario, topography classification was based on elevation. Subplots with elevations higher than the mean elevation of the Tiantong plot were categorized as high elevation, while those with elevations lower than the mean were classified as low elevation [36]. In the second scenario, topography classification was based on convexity using the multivariate regression tree [14,36]. Subplots were divided into three topographic categories: valley (convexity < −2), ridge (convexity ≥ 2), and slope (−2 ≤ convexity < 2).

2.3. Quantification of FM

The forest multifunctionality (FM) was assessed based on eight forest functions encompassing aspects of production, regeneration, nutrient cycling, and carbon stock [3,37]. These functions included aboveground biomass (AGB), woody productivity (Pro), litter production (Lit), sapling growth (Gro), sapling recruitment (Rec), soil nitrogen stock (SNstock), soil phosphorus stock (SPstock), and soil organic carbon stock (SOCstock). To estimate AGB, we used a general allometric equation [38], adjusted for site-specific environment stress factors, species' wood density (ρ, g·cm^{-3}), and tree DBH (cm). This equation has been validated by previous studies to reliably predict biomass in trees and shrubs within this site [39,40]. AGB was then calculated as the biomass of all trees with DBH ≥ 5 cm. Pro was determined as the mean annual AGB increments between 2010 and 2015 [41,42]. For litter production estimation, 187 traps were deployed in the 20 hm² plot at a spacing of >28.8 m in a relatively regular pattern in August 2008 [43]. Litter was collected from these traps twice per month, and litter production was quantified using data from January 2011 to December 2018. Gro was quantified by assessing the growth of trees with DBH ranging from 1 cm to 5 cm [3,16]. Similarly, Rec was determined by identifying saplings with DBH less than 1 cm in 2010 that exceeded 1 cm in the re-census conducted in 2015 [44,45]. SOCstock was estimated for the 0–10 cm soil layer [46–48] and calculated using measurements of soil organic carbon, soil bulk density, and soil sampling depth [49]. Similarly, SNstock and SPstock were expressed as nitrogen and phosphorus stock in the 0–10 cm soil layer [46,50]. After the removal of missing data and obvious outliers, our analysis focused on 166 subplots with litter traps positioned at their centers.

To quantify SOCstock, SNstock, and SPstock, we collected soil samples in 2011. Within each of the 500 subplots (20 × 20 m), soil sampling was conducted by collecting a sample from the southwest corner and two additional samples positioned at random compass directions, either 2 m, 5 m, or 8 m away [51]. This sampling strategy ensured comprehensive coverage of the entire plot, resulting in a total of 1292 soil samples collected. SOC content was determined using the potassium dichromate oxidation ($K_2Cr_2O_7$-H_2SO_4) method, while soil nitrogen (TN) was quantified using an Elemental Analyzer (vario MICRO cube, Elementar). Soil phosphorus (TP) levels were estimated using a flow-injection autoanalyzer (Skalar, The Netherlands), and soil bulk density was assessed using volumetric rings obtained during soil sample collection [47,52,53]. For the 166 subplots mentioned earlier, the actual soil conditions of each subplot were calculated using a standard block kriging approach based on the dataset of 1292 soil samples.

For the quantification of FM, we applied two widely used approaches: averaging and single threshold [5,54]. Integrating these methods allows for a more comprehensive understanding of FM due to their distinct strengths and weaknesses [33,55–57]. In the averaging approach, each individual ecosystem function was first standardized using Z-scores. Subsequently, the standardized Z-score variables of the eight forest functions were averaged, following the method outlined by Maestre et al. (2012) [58]. In the threshold-based index of multifunctionality, FM was determined by counting the number of ecosystem functions exceeding a predefined threshold [5]. Since there is no universal criterion for defining when an ecosystem function is functional, a 50% threshold, analogous to EC50 in ecotoxicology, was recommended [59]. EC50 represents the concentration that elicits a 50% maximum effect. Therefore, the eight forest functions mentioned earlier were initially scaled to values between 0 and 1 using the formula $f(x) = (x - min(x)) / (max(x) - min(x))$ [60]. Subsequently, the threshold value of 50% was applied to calculate FM in this study. For further details, refer to the works of Byrnes et al. (2014) and Wang et al. (2023) [5,33].

2.4. Statistical Analyses

All analyses were conducted in R 3.4.3 [61]. In this study, a non-parametric test was used to compare whether there were significant differences in individual forest functions and forest multifunctionality across different topographies. Pairwise Pearson correlations between forest functions were calculated as the relationships between functions. The cortest.jennrich function [62] in the psych package was employed to conduct a chi-square test, determining whether the correlation matrices (representing the overall degree of synergies or trade-offs between multiple functions) in different topographies exhibited significant differences. A significance level of $p < 0.05$ was used, indicating significant differences in overall synergies or trade-offs between multiple functions. Additionally, Fisher's test was conducted using the cocor function in the cocor package to assess significant differences between each pair of functions in different topographies [7,63]. A significance level of $p < 0.05$ in Fisher's test indicated significant differences between the two correlations.

3. Results

3.1. Topographic Variation in Individual Forest Functions

In the Tiantong plot, the average values of AGB, Pro, Lit, Gro, Rec, SNstock, SPstock, and SOCstock were 79.22 Mg/hm^2, 1.82 Mg/hm^2, 6.0 Mg/hm^2, 2180.32 cm^2/hm^2,

545.72 stems/hm^2, 2.59 Mg/hm^2, 0.22 Mg/hm^2, and 37.16 Mg/hm^2, respectively (Table 1). Correspondingly, the coefficients of variation among different subplots were 0.32, 0.43, 0.19, 0.53, 0.90, 0.30, 0.54, and 0.33, respectively, indicating considerable heterogeneity in individual functions overall. In addition, the multiple individual functions varied with topography (Figure 2). For example, the low elevation had visibly higher AGB, and the ridge had visibly higher SPstock. Figure 2 also shows that where there is lower AGB, there is more Gro and Rec. Where there is more Gro, there is more Rec. Subplots with higher SNstock often had higher SOCstock.

Table 1. Eight forest functions in the Tiantong plot.

Forest Functions	Mean Value	Maximum Value	Minimum Value	Coefficient of Variance
AGB (Mg/hm^2)	79.22	163.38	8.70	0.32
Pro (Mg/hm^2)	1.82	5.22	0.29	0.43
Lit (Mg/hm^2)	6.00	9.92	3.03	0.19
Gro (cm^2/hm^2)	2180.32	6040.16	350.55	0.53
Rec (stems/hm^2)	545.72	2605	50	0.90
SNstock (Mg/hm^2)	2.59	4.88	1.12	0.30
SPstock (Mg/hm^2)	0.22	0.63	0.06	0.54
SOCstock (Mg/hm^2)	37.16	76.42	15.15	0.33

AGB, aboveground biomass; Pro, woody productivity; Lit, litter production; Gro, sapling growth; Rec, sapling recruitment; SNstock, soil nitrogen stock; SPstock, soil phosphorus stock; SOCstock, soil organic carbon stock.

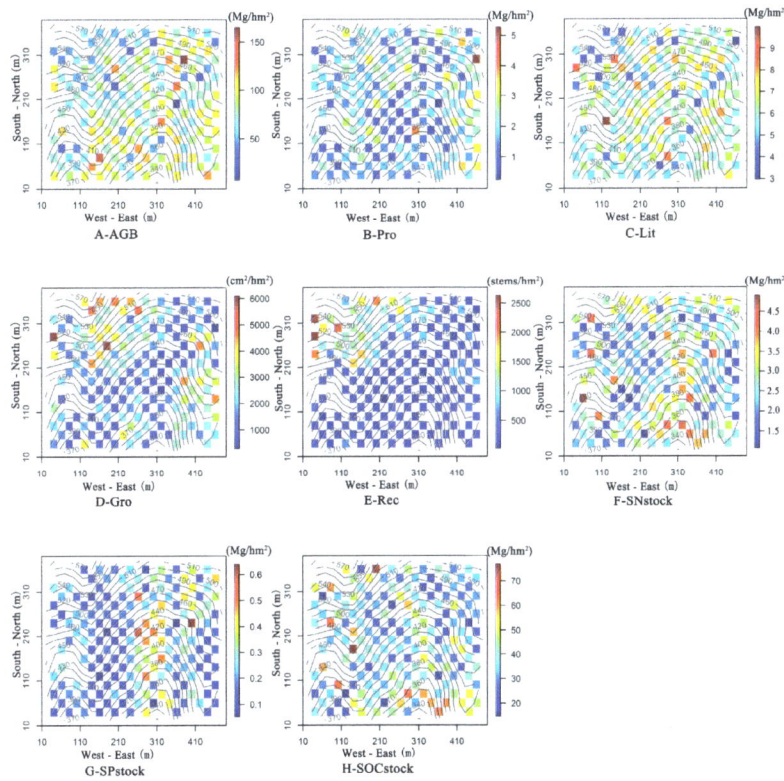

Figure 2. Spatial distribution of eight forest functions in the Tiantong 20 hm^2 plot. Contour lines in the map indicate elevation and the colors indicate biomass values. A, B, C, D, E, F, G, and H represent the spatial distribution of AGB, Pro, Lit, Gro, Rec, SNstock, SPstock, and SOCstock, respectively.

We compared the differences in eight forest functions between the low elevation and the high elevation. AGB was significantly higher at the low elevation compared to the high elevation ($p < 0.05$). However, other functions such as Pro, Gro, and Rec were significantly lower at the low elevation (Figure 3). When considering the differences in eight forest functions among valley, slope, and ridge, AGB was significantly higher on the ridge and the slope compared to the valley. Additionally, Pro, Rec, and SOCstock were significantly higher on the ridge compared to both the valley and the slope (Figure 4). Regarding Lit, there was no significant difference between the ridge and the other two topographies, but Lit on the slope was significantly higher than that in the valley. Gro was significantly higher on the ridge than in the valley, while there was no significant difference in the slope compared to either the valley or the ridge. SNstock in the valley was higher than that on the slope, although there was no significant difference between the ridge and the other two topographies. In terms of SPstock, the order was valley > slope > ridge.

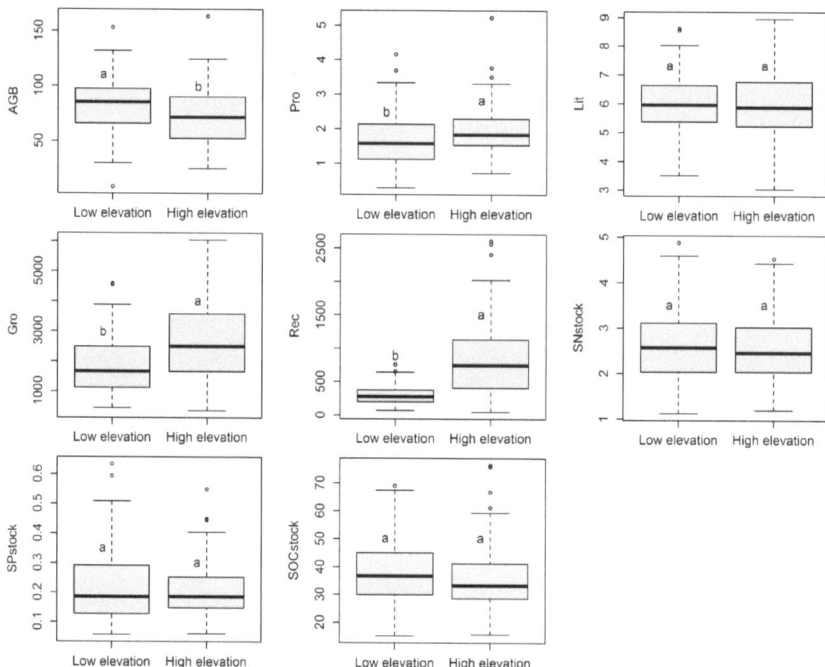

Figure 3. Differences in eight forest functions between low elevation and high elevation. Different lowercase letters indicate significant differences.

3.2. Topographic Variation in Forest Multifunctionality

In the Tiantong plot, FMA ranged from −0.96 to 1.19 while FMt50 ranged from 1 to 7 with a coefficient of variation of 0.35 (Figure 5). These results indicate spatial heterogeneity in forest multifunctionality. The distribution of the FM index (FMA and FMt50) illustrates variations in forest multifunctionality across different topographies. Specifically, compared to higher elevations, lower elevations exhibited higher FMA values, although FMt50 showed no significant difference between the two elevations (Figure 6A). Furthermore, when comparing valley, slope, and ridge, the ridge displayed higher FMA and FMt50 values (Figure 6B).

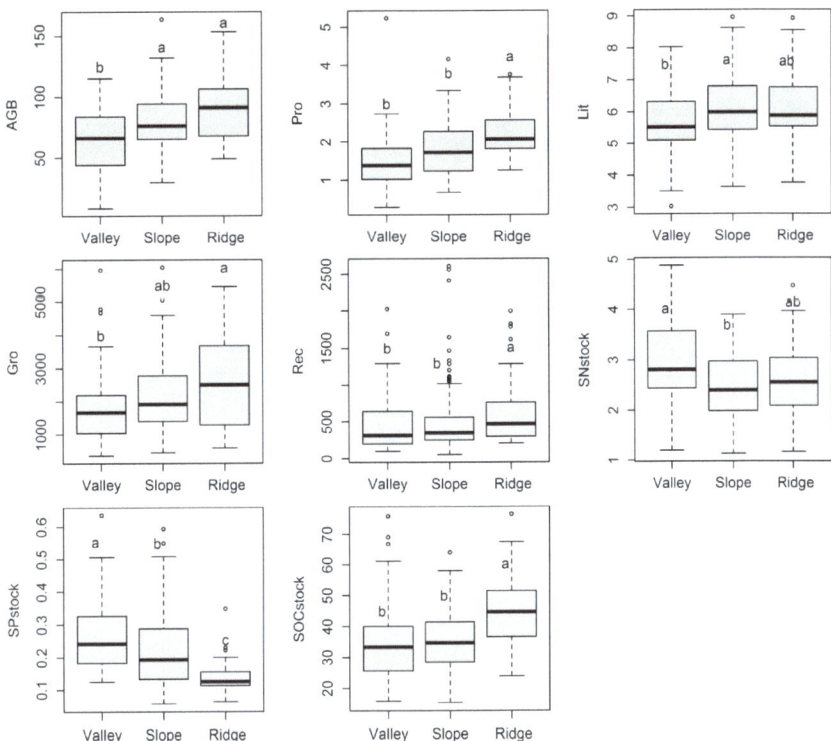

Figure 4. Differences in eight forest functions among valley, slope, and ridge. Different lowercase letters indicate significant differences.

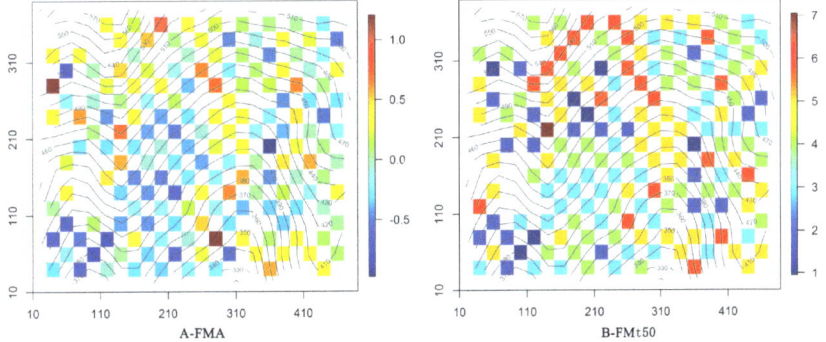

Figure 5. Spatial distribution of forest multifunctionality in the Tiantong 20 hm^2 plot. Contour lines in the map indicate elevation and the colors indicate forest multifunctionality values. A and B represent the spatial distribution of FMA (averaged forest multifunctionality) and FMt50 (forest multifunctionality at 50% threshold level), respectively.

3.3. Topographic Variation in Trade-Offs and Synergies between Functions

Compared to a high elevation, the mean value of relationships at a low elevation is significantly higher ($p < 0.05$), indicating that the overall degree of synergies between functions at a low elevation is significantly higher than that at a high elevation (Figure 7A). Among the relationships, the correlation between AGB and SOCstock changed from a synergistic relationship ($r = 0.21$) at a low elevation to a trade-off relationship ($r = -0.12$) at a high elevation (Figure 7B). The correlation between SPstock and SOCstock also changed from a synergy (r = 0.21) at a low elevation to a trade-off correlation (r = −0.11) at a high elevation. The correlation of soil N and P storage changed from a strong synergistic relationship (r = 0.51) at a low elevation to a weak synergy at a high elevation (r = 0.12). The correlation between Rec and SPstock changed from a weak trade-off at a low elevation (r = −0.11) to a strong trade-off relationship at a high elevation (r = −0.4).

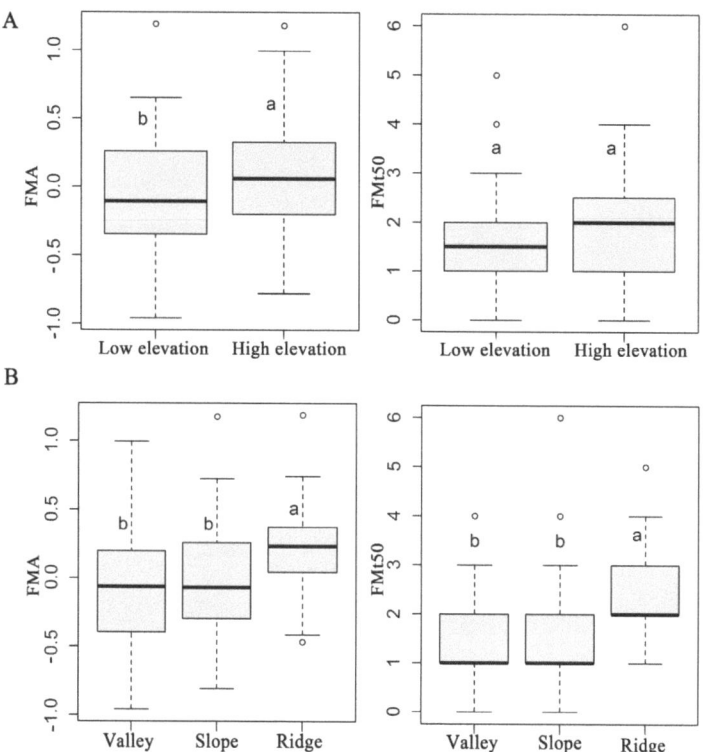

Figure 6. Differences in forest multifunctionality among different topographies in two scenarios. (**A**) Differences between low elevation and high elevation. (**B**) Differences among valley, slope and ridge. Different lowercase letters indicate significant differences.

When considering the differences between the valley, slope, and ridge, the mean value of relationships between functions in the ridge was significantly lower than that in the valley and the slope (Figure 8A). This indicates that the overall degree of synergies between functions in the ridge was significantly lower than that in the valley and the slope ($p < 0.05$). The trade-off degree between AGB and Gro in the ridge was stronger than that in the slope ($p < 0.05$) and the valley ($0.05 < p < 0.1$), while there was no significant difference between the valley and the slope (Figure 8B). The correlation between Pro and Lit was significantly different in the valley and in the slope, which was a synergistic relationship (r = 0.22) and a tradeoff relationship (r = −0.27), respectively. The correlation between Gro and SNstock changed from a synergistic relationship in the ridge (r = 0.32) to a trade-off relationship in the slope. The synergistic degree of SNstock and SOCstock (r = 0.64) in the ridge was significantly lower than that in the two other topographies.

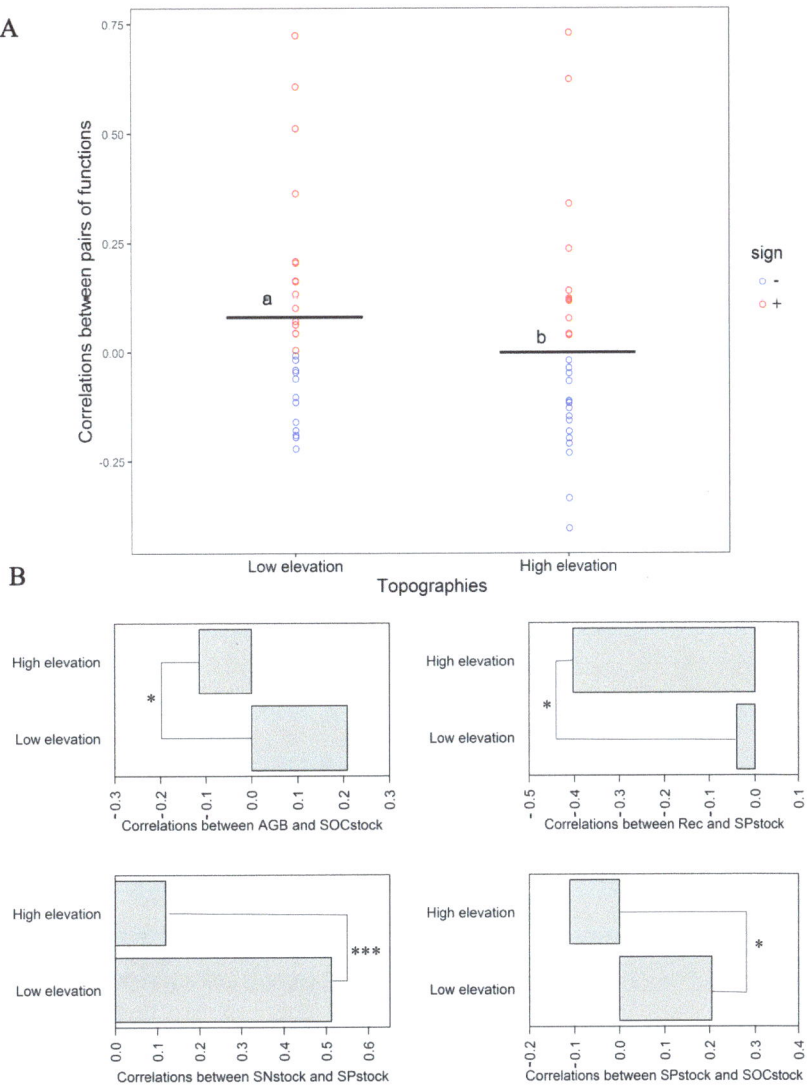

Figure 7. Differences in the correlations between forest functions between low elevation and high elevation. (**A**) A comprehensive picture of Pearson correlations between pairs of forest functions in two topographies. Red circles (+) mean synergies between pairs of forest functions, blue circles mean trade-off, and the long horizontal line represents the mean value of the correlations between pairs of functions. (**B**) Correlations between pairs of forest functions with significant differences between the two topographies. The significant differences in (**A**,**B**) are based on the Chi-square test and Fisher's test, respectively. Different lowercase letters indicate significant differences in the correlations between functions between low elevation and high elevation. * $p < 0.05$, *** $p < 0.001$.

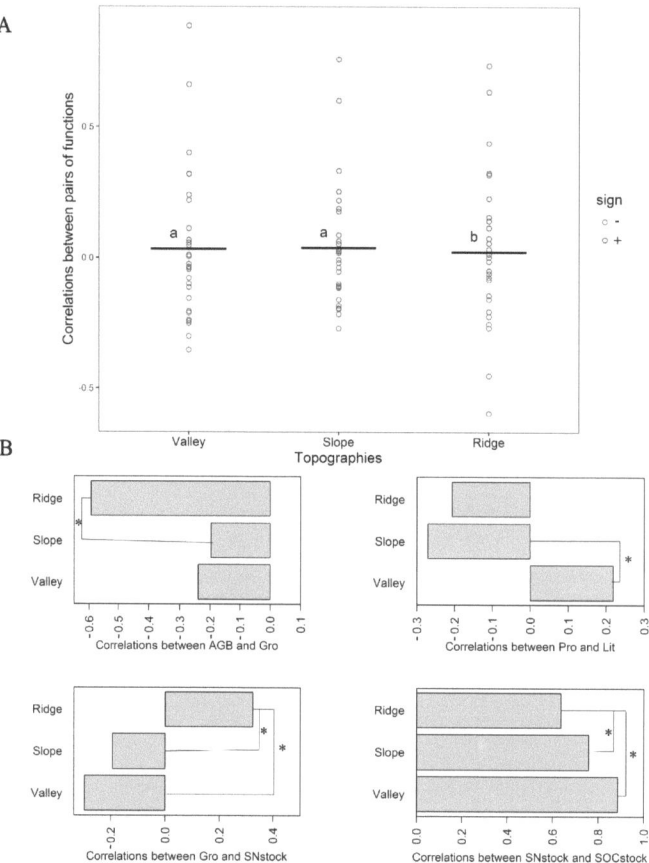

Figure 8. Differences in the correlations between forest functions among valley, slope, and ridge. (**A**) A comprehensive picture of Pearson correlations between pairs of forest functions among three topographies. Red circles (+) mean synergies between pairs of forest functions, blue circles (−) mean trade-offs, and the long horizontal line represents the mean value of the correlations between pairs of functions. (**B**) Correlations between pairs of forest functions with significant differences among the three topographies. The significant differences in (**A**,**B**) are based on the Chi-square test and Fisher's test, respectively. Different lowercase letters indicate significant differences in the correlations between functions among different topographies. * $p < 0.05$.

4. Discussion

4.1. Effects of Topography on Individual Forest Functions and Forest Multifunctionality

The occurrence of multiple individual functions and forest multifunctionality varied with topography. On the one hand, topography significantly influences the distribution and intensity of typhoons [12,14]. Previous studies have highlighted the frequent occurrence of typhoons in the study area, with their impact being notably more pronounced at higher elevations compared to lowlands [14,64]. The strong winds associated with typhoons directly constrained tree height and biomass [65] at higher elevations, resulting in simpler community structures and lower biomass. On the other hand, topography also played a role in redistributing the intensity of snow and ice disasters, with high elevations experiencing greater intensity of occasional snow and ice disasters [66]. For instance, the snow and ice disaster in southern China in 2008 had a significant impact on this plot, leading to increased forest gap density and gap fraction at higher elevations. Additionally, as elevation increased, temperatures decreased and ultraviolet radiation intensified [67], contributing to lower aboveground biomass and increased intra-forest light availability at higher elevations [68]. This increased light availability significantly promoted tree growth in the Tiantong plot [69]. Consequently, woody productivity, sapling growth, and recruitment were significantly higher at higher elevations compared to lower elevations, consistent with previous findings [70]. To adapt to the environment characterized by frequent typhoon disturbances, plant communities exhibited traits such as high growth rates, recruitment rates, and biomass turnover [70]. Overall, higher productivity, sapling growth, and sapling recruitment at high elevations drove higher FMA, but the results showed that FMt50 has no difference between the high elevation and the low elevation.

In this plot, valleys were more susceptible to flooding, river formation, and small-scale landslides [71], which reduced the living space available for plants. These disturbances, coupled with the thin soil layer, resulted in significantly lower AGB, reduced Pro, and hindered regeneration (Rec and Gro) in the valleys. However, in terms of SPstock, the order was valley > slope > ridge. This could be attributed to differences in community species composition among the different topographies. In valleys, small-scale landslides often led to the formation of forest gaps, which favored the survival of deciduous species. Deciduous species, with their higher leaf nutrient content, primarily rely on external nutrient cycling, contributing to decomposition and turnover, thus, benefiting the accumulation of soil nitrogen and phosphorus [72,73]. In turn, it also resulted in higher SNstock in the valley than that in the slope. Previous studies have demonstrated that natural ecosystems possess an inherent ability to accumulate nitrogen for carbon fixation [74]. The higher SOC accumulation in the ridge may coincide with increased soil nitrogen accumulation, placing SNstock levels between those of the valley and the slope. Overall, compared to the valley and the slope, we observed that higher Pro, Rec, and SOCstock contribute to higher forest multifunctionality index (FMA and FMt50) values in the ridge, designating it as a forest multifunctionality "hotspot" in the Tiantong plot. These results highlight the significance of topography as a key driver of forest multifunctionality in this plot, influencing tree composition and ecosystem processes.

4.2. Effects of Topography on Trade-Offs and Synergies between Functions

The overall degree of synergies between functions at high elevations is significantly lower than that at low elevations. For specific pairs of functions, the synergistic degree of the four pairs with significant changes is notably lower at high elevations, indicating a higher degree of trade-offs. This suggests that achieving high forest multifunctionality is more challenging at high elevations, potentially indicating that plant communities in stressed environments adopt more trade-off functional strategies, consistent with previous studies [75]. Scientists have observed that under water stress, plants develop more and longer roots to access water and accumulate more underground biomass, resulting in a greater trade-off between AGB and SOCstock [75,76]. Similarly, the greater trade-off observed between AGB and SOCstock in our study site's high elevation may be attributed to plants allocating more resources to the underground part to enhance resistance to strong typhoons, possibly leading to a decrease in AGB [77,78]. Additionally, the simpler forest structure at high elevations may contribute to weaker synergies between functions. Previous studies in this plot have demonstrated that structural diversity promotes multiple functions simultaneously and strengthens synergies between functions [33,79]. This suggests that forest management strategies should focus on regulating forest structure to build multifunctional forests.

The overall degree of synergies between functions in the ridge is significantly lower than in the valley and the slope. Regarding specific pairs of functions, there is no consistency observed from the ridge to the valley or the slope; some synergies weaken while others strengthen. The stronger trade-offs between AGB and Gro and the weaker synergy between SNstock and SPstock contribute to the lower overall synergies between functions in the ridge. The strong trade-off between AGB and Gro may be attributed to smaller forest gaps and higher canopy density in the ridge [14], where longer periods of photosynthetically active radiation [68] and lower intensity of water interference may benefit tree survival. Species composition also plays a role in regulating multiple ecosystem functions in forests [80]. Different species adapt to specific topography and thrive in it [81,82]. Therefore, topography can indirectly regulate the synergistic and trade-off relationships between functions by influencing species composition [79]. Our results suggest that forest management should prioritize topographic conditions. It is essential to consider different functions in various topographies and regulate the relationships between functions according to these conditions.

5. Conclusions

Our study demonstrates the critical role of topography in ecosystem multifunctionality within an old-growth subtropical evergreen broad-leaved forest. Topography not only influences the level of forest multifunctionality but also shapes the synergistic and trade-off relationships between functions. Our findings highlight the importance of considering topographic conditions in forest ecosystem management such as regulating the synergies and trade-offs between functions.

Author Contributions: Conceptualization, J.W. and X.W.; methodology, J.W. and Y.L. (Yepu Li); project administration, X.W.; resources, X.W.; data curation, Q.Y. (Qingsong Yang), Y.Q. and Q.Y. (Qiushi Yu); supervision, X.W.; validation, J.W. and Q.Y. (Qingsong Yang); writing—original draft preparation, J.W.; writing—review and editing, J.W., H.X., Y.L. (Yuying Li), M.J. and X.W.; visualization, J.W., Z.C. and X.W. All authors contributed to the article and agreed to the submitted version. All authors have read and agreed to the published version of the manuscript.

Funding: This research was funded by State Key Laboratory of Geo-Information Engineering and Key Laboratory of Surveying and Mapping Science and Geospatial Information Technology of MNR, CASM, and Science and Technology Program of Guangdong Province (grant number 2021B1212100003). Acknowledgement for the data support from "Guangdong Geographical Science Data Center".

Data Availability Statement: The data presented in this study are available on request from the corresponding author.

Conflicts of Interest: The authors declare no conflict of interest.

References

1. Costanza, R.; d'Arge, R.; de Groot, R.; Farber, S.; Grasso, M.; Hannon, B.; Limburg, K.; Naeem, S.; O'Neill, R.V.; Paruelo, J.; et al. The value of the world's ecosystem services and natural capital. *Nature* **1997**, *387*, 253–260. [CrossRef]
2. Manning, P.; Van Der Plas, F.; Soliveres, S.; Allan, E.; Maestre, F.T.; Mace, G.; Whittingham, M.J.; Fischer, M. Redefining ecosystem multifunctionality. *Nat. Ecol. Evol.* **2018**, *2*, 427–436. [CrossRef]
3. Ratcliffe, S.; Wirth, C.; Jucker, T.; van Der Plas, F.; Scherer-Lorenzen, M.; Verheyen, K.; Allan, E.; Benavides, R.; Bruelheide, H.; Ohse, B.; et al. Biodiversity and ecosystem functioning relations in European forests depend on environmental context. *Ecol. Lett.* **2017**, *20*, 1414–1426. [CrossRef] [PubMed]
4. Garland, G.; Banerjee, S.; Edlinger, A.; Oliveira, E.M.; Herzog, C.; Wittwer, R.; Philippot, L.; Maestre, F.; van der Heijden, M.G.A. A closer look at the functions behind ecosystem multifunctionality: A review. *J. Ecol.* **2021**, *109*, 600–613. [CrossRef]
5. Byrnes, J.E.K.; Gamfeldt, L.; Isbell, F.; Lefcheck, J.S.; Griffin, J.N.; Hector, A.; Cardinale, B.J.; Hooper, D.U.; Dee, L.E.; Duffy, J.E. Investigating the relationship between biodiversity and ecosystem multifunctionality: Challenges and solutions. *Methods Ecol. Evol.* **2014**, *5*, 111–124. [CrossRef]
6. Hector, A.; Bagchi, R. Biodiversity and ecosystem multifunctionality. *Nature* **2007**, *448*, 188–190. [CrossRef] [PubMed]
7. Felipe-Lucia, M.R.; Soliveres, S.; Penone, C.; Manning, P.; van der Plas, F.; Boch, S.; Prati, D.; Ammer, C.; Schall, P.; Gossner, M.M.; et al. Multiple forest attributes underpin the supply of multiple ecosystem services. *Nat. Commun.* **2018**, *9*, 4839. [CrossRef] [PubMed]
8. Jing, X.; He, J. Relationship between biodiversity, ecosystem multifunctionality and multiserviceability: Literature overview and research advances. *Chin. J. Plant Ecol.* **2021**, *45*, 1094–1111. [CrossRef]
9. Meyer, V.; Saatchi, S.S.; Chave, J.; Dalling, J.W.; Fricker, G.A.; Robinson, C.; Neumann, M.; Hubbel, S. Detecting tropical forest biomass dynamics from repeated airborne lidar measurements. *Biogeosciences* **2013**, *10*, 5421–5438. [CrossRef]
10. Xu, Y.; Franklin, S.B.; Wang, Q.; Shi, Z.; Luo, Y.; Lu, Z.; Zhang, J.; Qiao, X.; Jiang, M. Topographic and biotic factors determine forest biomass spatial distribution in a subtropical mountain moist forest. *For. Ecol. Manag.* **2015**, *357*, 95–103. [CrossRef]

11. Jin, Y.; Russo, S.E.; Yu, M. Effects of light and topography on regeneration and coexistence of evergreen and deciduous tree species in a Chinese subtropical forest. *J. Ecol.* **2018**, *106*, 1634–1645. [CrossRef]
12. Kubota, Y.; Murata, H.; Kikuzawa, K. Effects of topographic heterogeneity on tree species richness and stand dynamics in a subtropical forest in Okinawa Island, southern Japan. *J. Ecol.* **2004**, *92*, 230–240. [CrossRef]
13. Jucker, T.; Bongalov, B.; Burslem, D.F.R.P.; Nilus, R.; Dalponte, M.; Lewis, S.L.; Phillips, O.L.; Qie, L.; Coomes, D.A. Topography shapes the structure, composition and function of tropical forest landscapes. *Ecol. Lett.* **2018**, *21*, 989–1000. [CrossRef] [PubMed]
14. Zhang, Z.; Ma, Z.; Liu, H.; Zeng, Z.; Xie, Y.; Fang, X.; Wang, X. Topographic distribution patterns of forest gap within an evergreen broad-leaved forest in Tiantong region of Zhejiang Province, eastern China. *Chin. J. Appl. Ecol.* **2013**, *24*, 621–625.
15. Wang, J.; Xu, H.; Li, Y.; Zhou, Z.; Lin, M.; Luo, T.; Chen, D. Effects of topographic heterogeneity on community structure and diversity of woody plants in Jianfengling tropical montane rainforest. *Sci. Silvae Sin.* **2018**, *54*, 1–11. [CrossRef]
16. Van der Plas, F.; Ratcliffe, S.; Ruiz-Benito, P.; Scherer-Lorenzen, M.; Verheyen, K.; Wirth, C.; Zavala, M.A.; Ampoorter, E.; Baeten, L.; Barbaro, L.; et al. Continental mapping of forest ecosystem functions reveals a high but unrealised potential for forest multifunctionality. *Ecol. Lett.* **2018**, *21*, 31–42. [CrossRef] [PubMed]
17. Vinebrooke, R.D.; Cottingham, K.L.; Norberg, J.; Scheffer, M.; Dodson, S.; Maberly, S.; Sommer, U. Impacts of multiple stressors on biodiversity and ecosystem functioning: The role of species co-tolerance. *Oikos* **2004**, *104*, 451–457. [CrossRef]
18. Zavaleta, E.S.; Pasari, J.R.; Hulvey, K.B.; Tilman, G.D. Sustaining multiple ecosystem functions in grassland communities requires higher biodiversity. *Proc. Natl. Acad. Sci. USA* **2010**, *107*, 1443–1446. [CrossRef]
19. Dai, E.; Wang, X.; Zhu, J.; Gao, J. Progress and perspective on ecosystem services trade-offs. *Adv. Earth Sci.* **2015**, *30*, 1250–1259. [CrossRef]
20. Markesteijn, L. Drought Tolerance of Tropical Tree Species: Functional Traits, Trade-Offs and Species Distribution. Ph.D. Thesis, Wageningen University, Wageningen, The Netherlands, 2010.
21. Powers, B.F.; Ausseil, A.G.; Perry, G.L.W. Ecosystem service management and spatial prioritisation in a multifunctional landscape in the Bay of Plenty, New Zealand. *Australas. J. Environ. Manag.* **2020**, *27*, 275–293. [CrossRef]
22. Zhang, J. Quantification of Tradeoffs on Forest Ecosystem Services in MOUNT FUNIU REGION. Ph.D. Thesis, Henan University, Zhengzhou, China, 2019. [CrossRef]
23. Song, Y.; Chen, X.; Wang, X. Studies on evergreen broad-leaved forests of China: A retrospect and prospect. *J. East China Norm. Univ. Nat. Sci.* **2005**, *1*, 1–8. [CrossRef]
24. Ouyang, S.; Xiang, W.; Gou, M.; Chen, L.; Lei, P.; Xiao, W.; Deng, X.; Zeng, L.; Li, J.; Zhang, T.; et al. Stability in subtropical forests: The role of tree species diversity, stand structure, environmental and socio-economic conditions. *Glob. Ecol. Biogeogr.* **2021**, *30*, 500–513. [CrossRef]
25. Melillo, J.M.; McGuire, A.D.; Kicklighter, D.W.; Moore, B.; Vorosmarty, C.J.; Schloss, A.L. Global climate change and terrestrial net primary production. *Nature* **1993**, *363*, 234–240. [CrossRef]
26. Yu, G.; Chen, Z.; Piao, S.; Peng, C.; Ciais, P.; Wang, Q.; Li, X.; Zhu, X. High carbon dioxide uptake by subtropical forest ecosystems in the East Asian monsoon region. *Proc. Natl. Acad. Sci. USA* **2014**, *111*, 4910–4915. [CrossRef] [PubMed]
27. Zhang, Y.; Cristiano, P.M.; Zhang, Y.F.; Campanello, P.I.; Tan, Z.H.; Zhang, Y.P.; Cao, K.F.; Goldstein, G. Carbon Economy of Subtropical Forests. In *Tropical Tree Physiology*; Springer: Cham, Switzerland, 2016; pp. 337–355. [CrossRef]
28. Fan, D.; Hu, W.; Li, B.; Morris, A.B.; Zheng, M.; Soltis, D.E.; Soltis, P.S.; Zhang, Z. Idiosyncratic responses of evergreen broad-leaved forest constituents in China to the late Quaternary climate changes. *Sci. Rep.* **2016**, *6*, 31044. [CrossRef] [PubMed]
29. Yang, Q.; Shen, G.; Liu, H.; Wang, Z.; Ma, Z.; Fang, X.; Zhang, J.; Wang, X. Detangling the effects of environmental filtering and dispersal limitation on aggregated distributions of tree and shrub species: Life stage matters. *PLoS ONE* **2016**, *11*, e0156326. [CrossRef] [PubMed]
30. Song, Y.Q.; Wang, X.R. *Vegetation and Flora of Tiang Tong National Forest Park, Zhejiang Province, China*; Shanghai Science and Technical Literature Press: Shanghai, China, 1995.
31. Yang, Q.; Ma, Z.; Xie, Y.; Zhang, Z.; Wang, Z.; Liu, H.; Li, P.; Zhang, N.; Wang, D.; Yang, H.; et al. Community structure and species composition of an evergreen broadleaved forest in Tiantong's 20 ha dynamic plot, Zhejiang Province, eastern China. *Biodivers. Sci.* **2011**, *19*, 215–223. [CrossRef]
32. Condit, R. *Tropical Forest Census Plots: Methods and Results from Barro Colorado Island, Panama and a Comparison with Other Plots*; Springer Science & Business Media: Dordrecht, The Netherlands, 1998.
33. Wang, J.; Liu, H.; Yang, Q.; Shen, G.; Zhu, X.; Xu, Y.; Wang, X. Topography and structural diversity regulate ecosystem multifunctionality in a subtropical evergreen broadleaved forest. *Front. For. Glob. Chang.* **2023**, *6*, 1309660. [CrossRef]
34. Harms, K.E.; Condit, R.; Hubbell, S.P.; Foster, R.B. Habitat associations of trees and shrubs in a 50-ha neotropical forest plot. *J. Ecol.* **2001**, *89*, 947–959. [CrossRef]
35. Lai, J.; Mi, X.; Ren, H.; Ma, K. Species-habitat associations change in a subtropical forest of China. *J. Veg. Sci.* **2009**, *20*, 415–423. [CrossRef]
36. Xie, Y.; Ma, Z.; Yang, Q.; Fang, X.; Zhang, Z.; Yan, E.; Wang, X. Coexistence mechanisms of evergreen and deciduous trees based on topographic factors in Tiantong region, Zhejiang Province, eastern China. *Biodivers. Sci.* **2012**, *20*, 159–167. [CrossRef]
37. Li, S.; Huang, X.; Lang, X.; Shen, J.; Xu, F.; Su, J. Cumulative effects of multiple biodiversity attributes and abiotic factors on ecosystem multifunctionality in the Jinsha River valley of southwestern China. *For. Ecol. Manag.* **2020**, *472*, 118281. [CrossRef]

38. Chave, J.; Réjou-Méchain, M.; Búrquez, A.; Chidumayo, E.; Colgan, M.S.; Delitti, W.B.; Duque, A.; Eid, T.; Fearnside, P.M.; Goodman, R.C.; et al. Improved allometric models to estimate the aboveground biomass of tropical trees. *Glob. Chang. Biol.* **2014**, *20*, 3177–3190. [CrossRef] [PubMed]
39. Ali, A. The Forest Strata-Dependent Relationships among Environment, Biodiversity and Aboveground Biomass in a Subtropical Forest in Tiantong, Zhejiang Province. Ph.D. Thesis, East China Normal University, Shanghai, China, 2017.
40. Ali, A.; Chen, H.Y.; You, W.H.; Yan, E.R. Multiple abiotic and biotic drivers of aboveground biomass shift with forest stratum. *For. Ecol. Manag.* **2019**, *436*, 1–10. [CrossRef]
41. Ren, S.Y. Responses of Plant Diversity-Productivity Relationship to Canopy Gap Disturbance in Tiantong Evergreen Broad-Leaved Forest. Ph.D. Thesis, East China Normal University, Shanghai, China, 2021. [CrossRef]
42. Ren, S.; Yang, Q.; Liu, H.; Shen, G.; Zheng, Z.; Zhou, S.; Liang, M.; Yin, H.; Zhou, Z.; Wang, X. The driving factors of subtropical mature forest productivity: Stand structure matters. *Forests* **2021**, *12*, 998. [CrossRef]
43. Wang, Z.H. The Temporal and Spatial Distribution Characteristics of Litter Production in an Evergreen Broad-Leaved Forest in Tiantong, Zhejiang Province, Eastern China. Master's Thesis, East China Normal University, Shanghai, China, 2013.
44. Condit, R.; Ashton, P.S.; Manokaran, N.; LaFrankie, J.V.; Hubbell, S.P.; Foster, R.B. Dynamics of the forest communities at Pasoh and Barro Colorado: Comparing two 50-ha plots. *Philos. Trans. R. Soc. B-Biol. Sci.* **1999**, *354*, 1739–1748. [CrossRef] [PubMed]
45. Bin, Y.; Spence, J.; Wu, L.; Li, B.; Hao, Z.; Ye, W.; He, F. Species-habitat associations and demographic rates of forest trees. *Ecography* **2016**, *39*, 9–16. [CrossRef]
46. Guo, L.B.; Gifford, R.M. Soil carbon stocks and land use change: A meta analysis. *Glob. Change Biol.* **2002**, *8*, 345–360. [CrossRef]
47. Li, S.; Su, J.; Liu, W.; Lang, X.; Huang, X.; Jia, C.; Zhang, Z.; Tong, Q. Changes in biomass carbon and soil organic carbon stocks following the conversion from a secondary coniferous forest to a pine plantation. *PLoS ONE* **2015**, *10*, e0135946. [CrossRef]
48. Bleam, W.F. *Soil and Environmental Chemistry*; Academic Press: Pittsburgh, PA, America, 2016.
49. Mann, L.K. Changes in soil carbon storage after cultivation. *Soil Sci.* **1986**, *142*, 279–288. [CrossRef]
50. Tuo, B.; Tian, W.B.; Guo, C.; Xu, M.S.; Zheng, L.; Su, T.; Liu, X.; Yan, E. Latitudinal variation in soil carbon, nitrogen and phosphorus pools across island forests and shrublands in eastern China. *Chin. J. Appl. Ecol.* **2019**, *30*, 2631–2638. [CrossRef]
51. John, R.; Dalling, J.W.; Harms, K.E.; Yavitt, J.B.; Stallard, R.F.; Mirabello, M.; Hubbell, S.P.; Valencia, R.; Navarrete, H.; Vallejo, M.; et al. Soil nutrients influence spatial distributions of tropical tree species. *Proc. Natl. Acad. Sci. USA* **2007**, *104*, 864–869. [CrossRef] [PubMed]
52. Bao, S.D. *Agricultural and Chemistry Analysis of Soil*; China Agriculture Press: Beijing, China, 2000.
53. Lu, R.K. *Analytical Methods of Soil Agrochemistry*; China Agricultural Science and Technology Press: Beijing, China, 2000.
54. Hölting, L.; Beckmann, M.; Volk, M.; Cord, A.F. Multifunctionality assessments-more than assessing multiple ecosystem functions and services? A quantitative literature review. *Ecol. Indic.* **2019**, *103*, 226–235. [CrossRef]
55. Jing, X.; Sanders, N.J.; Shi, Y.; Chu, H.; Classen, A.T.; Zhao, K.; Chen, L.T.; Shi, Y.; Jiang, Y.X.; He, J.S. The links between ecosystem multifunctionality and above-and belowground biodiversity are mediated by climate. *Nat. Commun.* **2015**, *6*, 8159. [CrossRef]
56. Xu, W.; Ma, Z.; Jing, X.; He, J. Biodiversity and ecosystem multifunctionality: Advances and perspectives. *Biodivers. Sci.* **2016**, *24*, 55–71. [CrossRef]
57. Schuldt, A.; Assmann, T.; Brezzi, M.; Buscot, F.; Eichenberg, D.; Gutknecht, J.; Härdtle, W.; He, J.; Klein, A.M.; Kühn, P.; et al. Biodiversity across trophic levels drives multifunctionality in highly diverse forests. *Nat. Commun.* **2018**, *9*, 2989. [CrossRef]
58. Maestre, F.T.; Castillo-Monroy, A.P.; Bowker, M.A.; Ochoa-Hueso, R. Species richness effects on ecosystem multifunctionality depend on evenness, composition and spatial pattern. *J. Ecol.* **2012**, *100*, 317–330. [CrossRef]
59. Gamfeldt, L.; Hillebrand, H.; Jonsson, P.R. Multiple functions increase the importance of biodiversity for overall ecosystem functioning. *Ecology* **2008**, *89*, 1223–1231. [CrossRef]
60. Gamfeldt, L.; Roger, F. Revisiting the biodiversity-ecosystem multifunctionality relationship. *Nat. Ecol. Evol.* **2017**, *1*, 0168. [CrossRef]
61. R Development Core Team. *R Version 3.6.0*; R Foundation for Statistical Computing: Vienna, Austria, 2019.
62. Jennrich, R.I. An asymptotic χ2 test for the equality of two correlation matrices. *J. Am. Stat. Assoc.* **1970**, *65*, 904–912. [CrossRef]
63. Diedenhofen, B.; Musch, J. cocor: A comprehensive solution for the statistical comparison of correlations. *PLoS ONE* **2015**, *10*, e0121945. [CrossRef]
64. Yu, K.; Chen, D.; Xie, H.; Wan, X.; Wu, Y. Characters of typhoon affected Ningbo City and preventive measures. *China Water Resour.* **2017**, *11*, 11–13.
65. Chen, Z.; Hsieh, C.; Jiang, F.; Hsieh, T.; Sun, I. Relations of soil properties to topography and vegetation in a subtropical rain forest in southern Taiwan. *Plant Ecol.* **1997**, *132*, 229–241. [CrossRef]
66. Zhu, J.; Liu, S. *Ecological Research on Forest Disturbances*; China Forestry Publishing House: Beijing, China, 2007.
67. Girardin, C.A.J.; Farfan-Rios, W.; Garcia, K.; Feeley, K.J.; Jørgensen, P.M.; Murakami, A.A.; Pérez, L.C.; Seidel, R.; Paniagua, N.; Claros, A.F.F.; et al. Spatial patterns of above-ground structure, biomass and composition in a network of six Andean elevation transect. *Plant Ecol. Divers.* **2014**, *7*, 161–171. [CrossRef]
68. Ediriweera, S.; Singhakumara, B.M.P.; Ashton, M.S. Variation in canopy structure, light and soil nutrition across elevation of a Sri Lankan tropical rain forest. *For. Ecol. Manag.* **2008**, *256*, 1339–1349. [CrossRef]
69. Liu, H.; Ma, J.; Yang, Q.; Fang, X.; Lin, K.; Zong, Y.; Ar-dak, A.; Wang, X. Relationships between established seedling survival and growth in ever-green broad-leaved forest in Tiantong. *Biodivers. Sci.* **2017**, *25*, 11–22. [CrossRef]

70. Bellingham, P.J.; Tanner, E.V.J. The influence of topography on tree growth, mortality, and recruitment in a tropical montane forest. *Biotropica* **2000**, *32*, 378–384. [CrossRef]
71. Fei, X.Y. Study on Factors Affecting the Stem Growth in an Evergreen Broadleaved Forest in Zhejiang Tiantong. Master's Thesis, East China Normal University, Shanghai, China, 2016.
72. Aerts, R.; Chapin, F.S., III. The Mineral Nutrition of Wild Plants Revisited: A Re-Evaluation of Processes and Patterns. *Adv. Ecol. Res.* **1999**, *30*, 1–67. [CrossRef]
73. Cornwell, W.K.; Cornelissen, J.H.C.; Amatangelo, K.; Dorrepaal, E.; Eviner, V.T.; Godoy, O.; Hobbie, S.E.; Hoorens, B.; Kurokawa, H.; Perez-Harguindeguy, N.; et al. Plant species traits are the predominant control on litter decomposition rates within biomes worldwide. *Ecol. Lett.* **2008**, *11*, 1065–1071. [CrossRef]
74. Yang, Y.; Luo, Y.; Finzi, A.C. Carbon and nitrogen dynamics during forest stand development: A global synthesis. *New Phytol.* **2011**, *190*, 977–989. [CrossRef]
75. Lu, N.; Fu, B.; Jin, T.; Chang, R. Trade-off analyses of multiple ecosystem services by plantations along a precipitation gradient across Loess Plateau landscapes. *Landsc. Ecol.* **2014**, *29*, 1697–1708. [CrossRef]
76. Mokany, K.; Raison, R.J.; Prokushkin, A.S. Critical analysis of root: Shoot ratios in terrestrial biomes. *Glob. Chang. Biol.* **2006**, *12*, 84–96. [CrossRef]
77. Gewin, V. Food: An underground revolution. *Nat. News* **2010**, *466*, 552–553. [CrossRef]
78. Zhang, C.; Qian, W.; Song, L.; Zhao, Q. Growth patterns and environmental adaptions of the tree species planted for ecological remediation in typhoon-disturbed areas-A case study in Zhuhai, China. *Front. Sustain. Cities* **2022**, *4*, 1064525. [CrossRef]
79. Wang, J. Spatial Variation Pattern and Drivers of Multifunctionality of Subtropical Forests in Tiantong, Zhejiang. Ph.D. Thesis, East China Normal University, Shanghai, China, 2023.
80. Sanaei, A.; Ali, A.; Yuan, Z.; Liu, S.; Lin, F.; Fang, S.; Ye, J.; Hao, Z.Q.; Loreau, M.; Bai, E.D.; et al. Context-dependency of tree species diversity, trait composition and stand structural attributes regulate temperate forest multifunctionality. *Sci. Total Environ.* **2023**, *757*, 143724. [CrossRef]
81. Queenborough, S.A.; Burslem, D.F.R.P.; Garwood, N.C.; Valencia, R. Habitat niche partitioning by 16 species of Myristicaceae in Amazonian Ecuador. *Plant Ecol.* **2007**, *192*, 193–207. [CrossRef]
82. Punchi-Manage, R.; Getzin, S.; Wiegand, T.; Kanagaraj, R.; Gunatilleke, C.V.S.; Gunatilleke, I.A.U.N.; Wiegand, K.; Huth, A. Effects of topography on structuring local species assemblages in a Sri Lankan mixed dipterocarp forest. *J. Ecol.* **2013**, *101*, 149–160. [CrossRef]

Disclaimer/Publisher's Note: The statements, opinions and data contained in all publications are solely those of the individual author(s) and contributor(s) and not of MDPI and/or the editor(s). MDPI and/or the editor(s) disclaim responsibility for any injury to people or property resulting from any ideas, methods, instructions or products referred to in the content.

Article

Spatiotemporal Evolution and Prediction of Ecosystem Carbon Storage in the Yiluo River Basin Based on the PLUS-InVEST Model

Lei Li [1], Guangxing Ji [1], Qingsong Li [1], Jincai Zhang [1], Huishan Gao [1], Mengya Jia [2], Meng Li [1] and Genming Li [1,*]

[1] College of Resources and Environmental Sciences, Henan Agricultural University, Zhengzhou 450046, China; leili@stu.henau.edu.cn (L.L.); guangxingji@henau.edu.cn (G.J.); lqsdavid@126.com (Q.L.); zhangjincai3711@163.com (J.Z.); huishan@stu.henau.edu.cn (H.G.); limeng@stu.henau.edu.cn (M.L.)

[2] School of Hydraulic and Environmental Engineering, Changsha University of Science & Technology, Changsha 410114, China; mjia@live.esu.edu

* Correspondence: gm.li@henau.edu.cn

Abstract: Land-use change has a great impact on regional ecosystem balance and carbon storage, so it is of great significance to study future land-use types and carbon storage in a region to optimize the regional land-use structure. Based on the existing land-use data and the different scenarios of the shared socioeconomic pathway and the representative concentration pathway (SSP-RCP) provided by CMIP6, this study used the PLUS model to predict future land use and the InVEST model to predict the carbon storage in the study area in the historical period and under different scenarios in the future. The results show the following: (1) The change in land use will lead to a change in carbon storage. From 2000 to 2020, the conversion of cultivated land to construction land was the main transfer type, which was also an important reason for the decrease in regional carbon storage. (2) Under the three scenarios, the SSP126 scenario has the smallest share of arable land area, while this scenario has the largest share of woodland and grassland land area, and none of the three scenarios shows a significant decrease in woodland area. (3) From 2020 to 2050, the carbon stocks in the study area under the three scenarios, SSP126, SSP245, and SSP585, all show different degrees of decline, decreasing to $36,405.0204 \times 10^4$ t, $36,251.4402 \times 10^4$ t, and $36,190.4066 \times 10^4$ t, respectively. Restricting the conversion of land with a high carbon storage capacity to land with a low carbon storage capacity is conducive to the benign development of regional carbon storage. This study can provide a reference for the adjustment and management of future land-use structures in the region.

Keywords: land-use change; carbon stocks; CMIP6; PLUS model; InVEST model

1. Introduction

Since the second industrial revolution, a large amount of greenhouse gases, mainly carbon dioxide, have been emitted, resulting in rising global temperatures and frequent extreme weather [1–3]. Since China's reform and opening up, China's economy has developed rapidly, the proportion of construction land has risen rapidly, and China's carbon emissions have reached first place in the world [4]. In the face of common global challenges, General Secretary Xi Jinping made an important speech at the 75th session of the United Nations General Assembly on the carbon peak in 2030 and carbon neutrality in 2060 [5]. The process of the carbon cycle in terrestrial ecosystems is often accompanied by carbon exchange, and land-use change is the main factor affecting regional carbon balance [6]. Studying the intrinsic relationship between carbon storage and land use can provide a reference for regional development and even increase regional carbon storage under the premise of ensuring economic development.

The traditional methods of carbon storage assessment have shortcomings in research scale, temporal and spatial changes in carbon storage, and visual expression. Moreover, the

operation is complex and costly, and it is not suitable for large-scale carbon storage research, such as the biomass method, accumulation method, and field sampling method [7,8]. In recent years, InVEST has attracted the attention of scholars due to its simple parameters, small amount of data required, and high accuracy [9]. The InVEST model is a model developed by Stanford University in the United States, which can be applied to quantify ecosystem services [10]. Many scholars at home and abroad use the carbon storage plate in the InVEST model to predict carbon storage. For example, Tadese and Rajbanshi studied the relationship between land-use change and carbon storage in the Majang Forest Biosphere Reserve and the Konar catchment, India, respectively [11,12]. Xie [13], Wang [14], and Qing [15] estimated carbon storage and predicted different scenarios for the Huaihai Economic Zone, the Hubao and Yuyu urban agglomeration, and the Shihezi River Basin, respectively. Some scholars use land-use simulation models such as Dyna-CLUE, FLUS, CA-Markov, and other land-use simulation models coupled with InVEST models to predict future carbon stocks. Although the above methods can well simulate future land-use changes, they cannot find out the potential driving factors of land-use changes and the evolution of patches [8,16]. On the basis of these shortcomings, Liang [17] and other scholars proposed the PLUS model, which can improve the mining of transformation rules and the lack of landscape dynamic simulation and obtain higher simulation accuracy and more realistic landscape pattern indicators. At present, many scholars have coupled the InVEST and PLUS models to estimate and predict carbon stocks at provincial levels [18–20] and in urban agglomerations [21,22], cities [23,24], and counties [25]. At present, in the scenario provided by CMIP6, there are relatively few studies coupling the InVEST and PLUS models for research on basin. Therefore, this paper takes the Yiluo River Basin as the research object to evaluate the relationship between regional land-use change and carbon stock change.

The Yiluo River, composed of the Yi River and the Luo River, is one of the ten major tributaries of the Yellow River, and the Yellow River Basin is an indispensable ecological barrier in China [26,27]. In this paper, the PLUS model and InVEST model are coupled to simulate the prediction of carbon stocks in the study area for different periods in the future based on different scenarios provided by CMIP6. In this paper, the land-use change regulation between 2010 and 2020 is used to simulate and predict the land-use status of the study area in 2030, 2040, and 2050. Based on The InVEST model, this paper explores the relationship between land-use change and regional carbon storage in the Yiluo River Basin in different periods. It is expected to point out the trend of carbon stock changes in the basin in recent years and provide a reference for the future development of surrounding cities so as to promote the benign development of carbon storage in the basin.

2. Data Sources and Methodology

2.1. Overview of the Study Area

The Yiluo River Basin originates in Luanchuan County at the southern foot of the Xiong'er Mountain, with a total length of 974 km, passing through Shaanxi Province and Henan Province and mainly flowing through Shangluo City, Sanmenxia City, and Luoyang City. Most of them belong to the Henan boundary, of which the Luoyang section accounts for approximately 59.73% of the total area (Figure 1). The Yiluo River Basin covers an area of approximately 18,881 km^2, located between 109°43′~113°11′ E longitude and 33°39′~34°54′ N latitude, and it is located in the transition zone of the second and third tiers in China, with various landform types. The overall trend of the region is low in the east and high in the west, high in the north and south, and low in the middle. The region belongs to the warm temperate continental monsoon climate, summer and autumn are hot and rainy, and spring and winter are cold and dry [28]. The region is rich in mineral resources, and a series of enterprises such as mineral mining, processing, and transportation have been formed in the basin, which play a supporting role in local development.

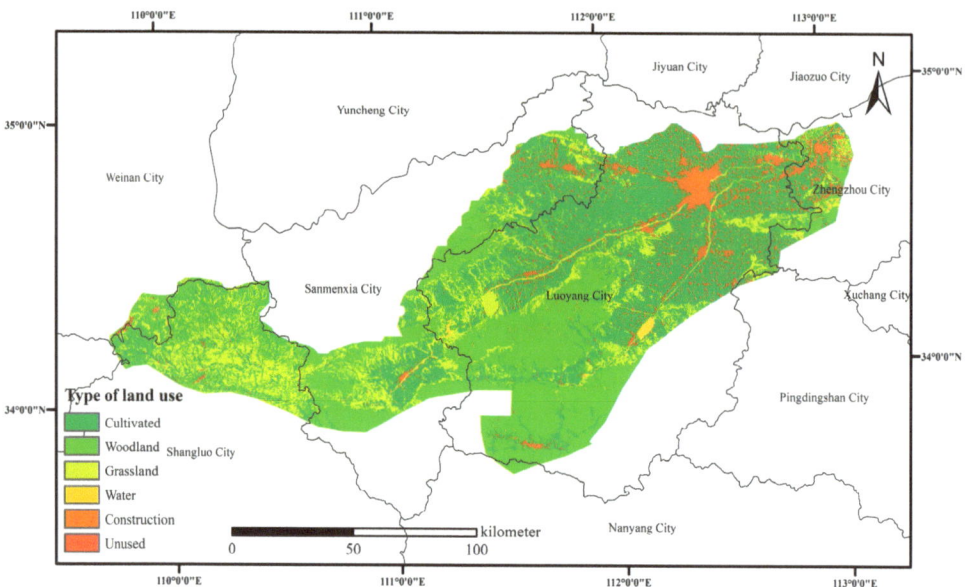

Figure 1. Overview of the Yiluo River Basin.

2.2. Data Sources

The 30 m land-use type, night light, population, GDP, and DEM data required for this study from 2000 to 2020 were obtained from the Resource and Environmental Science and Data Center of the Chinese Academy of Sciences (https://www.resdc.cn/ (accessed on 18 September 2023)). Road and town data were sourced from OpenStreetMap (https://www.openhistoricalmap.org/ (accessed on 20 September 2023)). Land-use data from 2030 to 2050 were sourced from the Global 0.25° × 0.25° Land-Use Harmonization (LUH2) dataset (https://luh.umd.edu/data.shtml (accessed on 23 September 2023)). Soil–root oxygen content data were sourced from the Harmonized World Soil Database (HWSD) (http://webarchive.iiasa.ac.at/Research/LUC/External-World-soil-database/ (accessed on 27 September 2023)). On the basis of the original data, ArcGIS was used to process the original data to ensure that the coordinate system was unified (Krasovsky_1940_Albers), the number of rows and columns was unified, and the accuracy of the land-use raster data was unified to 100 m × 100 m (Table 1).

Table 1. Sources of data.

Data Type	Data Name	Data Source
Social factors	Night lights	The Resource and Environmental Science and Data Center of Chinese Academy of Sciences
	Population	The Resource and Environmental Science and Data Center of Chinese Academy of Sciences
	GDP	The Resource and Environmental Science and Data Center of Chinese Academy of Sciences
Locational factors	Railway	OpenStreetMap
	Expressway	OpenStreetMap
	National highway	OpenStreetMap
	Provincial highway	OpenStreetMap
	Town	OpenStreetMap
	City	OpenStreetMap
Natural factors	Land-use data 2000–2020	The Resource and Environmental Science and Data Center of Chinese Academy of Sciences
	Land-use data 2030–2050	The Global 0.25° × 0.25° Land-Use Harmonization (LUH2) dataset
	Soil–root oxygen content	The Harmonized World Soil Database (HWSD)
	Soil types	The Resource and Environmental Science and Data Center of Chinese Academy of Sciences
	DEM	The Resource and Environmental Science and Data Center of Chinese Academy of Sciences
	Slope	Derived from extracting DEM data
	Slope orientation	Derived from extracting DEM data

2.3. Research Method

2.3.1. PLUS Model

The PLUS model is a model that generates land-use change simulations at the patch level and can better explore land-use drivers and sustainable landscape layouts. The PLUS model includes the Land Expansion Analysis Strategy (LEAS) and the CA Model (CARS) based on multi-type random plaque seeds [21,29]. Land expansion analysis strategy rule mining is used to obtain the development probabilities of various types of land use by extracting the parts of various types of land-use expansion in different time slices of land-use change and using the random forest algorithm to excavate the relationship between various types of land-use expansion and driving factors one by one. In the CA module, the expansion probability of each type of land based on the LEAS model was input, and the parameters of land-use conversion rules and domain weights were set to obtain the prediction results. The conversion rules and domain weights used in the CA module were set based on previous studies and the actual situation of the research area [30,31]. The Kappa coefficient was calculated by comparing the predicted land-use type with the real land-use type. If the Kappa coefficient was high, the land-use type under different scenarios in the future could be predicted.

2.3.2. InVEST Model

The InVEST model includes modules for assessing habitat quality, water supply, carbon stocks, and more [32]. In the assessment of carbon stocks, the carbon stocks in the ecosystem are divided into four basic carbon pools: above-ground biochar (C_{above}), below-ground biochar (C_{below}), soil carbon (C_{soil}), and dead organic carbon (C_{soil}). The formula for calculating total carbon stocks is:

$$C_i = C_{i-above} + C_{i-below} + C_{i-soil} + C_{i-dead}$$

$$C_{totali} = \sum_{i=1}^{n} C_i \times A_i$$

In the formula, i is a certain land-use type; C_i is the carbon density of land use in category i; and $C_{i-above}$, $C_{i-below}$, C_{i-soil}, and C_{i-dead} are the aboveground vegetation carbon density (t·hm^{-2}), belowground vegetation carbon density (t·hm^{-2}), soil carbon density (t·hm^{-2}), and dead organic carbon density (t·hm^{-2}) of type i land-use types, respectively. C_{total} is the total carbon stock of the ecosystem (t), A_i is the area of the type i land-use type (hm^2), and n is the number of land-use types.

The method of determining carbon density data is to use the average annual temperature and average annual precipitation in the study area and the nearby areas, and according to the carbon density correction formula, modify the carbon density of the nearby areas, and then obtain the carbon density of the study area. In this paper, the carbon density of the Yellow River Basin was selected to be corrected, and the average annual temperature and precipitation of the Yiluo River Basin and the Yellow River Basin were 680.1 mm/449.4 mm and 7.05 °C/13.1 °C, respectively [27,33]. The carbon density correction formula is [34–36]:

$$C_{SP} = 3.3968 \times P + 3996.1 \left(R^2 = 0.11 \right)$$

$$C_{BP} = 6.7981 e^{0.00541P} \left(R^2 = 0.70 \right)$$

$$C_{BT} = 28 \times T + 398 (R^2 = 0.47, P < 0.01)$$

In the formula, C_{sp} is the soil carbon density (kg·m^{-2}) obtained based on the average annual precipitation, and C_{bp} and C_{bt} are the biomass carbon density (kg·m^{-2}) obtained based on the average annual precipitation and average annual temperature, respectively. P is the average annual precipitation (mm), and T is the average annual temperature (°C).

$$K_{BP} = \frac{C'_{BP}}{C''_{BP}}$$

$$K_{BT} = \frac{C'_{BT}}{C''_{BT}}$$

$$K_B = K_{BT} \times K_{BP}$$

$$K_S = \frac{C'_{SP}}{C''_{SP}}$$

In the formula, K_{BP} and K_{BT} are the correction factors for the precipitation and temperature factors of the biomass carbon density, respectively, and C'_{BP} and C''_{BP} are the biomass carbon density data based on the average annual precipitation in the Yiluo River Basin and the Yellow River Basin, respectively. C'_{BT} and C''_{BT} are the biomass carbon density data of Yiluo River Basin and Yellow River Basin based on average annual temperature, respectively. C'_{SP} and C''_{SP} are the soil carbon density data of the Yiluo River Basin and Yellow River Basin based on average annual temperature, respectively. K_B and K_S are the correction coefficients of the biomass carbon density and the soil carbon density, respectively. According to the calculated carbon density correction coefficient, the carbon density data of the Yellow River Basin were corrected to obtain the carbon density data used in this paper (Table 2).

Table 2. Carbon density data of the study area.

Table	C_{above}	C_{below}	C_{soil}	C_{dead}
Cultivated	22.1	104.9	36.0	12.7
Woodland	55.1	150.7	52.7	18.3
Grassland	45.9	112.5	33.1	9.8
Water	0.4	0.0	0.0	0.0
Construction	3.3	35.8	0.0	0.0
Unused	1.7	0.0	7.2	0.0

3. Results and Analysis

3.1. Land-Use Change from 2000 to 2020

The construction land of the Yiluo River Basin is mainly concentrated in the northeast of the region, that is, the urban area of Luoyang. The cultivated land is mainly distributed in the relatively flat area in the lower reaches of the watershed, which envelops the urban area of Luoyang. Grassland and woodland are mainly concentrated in the middle and upper reaches of the watershed (Figure 2). From 2000 to 2010, 441.82 hectares of cultivated land was transferred to construction land in the Yiluo River Basin, accounting for 83.89% of the total amount of cultivated land transferred. The area of forest land and grassland decreased, mainly due to the conversion to cultivated land, which was 38.52 hectares and 69.06 hectares, respectively, and they accounted for 47.28% and 52.29% of the total transfers, respectively. By 2010, the construction land increased by 351.25 hectares, and the main reason for the increase was the encroachment of construction land on cultivated land. In the decade from 2010 to 2020, the loss of cultivated land mainly went to forest land and construction land, reaching 188.91 hectares and 210.51 hectares, respectively. At this stage, the area of cultivated land converted into construction land was only 47.65% of that in the previous decade. The total areas of forest land and grassland slightly fluctuated, which mainly showed the mutual transformations between cultivated land, forest land, and grassland. The changes in land-use types in the study area were mainly the transfer in and transfer out of cultivated land and construction land, and most of them occurred within the boundaries of Luoyang. The slowing down of the total conversion degree of various types of land to construction land was related to the entry of a new era and the implementation of water control and revitalization actions in Luoyang City. Luoyang City actively promotes comprehensive water environment management, systematic restoration, and the improvement of water ecology. Luoyang City implemented comprehensive management of the upstream and downstream and left and right banks of the "Four Rivers and Five Canals" in the Yiluo River Basin and carried out the construction of river composite ecological corridors and mountain ecological greening. The forest coverage rate has reached 45.8%, the wetland protection rate has reached 55%, and the soil and water conservation rate has reached 70%. Germplasm resource reserves have been designated, and biodiversity is increasing year by year.

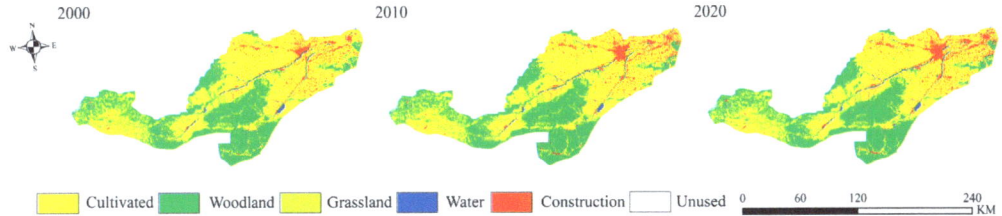

Figure 2. Spatiotemporal evolution of land use from 2000 to 2020.

Overall, cultivated land is the main type of land use in the Yiluo River Basin, and the cultivated land area reached 42.84% of the total area in 2020 (Table 3). The largest area change in the past 20 years has been in cultivated land occupied by construction land, which occupies a total of 593.879 hectares in the past 20 years. The construction land in 2020 was 1.59 times the area in 2000. The increase in the building area is concentrated in the Luoyang section of the Yiluo River Basin, which is related to the rapid economic development in Luoyang in the past 20 years. The other obvious ones are the mutual transformations between arable land, forest land, grassland, and water land. In 2020, in addition to the increase in the area of forest land, the land uses of cultivated land, grassland, and water area all showed different degrees of reduction.

Table 3. Land-use transfer matrix from 2000 to 2020 (km^2).

2000	2020						
	Cultivated	Woodland	Grassland	Water	Construction	Unused	Total
Cultivated	7518.743	198.940	132.989	30.293	593.879	0.176	8475.021
Woodland	164.991	5786.434	101.878	8.979	22.598	2.931	6087.812
Grassland	175.317	110.143	2792.163	5.427	48.448	1.419	3132.916
Water	50.240	5.189	6.235	252.246	17.275	0.010	331.194
Construction	178.962	2.444	1.916	1.647	654.486	0.062	839.517
Unused	0.352	2.610	0.145		0.663	10.771	14.541
Total	8088.605	6105.760	3035.325	298.591	1337.349	15.369	18,881.000

3.2. Multi-Scenario Land-Use Change Simulation Based on PLUS Model

According to the actual situation of the study area, 13 driving factors were selected from three aspects: social factors (population and GDP), location factors (distance to railway, distance to expressway, distance to national highway, distance to provincial highway, distance to city, and distance to town), and natural factors (DEM, slope, slope direction, soil type, and oxygen content in soil–roots). The driving factor data were input into the LEAS module of the PLUS model to obtain the contribution degrees of different driving factors to various land-use changes and the expansion probabilities of various land types. The land-use data of 2020 were predicted with the land-use data of 2010 and compared with the actual land-use data of 2020, and the comparison chart was finally obtained (Figure 3). The Kappa coefficient reached 0.896, and the overall accuracy was 0.929. The simulation results were more accurate, which can be used to predict future land use. Afterward, the land-use data for 2030, 2040, and 2050 were projected via the CA module in the PLUS model by combining the land-use data under different SSP-RCP scenarios.

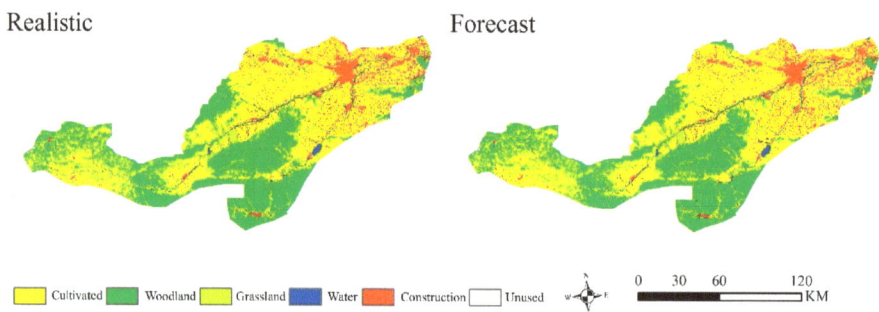

Figure 3. Chart comparison of reality and forecast of land-use types in 2020.

The CMIP emphasizes the impact of different development approaches on future climate change, combining different SSP-RCP scenarios. Five different scenarios are described for the future, depending on the rate at which greenhouse gases are emitted. SSP119 (the scenario combining SSP1 and RCP1.9) is the ideal scenario to reduce global carbon emissions to zero by approximately 2050. SSP126 (coupling SSP1 and RCP2.6) is a more moderate and eco-friendly sustainable development scenario with lower greenhouse gas emissions, with carbon emissions declining at a slower rate and reaching zero after 2050. SSP245 (the scenario that couples SSP2 and RCP4.5) is equivalent to a compromise scenario, representing the middle way for society, with moderate greenhouse gas emissions. Under the SSP370 (the scenario coupling SSP3 and RCP7.0), both carbon emissions and temperatures will rise, and carbon emissions will approximately double by the end of the century. SSP585 (coupling the SSP5 and RCP8.5 scenarios) is a high-speed development scenario dominated by fossil fuels, which is a barbaric development, the pursuit of development at all costs [37,38]. In this paper, three scenarios, SSP126, SSP245, and SSP585, were selected to predict carbon storage in the study area by considering the possibility of future development and the status of the study area [16].

Based on the current changes in land-use types, this paper predicted the land-use types under different scenarios in 2030, 2040, and 2050. In the forecast for the three different time periods, it was shown that under the SSP126 scenario, the cultivated land area will decrease to 8066.58 hectares by 2030, with a total decrease of 22.02 hectares, which is a small decrease. From 2030 to 2050, the cultivated land area shows a fluctuating trend of first increasing and then decreasing. At the end of the period, compared with 2020, the cultivated land area will decrease by 118.38 hectares, with a decrease of only 1.46%. In 30 years, forest land will increase significantly, and by 2050, the total area of this area will increase by 351.88 hectares, an increase of 5.76%. Grassland is decreasing year by year and will decrease by 455.08 hectares in 2050, with a change rate of 14.99%. The area of construction land will increase relatively rapidly before 2040, but there will be no significant change from 2040 to 2050, with a total increase of 220.63 hectares over the preceding 30 years.

Under the SSP245 scenario, the cultivated land area shows a trend of increasing year by year, reaching 8893.85 hectares by 2050, with a total increase of 805.24 hectares during the period, an increase of 9.96%. In this scenario, the change in forest area is relatively small, with a total increase of 75.51 hectares between 2020 and 2030, while the change is not significant in the following 20 years, with a total increase of 82.81 hectares by 2050. Grassland area will decrease rapidly between 2020 and 2050, with a total loss of 936.98 hectares in 30 years, accounting for 30.87% of the total area. The change trend of construction land area is similar to that under the SSP126 scenario, but the change amplitude is relatively small, with an increase of only 48.14 hectares in 30 years (Table 4).

Table 4. Areas of land-use types under different scenarios for the future period (km^2).

Year	SSP126			SSP245			SSP585		
	2030	2040	2050	2030	2040	2050	2030	2040	2050
Cultivated	8066.58	8097.63	7970.22	8337.60	8597.56	8893.85	8591.40	8771.78	8866.43
Woodland	6105.55	6237.27	6457.63	6181.06	6189.20	6188.37	6015.33	6046.18	6085.90
Grassland	2902.83	2674.57	2580.24	2679.77	2395.78	2098.31	2604.28	2380.52	2234.17
Water	298.56	298.56	298.56	298.56	298.56	298.56	298.56	298.56	298.56
Construction	1490.84	1556.50	1557.98	1367.40	1383.41	1385.49	1354.67	1367.23	1379.25
Unused	16.64	16.47	16.36	16.60	16.50	16.40	16.76	16.72	16.69

The largest change in the cultivated area occurs under the SSP585 scenario, while this scenario has the largest change in area between 2020 and 2030, with an increase of 6.22 percent. In the following 20 years, although the cultivated land area still increases, the increase rate is relatively lower, and the total area of cultivated land increases by 777.82 hectares in 30 years, with an increase rate of 9.62%. The area of forest land decreases in the first 10 years and then increases in the next 20 years. By 2050, the total amount of forest land will decrease by 19.86 hectares. The trend of grassland changes from 2020 to 2050 is similar to that of grassland area changes under the SSP245 scenario, but the decrease is relatively small, with a total reduction of 801.15 hectares, a decrease of 26.39%. The area of construction land will change minimally, with a total increase of 41.90 hectares by 2050.

The changes in the areas of water land and unused land in all three scenarios are relatively small. In the three scenarios of the 2050 node, the construction land under the SSP126 scenario is the largest, but at the same time, the forest and grassland areas under the SSP126 scenario are greater than those of the same land type areas under the SSP245 and SSP585 scenarios. The construction land area under the SSP126 scenario is relatively concentrated compared with SSP245 and SSP585. Under the three different scenarios, the increase in construction land mostly occurs around the urban area of Luoyang and the Zhengzhou section of the Yiluo River Basin (Figures 4 and 5).

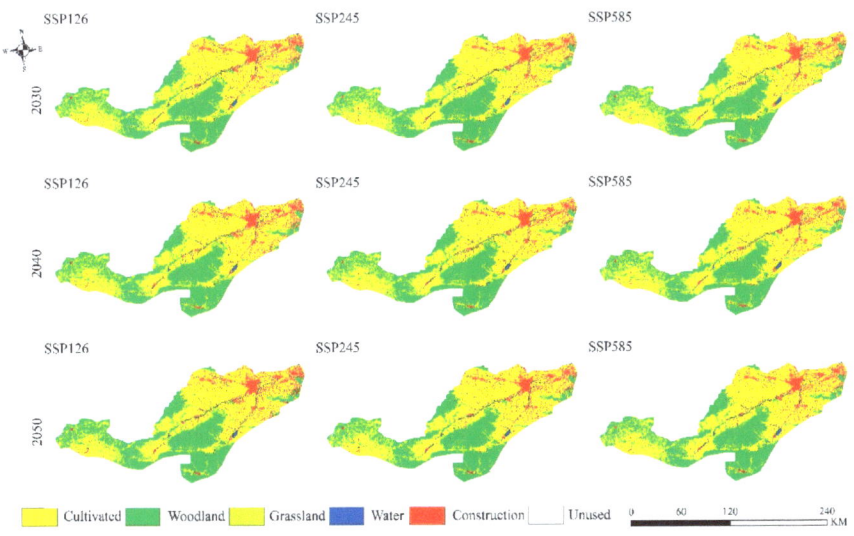

Figure 4. Spatiotemporal evolution of land use from 2030 to 2050.

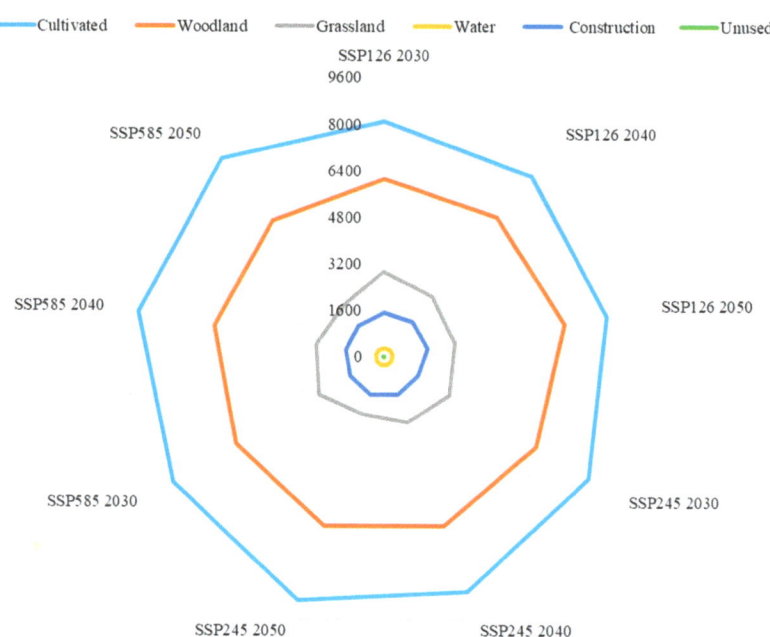

Figure 5. Overview of land-use types under different scenarios for the future period.

3.3. Changes in Carbon Storage in Historical and Future Scenarios Based on the InVEST Model

From 2000 to 2020, the overall carbon storage in the region showed a downward trend. The decrease in carbon storage from 2000 to 2010 was 399.0891×10^4 t, and from 2010 to 2020, the carbon storage decreased by 208.8340×10^4 t, with a total decrease of 607.9230×10^4 t over the past two decades. The intensity of change in the regional carbon stock coincided with the intensity of change in the built-up land, which shows that the change in land-use type affected the regional carbon stock. Limiting the transfer of land with a high carbon storage capacity and stabilizing or increasing the area of land types with a high carbon storage capacity is of great significance for the benign development of regional carbon storage capacity.

In this paper, the carbon storage values of the Yiluo River Basin were assigned to grids, and then ArcGIS was used to classify the carbon storage levels, resulting in Figure 6. From the perspective of spatial distribution, the overall distribution of carbon storage in the watershed shows a high level in the central and western regions and a low level in the eastern region. The distribution of carbon storage corresponds to the landforms of mountainous areas upstream of the watershed and hills and plains downstream. Low carbon density areas are mainly distributed near the main urban area of Luoyang City. This area is the economic center of the Yiluo River Basin, with a high population density and rapid urban development. The construction land area ratio is significant, and there is a trend of continued expansion. This phenomenon leads to land types with low carbon storage capacities constantly encroaching on land with a high carbon storage capacity, mainly around existing building areas. The areas with high carbon storage in the basin are mainly concentrated in the central and western parts of the Yiluo River Basin, with high vegetation coverage and primeval forests (Table 5).

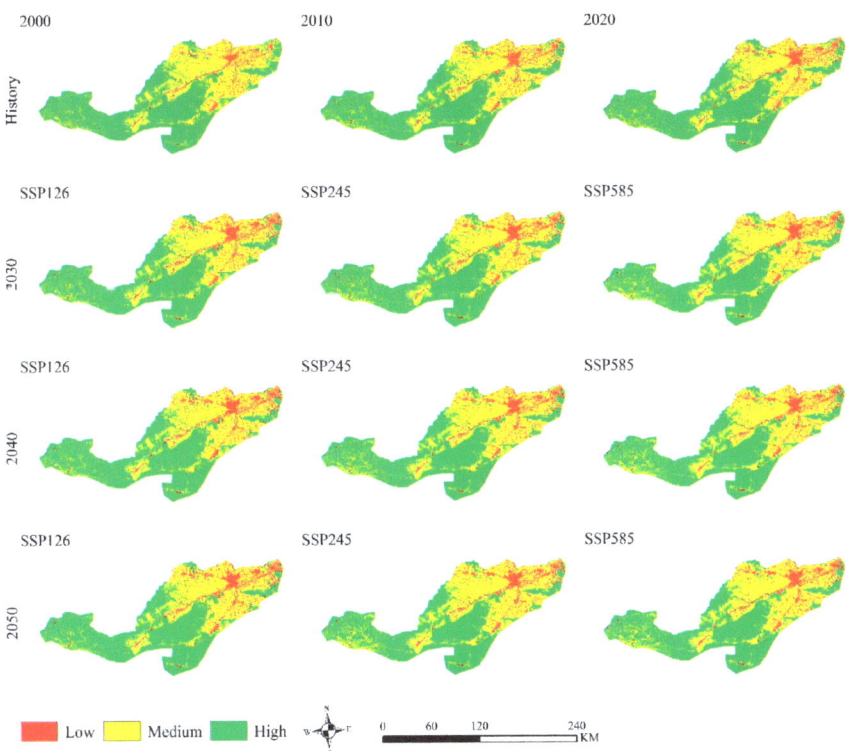

Figure 6. Distribution of carbon storage in the Yiluo River Basin in the historical and future periods.

Table 5. Carbon storage in the study area in different periods and scenarios (10^4 t).

	2000	2010	2020	2030	2040	2050
Reality	37,075.9030	36,676.8140	36,467.9800	—	—	—
SSP126	—	—	—	36,240.2765	36,227.6534	36,418.2218
SSP245	—	—	—	36,411.4745	36,327.9075	36,249.1117
SSP585	—	—	—	36,245.9717	36,204.1592	36,190.8492

Under the influence of future climate change, carbon storage in the study area under the three scenarios will decrease to different degrees compared with 2020. In the period from 2020 to 2050, the SSP126 scenario exhibits the least reduction in regional carbon storage, and in 2050, the regional carbon storage will be 36,418.2218 × 10^4 t, with a total reduction of 49.7582 × 10^4 t. In the SSP245 scenario, the regional carbon storage decreases the second most, and the carbon storage in 2050 will be 36,249.1117 × 10^4 t, with a total decrease of 218.8683 × 10^4 t. Under the SSP585 scenario, the regional carbon stock decreases the most, with a total decrease of 277.1308 × 10^4 t in 30 years, and the regional carbon stock in 2050 will be 36,190.8492 × 10^4 t (Table 5).

Figure 7 shows the spatiotemporal changes in carbon storage under multiple scenarios based on the carbon storage in the study area in 2020. The number of patches indicating an increase in carbon storage under the SSP126 scenario is much larger than that of the patches indicating increases in carbon storage under the SSP245 and SSP585 scenarios in the same period. Under the SSP126 scenario, the areas exhibiting carbon storage decreases in 2050 are mainly distributed downstream of the basin, around the main urban area of Luoyang city, and within the boundary of Zhengzhou. The main reason is that under urban development, land types with low carbon storage capacities continue to erode areas with high carbon storage capacities. The areas with increased carbon storage are mainly distributed in the upper reaches of the basin, where the landform is mostly mountainous and there are more forest lands. Furthermore, due to topography and other reasons, other land areas have the conditions for conversion to forest lands, and the forest area has a trend of expansion, which will increase the regional carbon storage. In the SSP245 scenario, the patches indicating a reduction in carbon storage are much larger than those in the SSP126 scenario, and the distribution is scattered throughout the basin. The patches of increased carbon storage in this scenario mainly exist in the middle and upper reaches of the basin. In the SSP585 scenario, the patches representing an increase in carbon storage are the least among the three and are far smaller than those representing a decrease in carbon storage in the region, and the distribution is similar to that in the SSP245 scenario.

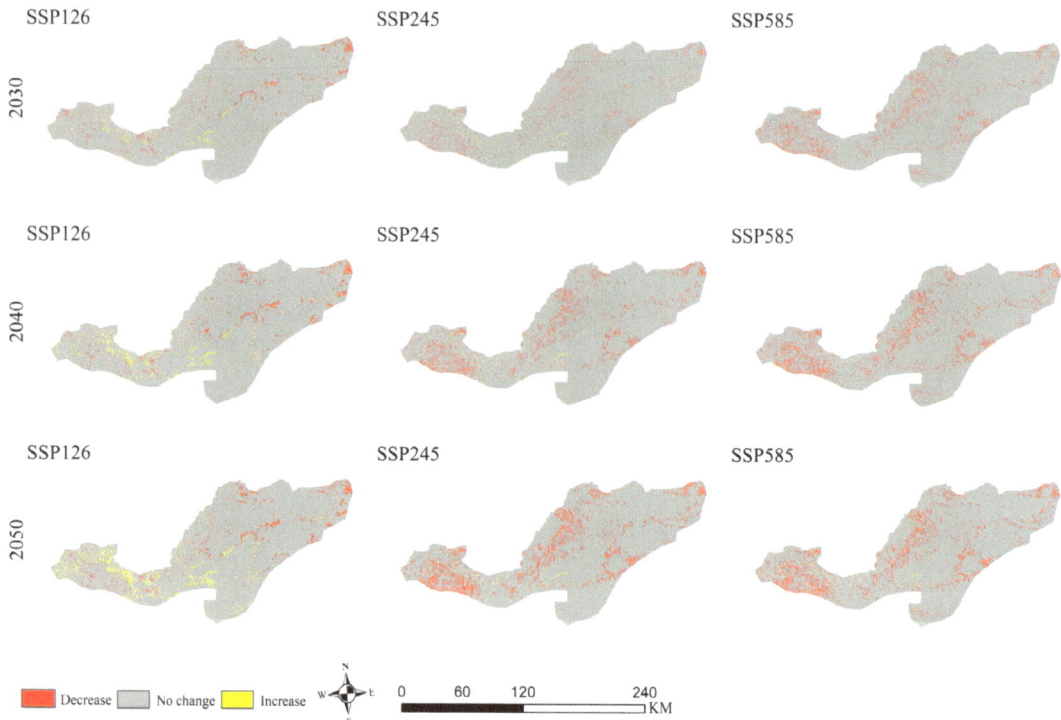

Figure 7. Changes in carbon storage in Yiluo River Basin in the future compared with 2020 under different scenarios.

4. Discussion

With the economic development of the study area, the change in land types with high carbon density values to land types with low carbon density values is the main reason for the decrease in regional carbon storage. In order to improve regional carbon storage, the conversion of cultivated land, forest land, and grassland into other land types should be controlled reasonably, and the area of land types with strong carbon storage capacities such as forest land should be appropriately increased. The future change trend of carbon storage in this paper is approximately similar to those in the studies by Yang [27] and Fan [39], and the distribution of carbon storage is similar to that in Yang's study [27]. In Yang's study, two scenarios were set up: an ecological protection scenario and a natural change scenario. In Fan's study, three scenarios were set up: business as usual, ecological conservation, and urban development scenarios. In this paper, three scenarios, SSP126, SSP245, and SSP585, were selected for research according to different paths provided by CMIP6. The land-use demand data of the study area in different periods in the future were also derived from CMIP6 rather than being predicted by the Markov chain.

In this paper, the average annual temperature and average annual precipitation in the Yellow River Basin and Yiluo River Basin were substituted into the carbon density correction formula, and then the K_B and K_S correction coefficients were obtained, respectively. The carbon density value of the Yellow River Basin was corrected with the correction coefficient, and the required carbon density value was obtained, which was similar to that in the study by Bian [40]. The carbon density will change due to environmental changes, human activities, and other factors. This study did not continuously track and record the carbon density values in the watershed, and there may be some differences between the carbon density used and the actual carbon density, leading to slight differences in carbon storage compared with the actual situation. This study was based on three different scenarios and the PLUS model to predict the land-use types at three time nodes in 2030, 2040, and 2050, respectively. Since the time intervals are all of ten years, this paper selected a decade closer to the future (from 2010 to 2020) for the simulation. Taking 2010 as the base period, the land-use type in 2020 was predicted, and the land-use types in 2030, 2040, and 2050 were predicted after passing the test. Luoyang City has carried out a series of ecological protection actions in the new era, resulting in changes in the intensity of land-use type changes from 2010 to 2020 compared with that from 2000 to 2010. If the year 2000 is used as the base period to simulate 2010 and the future land-use types are predicted based on this, will the transformation between land-use types be greater in the future? Based on this, the InVEST model was used to calculate the total carbon storage in the Yiluo River Basin under three scenarios. The magnitude of the change in total carbon stocks and whether and how the difference between different scenarios will change remain to be discussed.

5. Conclusions

Coupling the PLUS model and the InVEST model, on the basis of clarifying the land-use changes from 2010 to 2020, combined with three different scenarios provided by CMIP6, the land use and carbon storage in the study area in 2030, 2040, and 2050 were simulated and predicted, and the impact of land-use changes on the regional carbon storage were pointed out. The main conclusions are as follows:

(1) Land-use changes led to an increase in or loss of carbon storage. From 2000 to 2020, the areas of forest land and construction land in the Yiluo River Basin increased to varying degrees, while the areas of cultivated land, water area, grassland, and unused land decreased. The conversion of cultivated land to construction land was the main transfer type, which was also an important reason for the decrease in regional carbon storage.

(2) Under the three scenarios, the proportion of cultivated land area in the SSP126 scenario was the smallest, while the proportions of woodland and grassland areas in this scenario were the largest. All three scenarios had some protection of forest land area and none of them showed a significant reduction.

(3) From 2020 to 2050, the carbon storage in the study area under the three scenarios of SSP126, SSP245, and SSP585 all show varying degrees of decline, decreasing to $36,418.2218 \times 10^4$ t, $36,249.1117 \times 10^4$ t, and $36,190.8492 \times 10^4$ t, respectively. Forest land, grassland, and cultivated land have strong carbon storage capacities, and limiting the conversion of land with a high carbon storage capacity to land with a low carbon storage capacity is conducive to the benign development of regional carbon storage capacity.

Author Contributions: Conceptualization, G.L.; methodology, L.L.; software, J.Z. and L.L.; formal analysis, M.J.; data curation, L.L., M.L. and H.G.; writing—original draft, L.L.; writing—review and editing, L.L., G.J., Q.L. and M.L.; project administration, G.L., G.J. and Q.L.; funding acquisition, G.L., G.J. and Q.L. All authors have read and agreed to the published version of the manuscript.

Funding: This research was funded by the National Key R&D Program of China (2021YFD1700900), the Special Fund for Top Talents of Henan Agricultural University (30501031), the National Development and Reform Commission Energy Bureau Project ([2017]20-24), the Henan Agricultural University Graduate Education Reform Project (NDYJSJG2021-15), and the Study on High-Quality Development Path of Grain Production in Henan Province (SKL-2023-2727).

Data Availability Statement: The data presented in this study are available on request from the corresponding author.

Conflicts of Interest: The authors declare no conflict of interest.

References

1. Zou, S.; Zhang, T. CO_2 Emissions, Energy Consumption, and Economic Growth Nexus: Evidence from 30 Provinces in China. *Math. Probl. Eng.* **2020**, *2020*, 8842770. [CrossRef]
2. Siqin, Z.; Niu, D.; Li, M.; Zhen, H.; Yang, X. Carbon dioxide emissions, urbanization level, and industrial structure: Empirical evidence from North China. *Environ. Sci. Pollut. Res.* **2022**, *29*, 34528–34545. [CrossRef]
3. Giersch, J.J.; Hotaling, S.; Kovach, R.P.; Jones, L.A.; Muhlfeld, C.C. Climate-induced glacier and snow loss imperils alpine stream insects. *Glob. Chang. Biol.* **2017**, *23*, 2577–2589. [CrossRef]
4. Li, L.; Fu, W.; Luo, M. Spatial and Temporal Variation and Prediction of Ecosystem Carbon Stocks in Yunnan Province Based on Land Use Change. *Int. J. Environ. Res. Public Health* **2022**, *19*, 16059. [CrossRef]
5. Bo, C.; Chong, X.; Yin, W.; Zhi, L.; Ma, S.; Zhi, S. Spatiotemporal carbon emissions across the spectrum of Chinese cities: Insights from socioeconomic characteristics and ecological capacity. *J. Environ. Manag.* **2022**, *306*, 114510. [CrossRef]
6. Tao, Y.; Li, F.; Wang, R.; Zhao, D. Effects of land use and cover change on terrestrial carbon stocks in urbanized areas: A study from Changzhou, China. *J. Clean. Prod.* **2015**, *103*, 651–657. [CrossRef]
7. Du, S.; Zhou, Z.; Huang, D.; Zhang, F.; Deng, F.; Yang, Y. The Response of Carbon Stocks to Land Use/Cover Change and a Vulnerability Multi-Scenario Analysis of the Karst Region in Southern China Based on PLUS-InVEST. *Forests* **2023**, *14*, 2307. [CrossRef]
8. Sun, F.H.; Fang, F.M.; Hong, W.L.; Luo, H.; Yu, J.; Fang, L.; Miao, Y.Q. Evolution Analysis and Prediction of Carbon Storage in Anhui Province Based on PLUS and InVEST Model. *J. Soil Water Conserv.* **2023**, *37*, 151–158. (In Chinese) [CrossRef]
9. Zheng, H.; Zheng, H. Assessment and prediction of carbon storage based on land use/land cover dynamics in the coastal area of Shandong Province. *Ecol. Indic.* **2023**, *153*, 110474. [CrossRef]
10. Piyathilake, I.D.U.H.; Udayakumara, E.P.N.; Ranaweera, L.V.; Gunatilake, S.K. Modeling predictive assessment of carbon storage using InVEST model in Uva province, Sri Lanka. *Model. Earth Syst. Environ.* **2022**, *8*, 2213–2223. [CrossRef]
11. Tadese, S.; Soromessa, T.; Aneseye, A.B.; Gebreyehu, G.; Noszczyk, T.; Kindu, M. The impact of land cover change on the carbon stock of moist afromontane forests in the Majang Forest Biosphere Reserve. *Carbon Balance Manag.* **2023**, *18*, 24. [CrossRef] [PubMed]
12. Rajbanshi, J.; Das, S. Changes in carbon stocks and its economic valuation under a changing land use pattern—A multitemporal study in Konar catchment, India. *Land Degrad. Dev.* **2021**, *32*, 3573–3587. [CrossRef]
13. Xie, T.Q.; Li, L.; Chen, X.; Bai, L.; Xia, L.; Wang, C.; Li, T. Estimation and prediction of carbon storage based on land use in the Huaihai Economic Zone. *J. China Agric. Univ.* **2021**, *26*, 131–142. (In Chinese)
14. Wang, C.Y.; Gao, X.H.; Guo, L.; Bai, L.F.; Xia, L.L.; Wang, C.B.; Li, T.Z. Land Use Change and Its Impact on Carbon Storage in Northwest China Based on FLUS-Invest: A Case Study of Hu-Bao-Er-Yu Urban Agglomeration. *Ecol. Environ. Sci.* **2022**, *31*, 1667–1679. (In Chinese) [CrossRef]
15. Qing, M.; Zhao, J.; Feng, C.; Huang, Z.H.; Wen, Y.Y.; Zhang, W.J. Response of ecosystem carbon storage service to land-use change in Shiyang River Basin from 1980 to 2030. *Acta Ecol. Sin.* **2022**, *42*, 9525–9536. (In Chinese)
16. Wang, R.; Zhao, J.; Chen, G.; Lin, Y.; Yang, A.; Cheng, J. Coupling PLUS–InVEST Model for Ecosystem Service Research in Yunnan Province, China. *Sustainability* **2023**, *15*, 271. [CrossRef]
17. Liang, X.; Guan, Q.; Clarke, K.C.; Liu, S.; Wang, B.; Yao, Y. Understanding the drivers of sustainable land expansion using a patch-generating land use simulation (PLUS) model: A case study in Wuhan, China. *Comput. Environ. Urban Syst.* **2021**, *85*, 101569. [CrossRef]
18. Li, P.; Chen, J.; Li, Y.; Wu, W. Using the InVEST-PLUS Model to Predict and Analyze the Pattern of Ecosystem Carbon storage in Liaoning Province, China. *Remote Sens.* **2023**, *15*, 4050. [CrossRef]
19. Yue, S.; Ji, G.; Chen, W.; Huang, J.; Guo, Y.; Cheng, M. Spatial and Temporal Variability Characteristics of Future Carbon Stocks in Anhui Province under Different SSP Scenarios Based on PLUS and InVEST Models. *Land* **2023**, *12*, 1668. [CrossRef]

20. Huang, Y.; Xie, F.; Song, Z.; Zhu, S. Evolution and Multi-Scenario Prediction of Land Use and Carbon Storage in Jiangxi Province. *Forests* **2023**, *14*, 1933. [CrossRef]
21. Wu, F.; Wang, Z. Assessing the impact of urban land expansion on ecosystem carbon storage: A case study of the Changzhutan metropolitan area, China. *Ecol. Indic.* **2023**, *154*, 110688. [CrossRef]
22. Sun, X.-X.; Wang, S.-G.; Xue, J.-H.; Dong, L.-N. Assessment and simulation of ecosystem carbon storage in rapidly urbanizing areas based on land use cover: A case study of the southern Jiangsu urban agglomeration, China. *Front. Ecol. Evol.* **2023**, *11*, 1197548. [CrossRef]
23. Wang, Y.; Liang, D.; Wang, J.; Zhang, Y.; Chen, F.; Ma, X. An analysis of regional carbon stock response under land use structure change and multi-scenario prediction, a case study of Hefei, China. *Ecol. Indic.* **2023**, *151*, 110293. [CrossRef]
24. Chen, L.; Ma, Y. Exploring the Spatial and Temporal Changes of Carbon Storage in Different Development Scenarios in Foshan, China. *Forests* **2022**, *13*, 2177. [CrossRef]
25. Wang, X.; Wang, C.Y.; Lv, F.N.; Chen, S.L.; Yu, Z.R. Temporal and spatial carbon storage change and carbon sink improvement strategy of district and county level based on PLUS-InVEST model: Taking Yanging District as an example. *Chin. J. Appl. Ecol.* **2023**, in press. (In Chinese) [CrossRef]
26. Huang, Y.; Li, X.P.; Zhao, N.; Niu, X.L.; Yin, D.X.; Qin, D. Analysis on Characteristics of Temporal and Spatial Changes of Land Use in the Yiluo River Basin. *Spectrosc. Spectr. Anal.* **2022**, *42*, 3180–3186. (In Chinese)
27. Yang, J.; Xie, B.P.; Zhang, D.G. Spatio-temporal evolution of carbon stocks in the Yellow River Basin based on InVEST and CA-Markov models. *Chin. J. Eco-Agric.* **2021**, *29*, 1018–1029. (In Chinese) [CrossRef]
28. Ling, M.; Yang, Y.; Xu, C.; Yu, L.; Xia, Q.; Guo, X. Temporal and Spatial Variation Characteristics of Actual Evapotranspiration in the Yiluo River Basin Based on the Priestley–Taylor Jet Propulsion Laboratory Model. *Appl. Sci.* **2022**, *12*, 9784. [CrossRef]
29. Li, Q.; Pu, Y.; Gao, W. Spatial correlation analysis and prediction of carbon stock of "Production-living-ecological spaces" in the three northeastern provinces, China. *Heliyon* **2023**, *9*, e18923. [CrossRef]
30. Luo, S.Q.; Hu, X.M.; Sun, Y.; Yan, C.; Zhang, X. Multi-scenario land use change and its impact on carbon storage based on coupled Plus-Invest model. *Chin. J. Eco-Agric.* **2023**, *31*, 300–314. (In Chinese) [CrossRef]
31. Zhu, Z.Q.; Ma, X.S.; Hu, H. Spatio-temporal Evolution and Prediction of Ecosystem Carbon Stocks in Guangzhou City by Coupling FLUS-InVEST Models. *Bull. Soil Water Conserv.* **2021**, *41*, 222–229+239. (In Chinese) [CrossRef]
32. Zhu, J.; Hu, X.; Xu, W.; Shi, J.; Huang, Y.; Yan, B. Regional Carbon Stock Response to Land Use Structure Change and Multi-Scenario Prediction: A Case Study of Hunan Province, China. *Sustainability* **2023**, *15*, 12178. [CrossRef]
33. Hou, J.; Qin, T.; Yan, D.; Feng, J.; Liu, S.; Zhang, X.; Li, C. Evaluation of water-land resources regulation potential in the Yiluo River Basin, China. *Ecol. Indic.* **2023**, *153*, 110410. [CrossRef]
34. Alam, S.; Starr, M.; Clark, B. Tree biomass and soil organic carbon densities across the Sudanese woodland savannah: A regional carbon sequestration study. *J. Arid. Environ.* **2013**, *89*, 67–76. [CrossRef]
35. Giardina, C.; Ryan, M. Evidence that decomposition rates of organic carbon in mineral soil do not vary with temperature. *Nature* **2000**, *404*, 858–861. [CrossRef]
36. Chen, G.S.; Yang, Y.S.; Liu, L.Z.; Li, X.B.; Zhao, Y.C.; Yuan, Y.D. Research Review on Total Belowground Carbon Allocation in Forest Ecosystems. *J. Subtrop. Resour. Environ.* **2007**, *2*, 34–42. (In Chinese) [CrossRef]
37. Hurtt, G.C.; Chini, L.P.; Sahajpal, R.; Frolking, S.; Bodirsky, B.L.; Calvin, K.V.; Doelman, J.C.; Fisk, J.P.; Fujimori, S.; Klein Goldewijk, K.; et al. Harmonization of global land use change and management for the period 850–2100 (LUH2) for CMIP6. *Geosci. Model Dev.* **2020**, *13*, 11. [CrossRef]
38. O'Neill, B.C.; Tebaldi, C.; van Vuuren, D.P.; Eyring, V.; Friedlingstein, P.; Hurtt, G.; Knutti, R.; Kriegler, E.; Lamarque, J.-F.; Lowe, J.; et al. The Scenario Model Intercomparison Project (ScenarioMIP) for CMIP6. *Geosci. Model Dev.* **2016**, *9*, 3461–3482. [CrossRef]
39. Fan, L.; Cai, T.; Wen, Q.; Han, J.; Wang, S.; Wang, J.; Yin, C. Scenario simulation of land use change and carbon storage response in Henan Province, China: 1990–2050. *Ecol. Indic.* **2023**, *154*, 110660. [CrossRef]
40. Bian, R.; Zhao, A.Z.; Liu, X.F.; Xu, R.H.; Li, Z.Y. Impact of Land Use Change on Carbon Storage in Urban Agglomerations in the Guanzhong Plain. *Environ. Sci.* **2023**, in press. (In Chinese) [CrossRef]

Disclaimer/Publisher's Note: The statements, opinions and data contained in all publications are solely those of the individual author(s) and contributor(s) and not of MDPI and/or the editor(s). MDPI and/or the editor(s) disclaim responsibility for any injury to people or property resulting from any ideas, methods, instructions or products referred to in the content.

Article

Response of Vegetation Coverage to Climate Changes in the Qinling-Daba Mountains of China

Han Ren, Chaonan Chen, Yanhong Li, Wenbo Zhu *, Lijuan Zhang, Liyuan Wang and Lianqi Zhu

College of Geography and Environment, Henan University, Kaifeng 475004, China
* Correspondence: zhuwb517@163.com

Abstract: As a major component of the north–south transition zone in China, the vegetation ecosystem of the Qinling-Daba Mountains (QBM) is highly sensitive to climate change. However, the impact of sunshine duration, specifically, on regional vegetation remains unclear. By using linear trend, correlation, and multiple regression analyses, this study systematically analyzed the spatiotemporal characteristics and trend changes of the vegetation coverage in the QBM from 2000–2020. Changes in the main climate elements in different periods and the responses to them are also discussed. Over the past 21 years, the vegetation coverage on the east and west sides of the QBM has been lower than that in the central areas. However, it is showing a continuously improving trend, especially in winters and springs. The findings indicate that change of FVC in the QBM exhibited a positive correlation with temperature, a negative correlation with sunshine hours, and both positive and negative correlation with precipitation. On an annual scale, average temperature was the main controlling climatic factor. On a seasonal scale, the area dominated by precipitation in spring was larger. In summer, the relative importance of the three was weak. In autumn and winter, sunshine duration became the main factor affecting vegetation coverage in most areas.

Keywords: the Qinling-Daba Mountains; vegetation coverage; climatic factors; main control factor

1. Introduction

Terrestrial ecosystems are the basis for human survival and sustainable development. As an important part of the terrestrial ecosystem, vegetation is a link among ecological elements such as soil, hydrology, and atmosphere, and plays an important role in improving regional microclimate, purifying air, containing water, maintaining soil and water, and in the process of ecosystem evolution [1]. Vegetation change is a concrete manifestation of the change of human living environment, and plays the role of "indicator" in the study of global or regional environmental change. Therefore, monitoring and attribution analysis of regional vegetation cover dynamics has become an important part of global change research.

Climate change has a major impact on the structure and function of global ecosystems [2]. Among the different climate factors, temperature and precipitation are generally considered to be the key factors affecting vegetation growth and development. Temperature is a regulator of vegetation growth, especially at high latitudes and high altitudes. In recent decades, increased vegetation activity in the Northern Hemisphere has been related to an increase in temperature [3–5]. This is because warmer temperatures extend the vegetation growing season and increase the efficiency of photosynthesis and water use for vegetation growth [6]. However, temperature can also have a negative impact on vegetation, as exceeding the temperature required for optimal vegetation growth can lead to inhibition of photosynthesis. At higher temperatures, nutrient consumption due to respiration increases, thus limiting the growth of vegetation [7]. Precipitation is another key climatic factor regulating vegetation growth, as it increases soil moisture, which is essential to promote plant root activity and the water status of the vegetation. Several

Citation: Ren, H.; Chen, C.; Li, Y.; Zhu, W.; Zhang, L.; Wang, L.; Zhu, L. Response of Vegetation Coverage to Climate Changes in the Qinling-Daba Mountains of China. Forests 2023, 14, 425. https://doi.org/10.3390/f14020425

Academic Editor: Romà Ogaya

Received: 17 January 2023
Revised: 10 February 2023
Accepted: 15 February 2023
Published: 19 February 2023

Copyright: © 2023 by the authors. Licensee MDPI, Basel, Switzerland. This article is an open access article distributed under the terms and conditions of the Creative Commons Attribution (CC BY) license (https://creativecommons.org/licenses/by/4.0/).

studies have shown that the correlation between vegetation change and precipitation in the Northern Hemisphere has increased in recent years [8,9]. In addition, some studies have confirmed that sunshine duration, relative humidity, and wind speed are also important factors affecting vegetation growth. For example, results showed that the annual sunshine hours in the Qilian Mountains had the greatest explanatory power for regional vegetation changes from 2000 to 2020 [10]. Throughout the Tibetan Plateau region, relative humidity and water vapor pressure play a dominant role in the variation of vegetation during the growing season [11].

Mountains are the most active interface and the most vulnerable geographical unit in terrestrial ecosystems and are the drivers and amplifiers of environmental change. Therefore, mountain vegetation ecosystems are more sensitive to global changes. The Qiling and Daba Mountains (QBM) constitute a complete geographical unit in the center of China's interior. Not only are they a main part of China's north–south transition zone, but they also provide an important ecological channel connecting China's eastern plains and the Qinghai Tibet Plateau. The special geographical location and complex landform conditions render the vegetation ecosystem in the QBM highly sensitive to climate change. Monitoring the dynamics of vegetation cover in the QBM and studying their relationship with climate elements are crucial for assessing the environmental quality of regional ecosystems and maintaining optimal ecosystem functions.

Research has covered the dynamic changes in vegetation [12,13] and its driving factors in some areas of the QBM [14–18]. These studies are valuable for understanding local vegetation-climate relationships in the QBM. However, current studies on the relationship between vegetation pattern evolution and climatic factors in the QBM mostly focus on two factors, temperature, and precipitation, ignoring the impact of sunshine duration on regional vegetation change, and the relationship between them is not clear. In addition, previous studies have usually used simple correlation analysis methods to investigate the response of vegetation change to changes in a single climatic factor, but rarely have multiple climatic elements been integrated to identify the main controlling climatic factors affecting regional vegetation cover change, ignoring the spatial variation characteristics of vegetation change response to climate change. Therefore, the purpose of the present study was to use MODIS-NDVI and meteorological station data of long time series to identify the characteristics of vegetation coverage distribution and spatiotemporal changes in the QBM from 2000–2020 and explore the response mechanism to climate factor changes. Specifically, the main contents of this study are as follows: (1) the spatiotemporal variation characteristics of vegetation coverage and main climate elements during 2000–2020; (2) the response mechanism of vegetation coverage change to a single climate factor; and (3) the main climate factor changes of regional vegetation coverage during different periods.

2. Data and Methods

2.1. Study Area

The QBM is in central China between 102° E–114° E and 30° N–36° N (Figure 1a). It stretches across Gansu, Sichuan, Shaanxi, Chongqing, Hubei, and Henan from west to east, with a total area of approximately 3.0×10^5 km^2. The altitude gradually rises from east to west. The QBM includes three major geomorphic units: Qinling Mountains, Daba Mountains and Hanjiang Valley [19] (Figure 1b). As the main body of the north–south transition zone in China, the climate types in the study area are diverse and exhibit significant vertical changes. The area to the north of Qinling Mountains is mainly affected by the continental climate of the warm temperate zone, which is cool in summer and dry and cold in winter. The area south of Qinling Mountains is mainly affected by the subtropical monsoon climate, which is humid and has four distinct seasons. In addition, the study area includes three types of ecological function protection areas for water conservation, water and soil conservation, and species resources in China (Figure 1c), as well as four forestry projects: the Three North Shelterbelt, the Middle Yellow River Shelterbelt, the Taihang Mountain Greening Project, and the Middle and Upper Yangtze River Shelterbelt (Figure 1d).

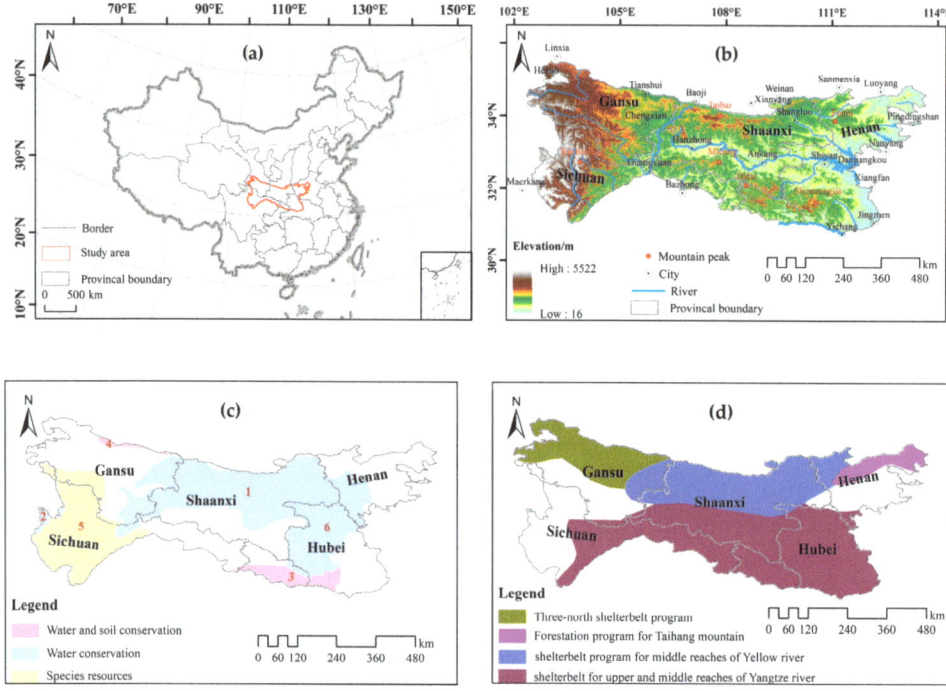

Figure 1. Overview of the QBM: (**a**) Geographical position; (**b**) Basic Elements: elevation, mountain peak, city location, and rivers; (**c**) Ecological Function Reserve (Number 1–6 represent the six ecological function reserves); (**d**) Forestry Engineering.

2.2. Data

The data sources used in this study mainly included MODIS satellite normalized difference vegetation index (MODIS-*NDVI*) data and meteorological station data (temperature, precipitation, and sunshine hours).

2.2.1. NDVI

Remote sensing satellite images provide more possibilities for monitoring vegetation changes in large-scale and long-term time series. The *NDVI* time series data used in this study were obtained from MOD13Q1-*NDVI* data provided by the NASA Land Process Distribution Dynamic Data Center (https://ladsweb.modaps.eosdis.nasa.gov, accessed on 1 September 2020), namely the normalized vegetation index dataset. First, we used the MODIS Reprojection Tool (MRT) to batch extract *NDVI* data and perform splicing, resampling, projection conversion, and other processes to convert them into Tiff images. Second, based on the ENVI platform, the maximum value composites (MVC) method was used to obtain monthly *NDVI* data, and *FVC* data were calculated through the pixel dichotomy model.

2.2.2. Climatic Data

The meteorological data used in this study were obtained from the Daily Data Set of China Surface Climate Data (V3.0) provided by the China Meteorological Data Sharing Service Network (http://cdc.cma.gov.cn, accessed on 1 September 2020). A 50 km buffer zone was created at the boundary of the study area, and the daily average temperature (TEM), precipitation (PRE), and sunshine duration (SSD) data of 100 meteorological stations in the study area and the buffer zone from 2000–2020 were used (Figure 2). Based on the observation data, ANUSPLIN software was used to conduct spatial interpolation processing on the station data, and grid data of monthly average *TEM*, monthly *PRE*, and monthly *SSD* with a spatial resolution of 250 m were obtained.

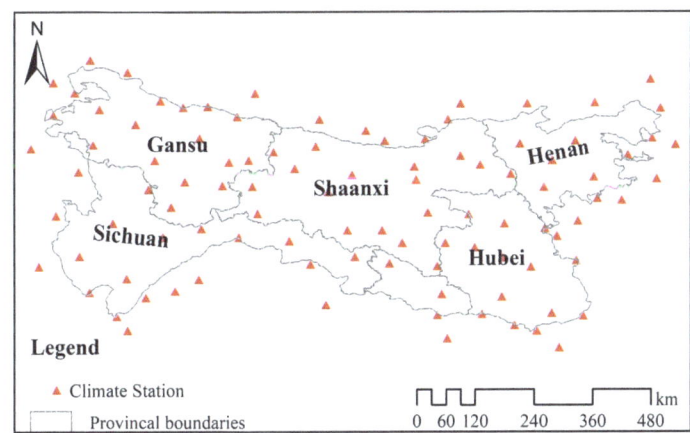

Figure 2. Spatial Distribution of Selected Meteorological Stations in the Study Area.

2.3. Methods

2.3.1. Estimation of Vegetation Coverage

The normalized difference vegetation index (*NDVI*) is the most commonly used vegetation index, and the pixel dichotomy model is the most common linear model used for calculation of vegetation coverage. This model assumes that the information contained in one pixel of a remote sensing image is composed only of vegetation and bare soil [20]. Pixel information includes pure vegetation composition information and pure soil composition information; therefore, the mixed pixel *S* can be expressed as:

$$S = S_v + S_s \quad (1)$$

where S_v is the vegetation information in the pixel and S_s is the information of bare soil.

The fractional vegetation cover (*FVC*) of a pixel is the area ratio of the vegetation in the pixel; therefore, the bare soil coverage in the pixel can be expressed as $(1 - FVC)$. Assuming that a pixel with pure vegetation coverage is S_{veg}, the pixel covered by pure bare soil is S_{soil}, then:

$$S_v = FVC \times S_{veg} \quad (2)$$

$$S_s = (1 - FVC) \times S_{soil} \quad (3)$$

If Equations (2) and (3) are introduced into Equation (1), we obtain:

$$FVC = \frac{S - S_{soil}}{S_{veg} - S_{soil}} \quad (4)$$

Therefore, the binary model expression of vegetation coverage pixel based on *NDVI* is as follows:

$$FVC = \frac{NDVI - NDVI_{soil}}{NDVI_{veg} - NDVI_{soil}} \quad (5)$$

The range of *NDVI* is between $[-1,1]$, and negative values indicate that the ground cover is the reflection of visible light by clouds, water, snow. The value 0 indicates the presence of rocks or bare soil. Positive values indicate that there is vegetation cover and they increase with the increase in cover. Therefore, in the calculation of *FVC*, we set the pixels with *NDVI* less than 0 as null values. Theoretically, the value of $NDVI_{soil}$ should be close to 0, but in fact, it fluctuates within the range of –0.1 to 0.2 owing to different research areas or surface environments. Due to the lack of systematically measured surface data for reference in this study, the NDVI statistical histogram is usually given a confidence interval, and the minimum and maximum values within this interval are considered as $NDVI_{soil}$ and $NDVI_{veg}$, or the NDVI value of the cumulative frequency [21]. According to the situation of the study area, NDVI values of 5% and 95% of the cumulative frequency were considered as $NDVI_{soil}$ and $NDVI_{veg}$. We assigned 0 to values less than 5%, and 1 to values greater than 95%.

To better reflect the distribution and changes in vegetation coverage in the study area, the vegetation coverage was divided into five grades according to the Classification and Grading Standards for Soil Erosion and the specific situation. The results are presented in Table 1.

Table 1. FVC Level Classification.

Class	FVC	Description
1	≤0.30	Low vegetation coverage
2	(0.30, 0.45]	Sub-low vegetation coverage
3	(0.45, 0.60]	Middle vegetation coverage
4	(0.60, 0.75]	Sub-high vegetation coverage
5	>0.75	High vegetation coverage

2.3.2. Change Trend

Change trend analysis refers to changes in a particular element of the time series (such as FVC, TEM, PRE, and SSD) that continuously increase or decrease over a certain period of time [22]. In the present study, the trend analysis method (i.e., least squares method) was used to calculate the interannual change trend of climate elements and FVC at different spatial scales, and the slope of the linear regression equation is defined as the interannual change trend rate of elements (slope) [23]. The calculation equation for slope is as follows:

$$Slope = \frac{n \times \sum_{i=1}^{n}(i \times p_i) - \sum_{i=1}^{n} i \times \sum_{i=1}^{n} p_i}{n \times \sum_{i=1}^{n} i^2 - (\sum_{i=1}^{n} i)^2} \quad (6)$$

where n is the total year of the study period, i is the serial number of each year, p_i is the pixel value of the i-th year, and Slope is the change in slope of the data on each pixel time series. When Slope > 0, it indicates that the pixel value increases with time; that is, it shows an improvement trend during the study period. When Slope = 0 indicates no change, and when slope < 0, the pixel value decreases with time; that is, it shows a degradation trend. The greater the absolute value of Slope, the greater is the change rate of the elements. Combined with the significant results of the t test, the change trend was divided into the following five levels, as shown in Table 2.

Table 2. Classification of change trend level.

Class	Slope	p Value	Description
1	>0	<0.05	Significant increase
2	>0	>0.05	No significant increase
3	<0	<0.05	Significant decrease
4	<0	>0.05	No significant decrease
5	=0	-	No changed

2.3.3. Correlation Analysis

To analyze the relationship between FVC and climate elements, it is necessary to establish a simple correlation coefficient between them on a pixel scale. This correlation coefficient, also known as the Pearson correlation coefficient, is widely used to measure the correlation between two variables, and its value is between −1 and 1. The calculation equation is as follows:

$$R = \frac{\sum_{i=1}^{n}\left[(x_i - \overline{X})(y_i - \overline{Y})\right]}{\sqrt{\sum_{i=1}^{n}(x_i - \overline{X})^2 \sum_{i=1}^{n}(y_i - \overline{Y})^2}} \quad (7)$$

where R is the correlation coefficient; n is the number of samples; x_i, y_i are the variables to be evaluated; and $\overline{X}$ and $\overline{Y}$ are, respectively, the mean of x_i and y_i. If $R > 0$, there is a positive correlation between the two; otherwise, it indicates a negative correlation. The closer the absolute value of R is to 1, the closer is the correlation between x and y; the closer the absolute value of R is to 0, the less close is the correlation between them. The correlation level can be classified into five types (Table 3).

Table 3. Classification of correlation level.

Class	R	p Value	Description
1	>0	<0.05	Significant positive correlation
2		>0.05	No significant positive correlation
3	<0	<0.05	Significant negative correlation
4		>0.05	No significant negative correlation
5	=0	-	No correlation

2.3.4. Relative Importance

When analyzing the impact of multiple elements on a single element, we must pay attention to the relative importance of each element. The calculation of relative importance was based on standardized coefficients or variance interpretation. Based on raster data from a long time series, the present study uses a multiple regression analysis method. First, a multiple linear regression model between FVC and climate factors was developed to explain the influence of multiple climate factors changes on FVC using TEM, PRE, and SSD as independent variables and FVC as a dependent variable.

The equation used was as follows:

$$FVC = A \times TEM + B \times PRE + C \times SSD + d \quad (8)$$

where A, B and C are regression coefficients of the three climatic elements, and d is a constant.

Secondly, in order to better identify the most important climate factors affecting the FVC variation in each pixel, we standardized the coefficients of the multiple regression model. The maximum absolute value of the standardized regression coefficient is considered as the most important variable. The formula for standardization was as follows:

$$\begin{aligned} A' &= A \times \frac{std(TEM)}{std(FVC)} \\ B' &= B \times \frac{std(PRE)}{std(FVC)} \\ C' &= C \times \frac{std(SSD)}{std(FVC)} \end{aligned} \quad (9)$$

where A', B', C' represent the normalization factors of TEM, PRE, and SSD.

3. Results

3.1. Spatiotemporal Changes of FVC

3.1.1. Time Variation

Figure 3a shows the overall trend of annual average FVC in the QBM from 2000–2020. In the past 21 years, the vegetation coverage of the QBM improved continuously, and increased significantly at a rate of 0.058/10a ($p < 0.01$). From 2011–2020, the annual growth rate of FVC was faster than that of the previous 11 years.

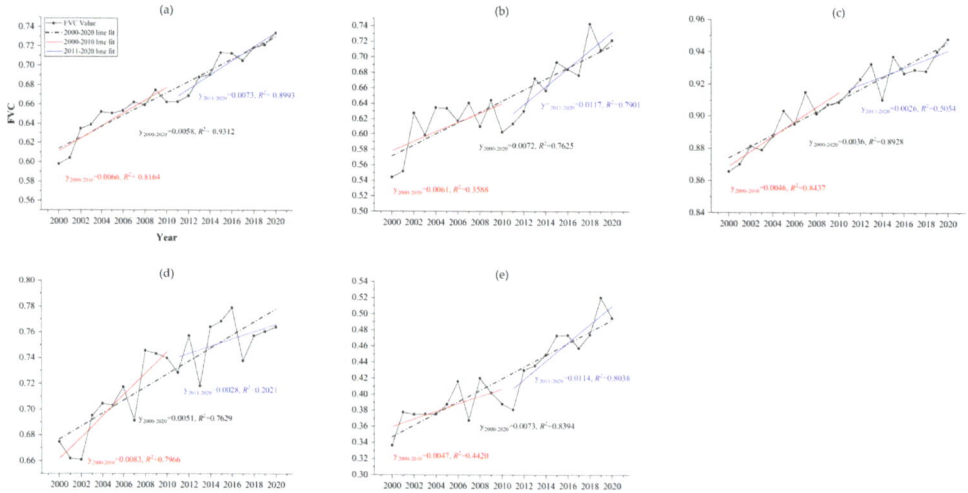

Figure 3. Change of *FVC* average value in the QBM from 2000–2020. (**a**) Year; (**b**) Spring; (**c**) Summer; (**d**) Autumn; (**e**) Winter.

The growth of surface vegetation generally exhibits a certain periodicity (Figure 3b,e). The growth rate in summer was the slowest, but the fastest in winter, and the growth rate of the *FVC* average value in spring was faster than that in autumn.

3.1.2. Spatial Pattern Change

In terms of spatial distribution, the annual average *FVC* value of the QBM was higher in the south, lower in the north, and lower in the east and west than in the middle (Figure 4a). Most areas are at a sub-high or high vegetation coverage level (Figure 4b). The high-value (>0.75) was mainly distributed in Daba Mountains and Qinling Mountains in Shaanxi, whereas the low-value (<0.30) area is very small and scattered in northwest Sichuan. The average values of *FVC* in Chongqing, Hubei, and Shaanxi were larger (>0.70), and in Henan and Gansu they were the smallest. This shows that the vegetation coverage in most areas has been in a good state over the past 21 years, especially in Daba Mountains and surrounding areas.

Spatially, the annual change trend rate generally showed increasing distribution characteristics from southwest to northeast (Figure 4c). The vegetation coverage in most regions of the QBM has significantly improved over the past 21 years, and the regions with faster growth rate (>0.06/10a) were concentrated in Shaanxi, Gansu, and Chongqing (Table 4). From the perspective of space, these areas have a good basis for vegetation coverage and are also key implementation areas for forestry projects, such as returning farmland to forests and grasslands. Less than 3% of the regional average annual *FVC* showed a downward trend, scattered across in Chengdu, Aba, Hanzhong, Shiyan, and Luoyang. The growth and development of vegetation in these areas may be affected by altitude or population distribution.

The spatial pattern of the *FVC* average in each season was similar to the annual average (Figure 5a); however, there were regional differences in the *FVC* trend in different seasons (Figure 5c). The average *FVC* in spring at the sub-high vegetation coverage level (Figure 5b) and the vegetation cover were significantly improved. The significantly improved area was distributed within Shaanxi, especially in Ankang and Shangluo; the decrease trends were scattered across Tianshui and Hanzhong. Spatially, the trend rate showed a distribution feature of increasing from both sides to the central region. The average change trend rate of each region was between 0.060/10a and 0.087/10a, with Chongqing and Shaanxi being the fastest and Sichuan the slowest.

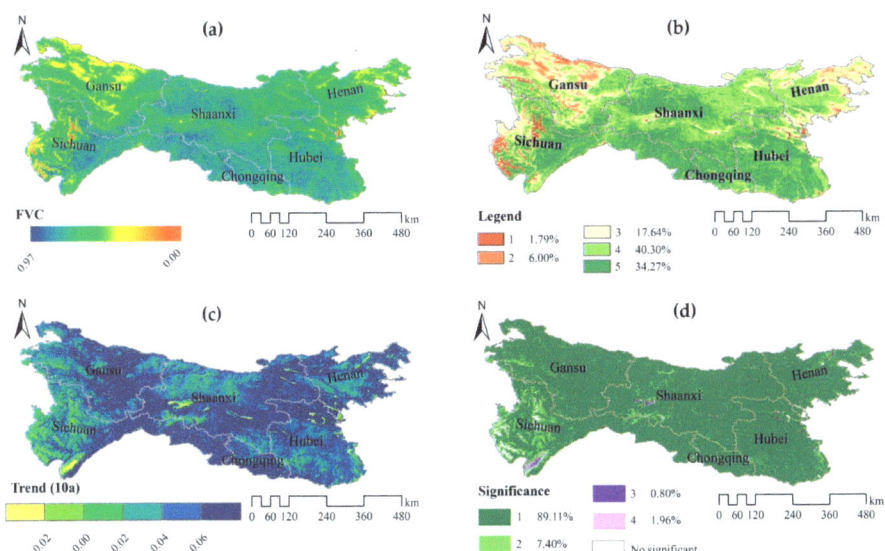

Figure 4. Changes of *FVC* spatial pattern in the QBM from 2000–2020. (**a**) Average value; (**b**) *FVC* level; (**c**) Change trend rate; (**d**) Significance level.

Table 4. Annual average vegetation coverage and trend rate in the QBM.

	FVC					Slope/10a				
	Year	Spring	Summer	Autumn	Winter	Year	Spring	Summer	Autumn	Winter
QB	0.67	0.64	0.91	0.73	0.42	0.058	0.072	0.036	0.051	0.073
SC	0.69	0.63	0.91	0.74	0.48	0.050	0.060	0.019	0.040	0.075
SX	0.72	0.71	0.95	0.78	0.44	0.060	0.081	0.031	0.045	0.081
HB	0.73	0.72	0.94	0.78	0.47	0.056	0.070	0.026	0.051	0.074
HN	0.59	0.58	0.85	0.64	0.29	0.057	0.073	0.043	0.058	0.057
GS	0.58	0.52	0.86	0.64	0.32	0.065	0.070	0.065	0.065	0.058
CQ	0.76	0.73	0.94	0.82	0.55	0.067	0.087	0.028	0.055	0.095

(SC, SX, HB, HN, GS and CQ are abbreviation for Sichuan, Shaanxi, Hubei, Hunan, Gansu, and Chongqing provinces, respectively).

Most areas in summer had high vegetation coverage. A small number of low-value areas were distributed in Aba, Sichuan. In 21 years, the area with a positive trend rate of average *FVC* change in summer accounted for the least (84.93%), and the faster increase rate were mainly in the east and west of the Qinling Mountains. Compared with other seasons, the area proportion of *FVC* showing a downward trend in summer was larger (10.62%) and was mainly distributed in western Sichuan and southern Shaanxi, but the significance of this decline in most regions was weak. The average change trend rate of *FVC* in each region in summer was between 0.019/10a and 0.065/10a, and the fastest and slowest were Gansu and Sichuan, respectively.

The spatial distribution of *FVC* in autumn was consistent with that in spring, and it was also at the sub-high vegetation coverage level overall. Approximately 7% of regions with a decreasing trend were concentrated in Deyang and Aba. The regional proportion of *FVC* showing a very significant upward trend was 62.53%, mainly distributed in Gansu and eastern regions. The average change trend rate of regional *FVC* was within the range of 0.040/10a–0.065/10a, and the fastest and slowest rates remained in Gansu and Sichuan, respectively.

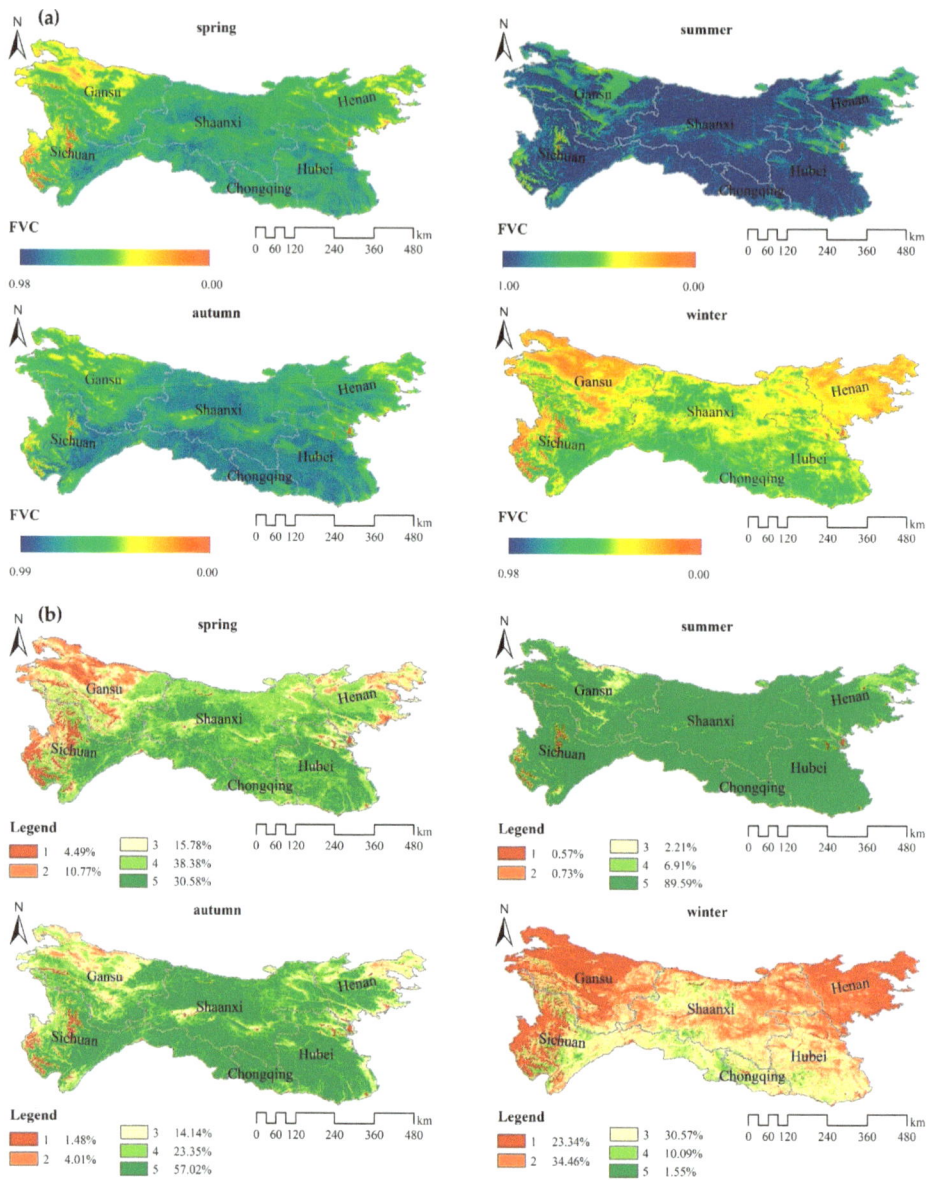

Figure 5. Cont.

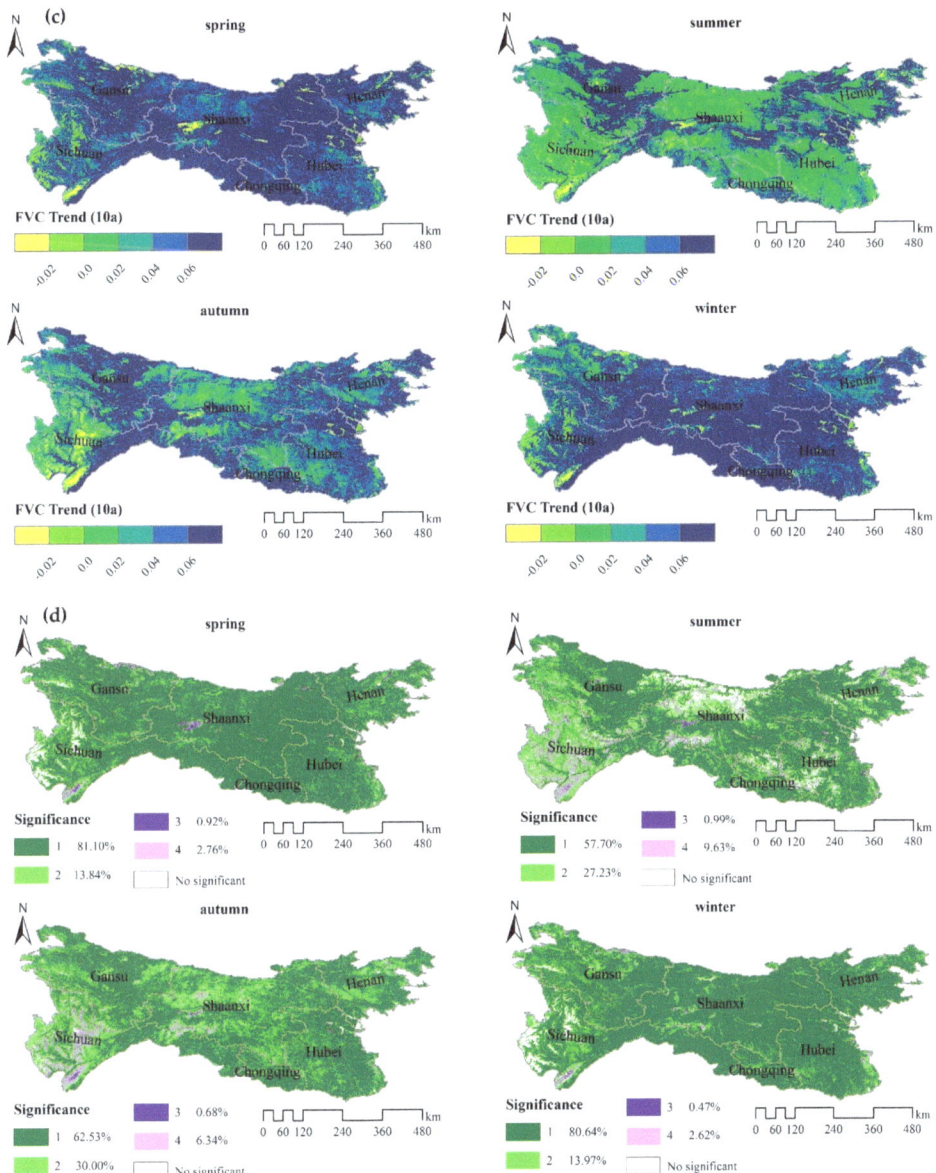

Figure 5. Change of seasonal average *FVC* spatial pattern in the QBM from 2000–2020. (**a**) Average value; (**b**) *FVC* level; (**c**) Change trend rate; (**d**) Significance level.

The average *FVC* value in winter was the lowest, which is generally at the sub-low vegetation coverage level. The low-value areas were concentrated in the west and northeast, while the high-value areas were few and scattered around Daba Mountains. The change trend rate of *FVC* in winter is close to that in spring; Chongqing and Henan had the fastest and slowest values, respectively. In addition, the area proportion of *FVC* increasing or decreasing was also similar to that in spring. The difference was that in winter, the *FVC* in the western region showed a downward trend, whereas that in Hanzhong showed an upward trend.

In general, vegetation coverage in the QBM was low only in winter and high in the other seasons. During the 21 years, the vegetation coverage in most areas improved in each season, but the improvement rate had a certain heterogeneity. The seasonal average *FVC* change trend rate in Sichuan, Shaanxi, Hubei, and Chongqing decreased and then increased. The minimum trend rate appeared in summer and, the maximum in winter; while in spring, it was faster than that in autumn. This trend was contrary to that of the average *FVC* in each season. In contrast, the trend rate of seasonal average *FVC* in Henan and Gansu showed a trend to "decrease-increase-decrease." The minimum value appeared in summer, while the maximum value occurred in spring.

3.2. Spatiotemporal Changes of Climate Factors

3.2.1. Time Variation

In the last 21 years, the annual average *TEM* in the QBM has been 10.12 °C, rising at a rate of 0.175 °C/10a (Figure 6a). However, since 2016, the average annual *TEM* has decreased by approximately 0.3 °C. This indicates that during the study period, the QBM also experienced a "warming" trend, especially in the past 10 years, when the *TEM* increased significantly; but in the past five years, "cooling" was observed. The growth rate of the average *TEM* in summer and autumn was higher, whereas that in winter was the slowest and showed a weak downward trend. The growth rate/deceleration was fastest in autumn/winter before 2010, and faster in winter and spring after 2010 (Figure 7a).

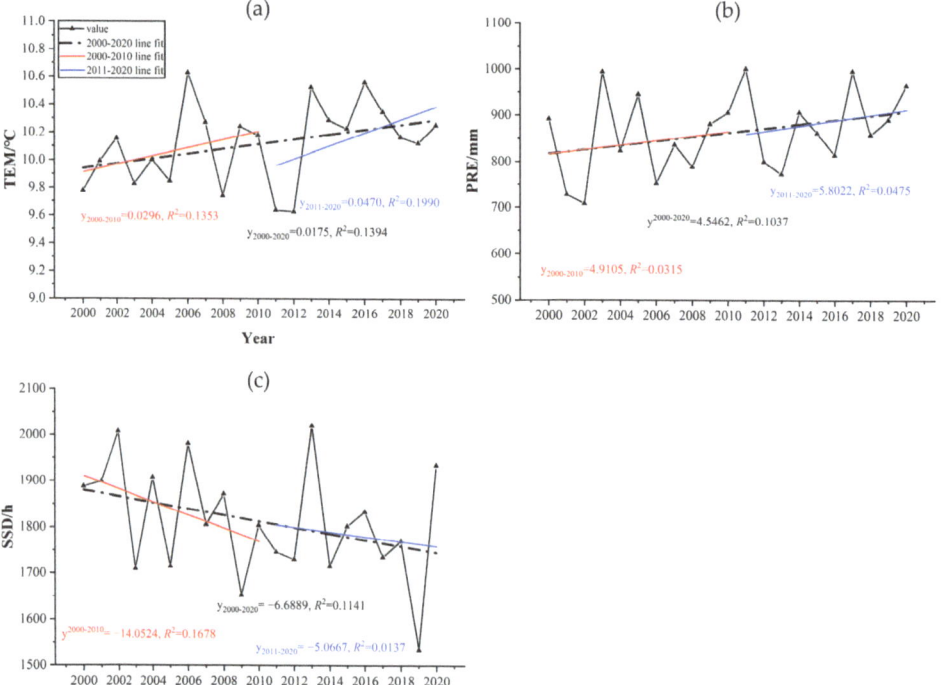

Figure 6. Annual mean change of climate factors in the QBM from 2000–2020: (**a**) *TEM*; (**b**) *PRE*; (**c**) *SSD*.

Figure 6b shows that the annual *PRE* has increased at a rate of 45.46 mm/10a in the past 21 years. Similar to the annual average *TEM*, the average value and increase rate of annual *PRE* before 2010 were smaller than those after 2010. The *PRE* in summer was the largest, slightly lower in spring than in autumn, and the lowest in winter (Figure 7b). In all four seasons, the growth rate was fastest in spring, particularly before 2010; the next was autumn and summer, and their rate of change was faster after 2010. Generally, there was a slight downward trend in winter.

In contrast to the *TEM* and *PRE*, the annual *SSD* showed a gradual downward trend at a rate of 66.89 h/10a during the study period (Figure 6c), and the decline rate in the first 11 years was 2.8 times that in the next 10 years. According to the seasonal average, the *SSD* in summer and spring was longer, followed by winter and autumn. Seasonally, spring and autumn showed a decreasing trend, summer and winter showed an increasing trend, and the fall/rise speed was faster in autumn/winter. However, it was found that the rate of decline in spring and summer was faster in 2000–2010, being fastest in autumn, whereas it was the fastest in winter in 2011–2020.

3.2.2. Spatial Pattern Change

From 2000–2020, the annual average *TEM* of the QBM was generally distributed in the zonal direction, that is, the spatial distribution characteristics of gradual increase from northwest to southeast (Figure 8(a1)). The low-value was concentrated in the western Songpan Plateau, whereas the high-value was mainly distributed in the eastern plain. Among them, Hubei and Henan had the highest, Gansu had the lowest (Table 5). Spatially, the heating rate generally increased from the central regions to both sides (Figure 8(a2)). The cooling areas were mainly distributed in the south, Guangyuan and Yichang, but the significance was weak. The regions that warmed faster (>0.60 °C/10a) and passes the significance level test were located mainly in Gannan, Luoyang, and Zhengzhou. In all regions, the fastest average warming rate was in Henan, and the slowest was in Shaanxi.

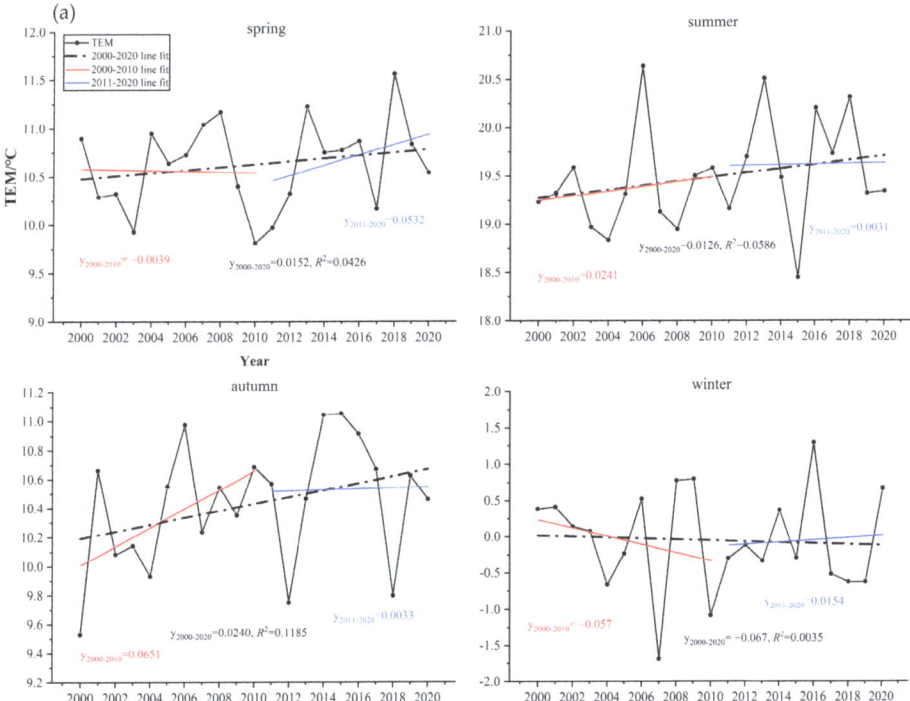

Figure 7. *Cont.*

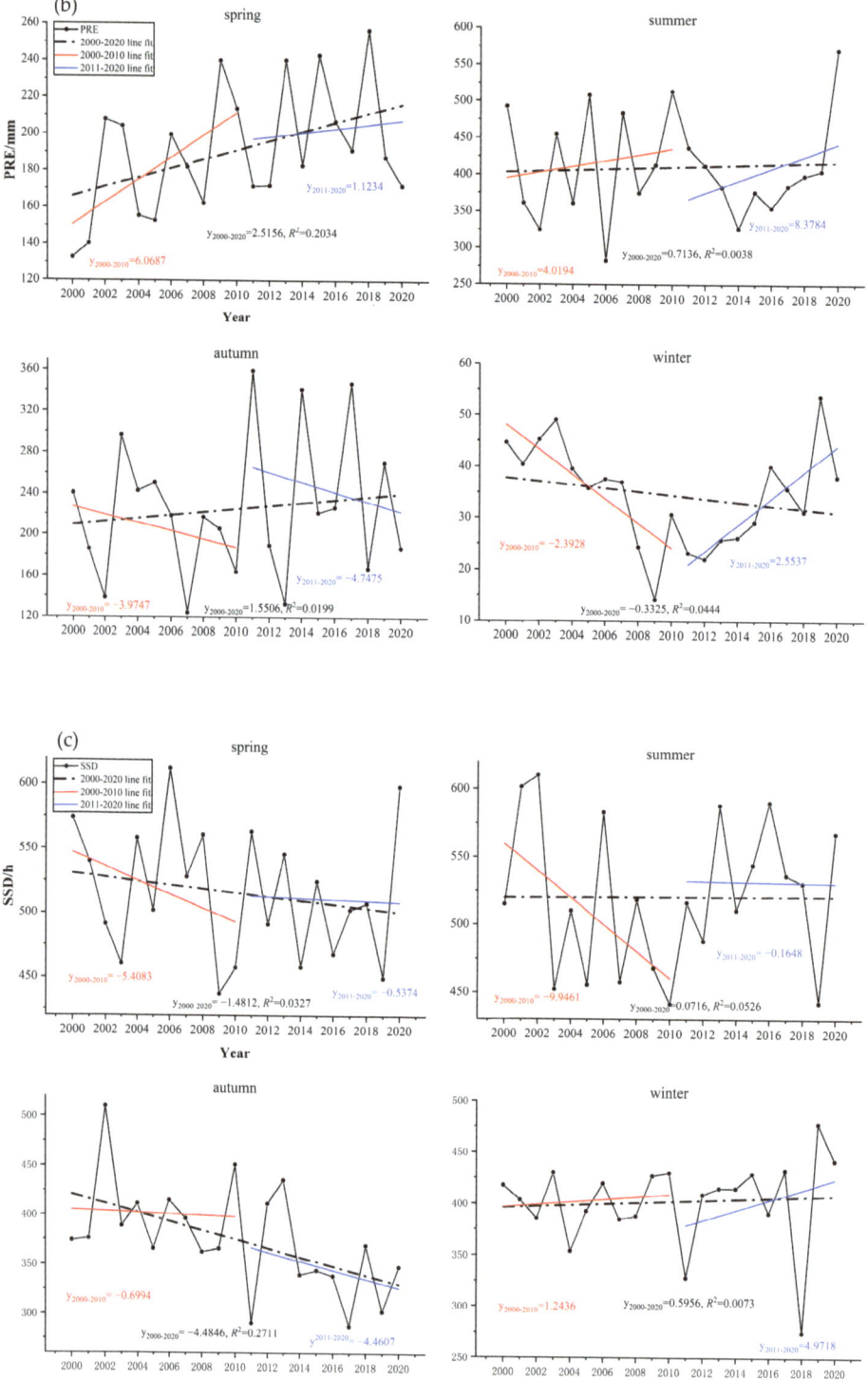

Figure 7. Seasonal mean changes of climate factors in the QBM from 2000–2020: (**a**) *TEM*; (**b**) *PRE*; (**c**) *SSD*.

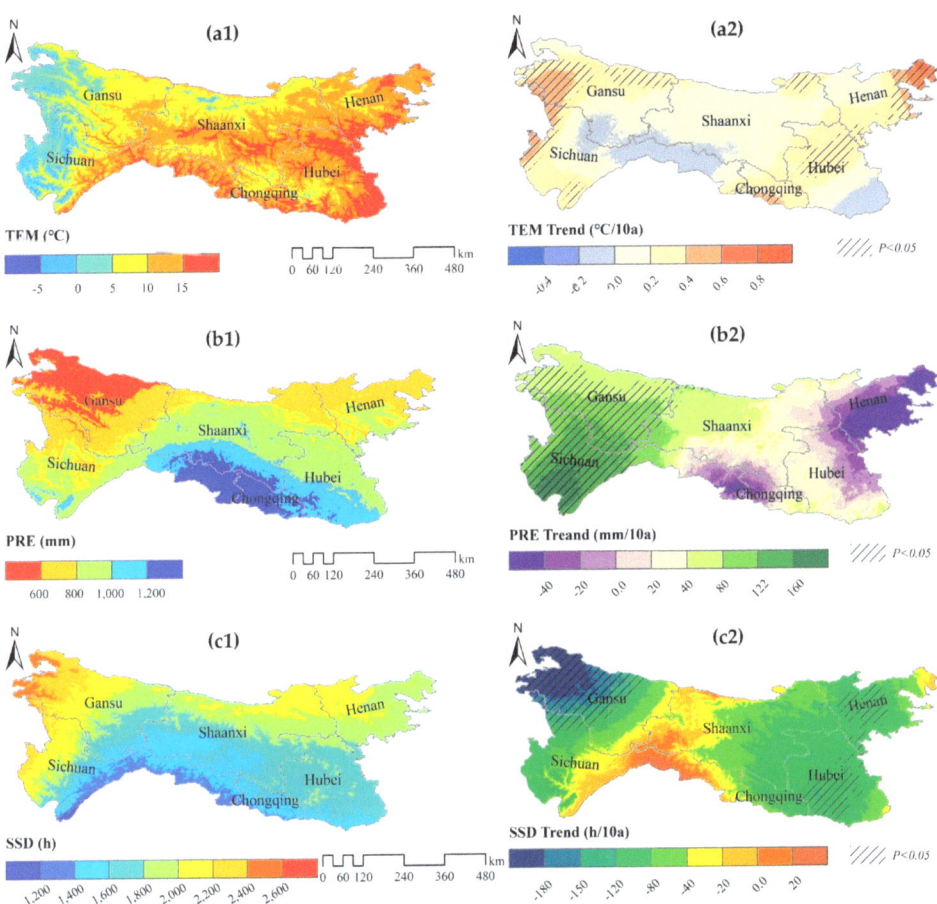

Figure 8. The spatial pattern changes of climate factors in the QBM from 2000–2020: (**a**) *TEM*; (**b**) *PRE*; (**c**) *SSD*. (1. Annual average; 2. Change trend rate).

Table 5. Annual average and change trend rate of climate factors in the QBM.

	Average Value			Slope/10a		
	TEM/°C	*PRE*/mm	*SSD*/h	*TEM*/°C	*PRE*/mm	*SSD*/h
QB	10.12	863.26	1812.51	0.175	45.462	−66.885
SC	7.50	925.20	1702.18	0.139	103.575	−48.757
SX	11.01	916.69	1779.01	0.118	39.114	−56.645
HB	13.54	976.99	1707.04	0.150	9.929	−101.508
HN	13.11	754.91	1949.17	0.307	−36.342	−87.147
GS	6.81	626.26	2028.31	0.208	81.862	−124.771
CQ	11.97	1255.14	1526.85	0.292	1.232	−83.292

Figure 8(b1) shows that the annual *PRE* was generally characterized by a gradual decrease in spatial distribution from southeast to northwest. The average annual *PRE* in Chongqing was the largest, whereas that in Gansu was the smallest. The annual *PRE* change rate of increase is generally characterized by its distribution increasing from east to west (Figure 8(b2)). In the past 21 years, the regions with reduced *PRE* were concentrated around the mountains of western Henan and Daba Mountains but did not pass the significance test. The faster (>160 mm/10a) increase trend rate was mainly distributed in the southwest, such as in Maoxian, Sichuan. The regional average annual *PRE* growth rate was the largest in Sichuan and the smallest in Henan.

The spatial distribution of annual *SSD* generally increases with increasing latitude, that is, it is low in the south and high in the north. High values of annual *SSD* are mainly distributed in the northwest, whereas the low-value area was concentrated in the northern edge of Sichuan Basin. Gansu and Henan have the largest value, and Chongqing has the smallest value. Figure 8(c1) shows that the annual *SSD* in most regions is decreasing. The increasing trend was concentrated in the northern edge of Sichuan Basin, while the decreasing trend was in the northwest, including Dingxi, Gannan and Linxia.

In summary, the west of the QBM was cold and dry with ample sunshine. The south was warm and humid, but the *SSD* was short. Most areas in the north were warm and dry with long *SSD*. However, these characteristics have changed over the past 21 years. The western region is gradually warming and humidifying, whereas the *SSD* is significantly reduced. In the southern region, *TEM* and humidity have decreased, but *SSD* has begun to increase. *TEM*, *PRE*, and *SSD* in the north show a weak increase/decrease trend.

In the past 21 years, the spatial distribution characteristics of the average *TEM* in each season showed a gradual increase from west to east. The spatial difference of average *TEM* in summer was the largest, the maximum value between east and west was more than 25 °C (Figure 9a). The annual average *TEM* in each season was 10.63 °C, 19.49 °C, 10.43 °C and −0.05 °C in spring, summer, autumn, and winter, respectively (Table 6). Gansu had the lowest average *TEM* in spring, autumn and winter, and Shaanxi had the lowest average *TEM* in summer, whereas Henan had the highest in spring and summer, and Hubei and Chongqing had the highest in autumn and winter.

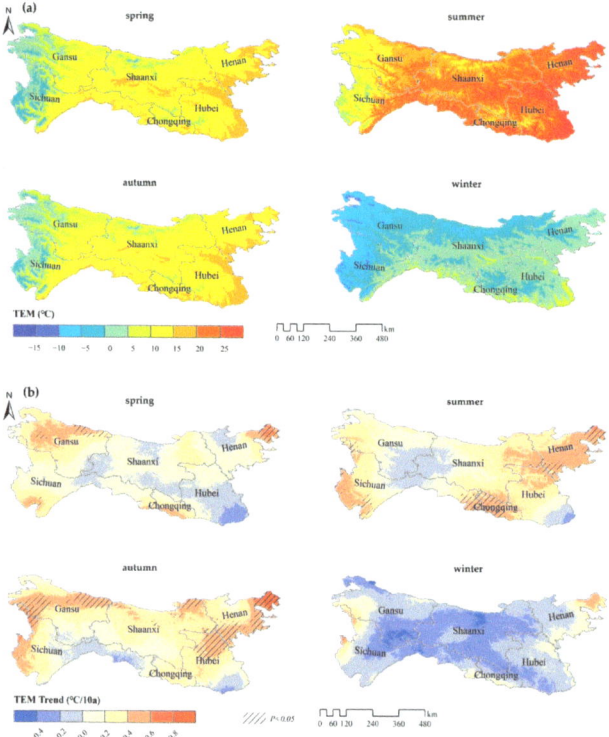

Figure 9. Change of seasonal average *TEM* spatial pattern in the QBM from 2000–2020. (**a**) Seasonal average; (**b**) Seasonal change trend rate.

Table 6. Seasonal average and change trend rate of TEM in the QBM.

		Average Value				Slope (°C/10a)			
		Spring	Summer	Autumn	Winter	Spring	Summer	Autumn	Winter
TEM	QB	10.63	19.49	10.43	−0.05	0.15	0.22	0.24	−0.07
	SC	7.75	15.55	8.04	−1.32	0.17	0.26	0.16	−0.10
	SX	11.62	20.74	11.12	0.64	0.07	0.20	0.25	−0.25
	HB	13.87	23.32	14.06	2.98	−0.04	0.23	0.23	−0.10
	HN	14.03	23.82	13.33	1.30	0.18	0.45	0.42	0.05
	GS	7.49	16.07	7.03	−3.35	0.29	0.11	0.30	−0.14
	CQ	11.95	20.79	12.59	2.72	0.27	0.47	0.23	−0.09

The variation in average TEM in each season exhibited spatial heterogeneity. Most areas showed a warming trend in all seasons except for winter (Figure 9b). The TEM dropped in spring, mainly in the junction of Gansu, Shaanxi, and Sichuan, and south of the Han River. In summer, the cooling area was still concentrated at the junction of Gansu, Shaanxi, and Sichuan, the trend rate changed from negative to positive, and the diffusion increased from this central area. The autumn cooling areas were concentrated in the south, especially in Sichuan. The average TEM in winter showed a downward trend overall, and the warming trend was mainly to the east and west.

From a regional perspective, the seasonal average TEM in Sichuan, Henan, Hubei, and Chongqing, as for FVC, initially experienced an increasing but then decreased. The maximum and minimum trend rates occurred in summer and winter, respectively. Although Shaanxi initially increased and then decreased, the maximum trend occurred in autumn. The change in trend rate in Gansu alternated from a change of decrease to increase to decrease, with the maximum appearing in autumn and the minimum in winter.

The seasonal PRE in the QBM generally shows a decreasing spatial distribution from southeast to northwest (Figure 10a). The PRE values were the most abundant in summer and the spatial difference was the largest, reaching more than 400 mm. In contrast, the PRE was rare in winter, and the spatial difference was the least. The distribution of PRE in spring and autumn was similar in the western region, but in the central and eastern regions, it was lower in spring than in autumn. Chongqing had the largest regional PRE in each season, Henan had the lowest PRE in spring, and Gansu had the lowest PRE in other seasons (Table 7).

Figure 10b shows that the trend rate of the PRE in spring was positive and gradually increased from four sides to the central region. The regions with increased and decreased summer PRE presented a spatial pattern of opposite distributions. In the area west of Shangluo-Ankang-Bazhong, the summer PRE showed an increasing trend, and the growth rate in the south was higher than that in the north. The area to the east showed a decreasing trend, and the rate of decrease shrank from the center to the northwest and southeast. In autumn, the PRE shows an increasing trend in most areas. The downward trend was mainly distributed in Gansu, central Sichuan, western Shaanxi, and central Henan. The rate of change of winter PRE gradually decreased from northwest to southeast. Only the northwest region exhibited a weak increasing trend.

From a regional perspective, the trend rate of seasonal PRE change in Shaanxi, Hubei, Henan, and Chongqing was characterized by a decrease-increase-decrease. The minimum value appeared in summer, and the maximum value in autumn. Sichuan and Gansu were characterized by first increasing and then decreasing, with the maximum in summer and the minimum in winter.

In the past 21 years, the SSD in the QBM has been long in spring and summer; and short in autumn and winter (Table 8). Spatially, the change along the latitude generally decreased from north to south (Figure 11a). The spatial difference was the largest in winter, reaching more than 450 h, and the smallest in summer. In terms of different regions, in Sichuan and Chongqing, SSD was relatively short, and in Henan and Gansu it was relatively long.

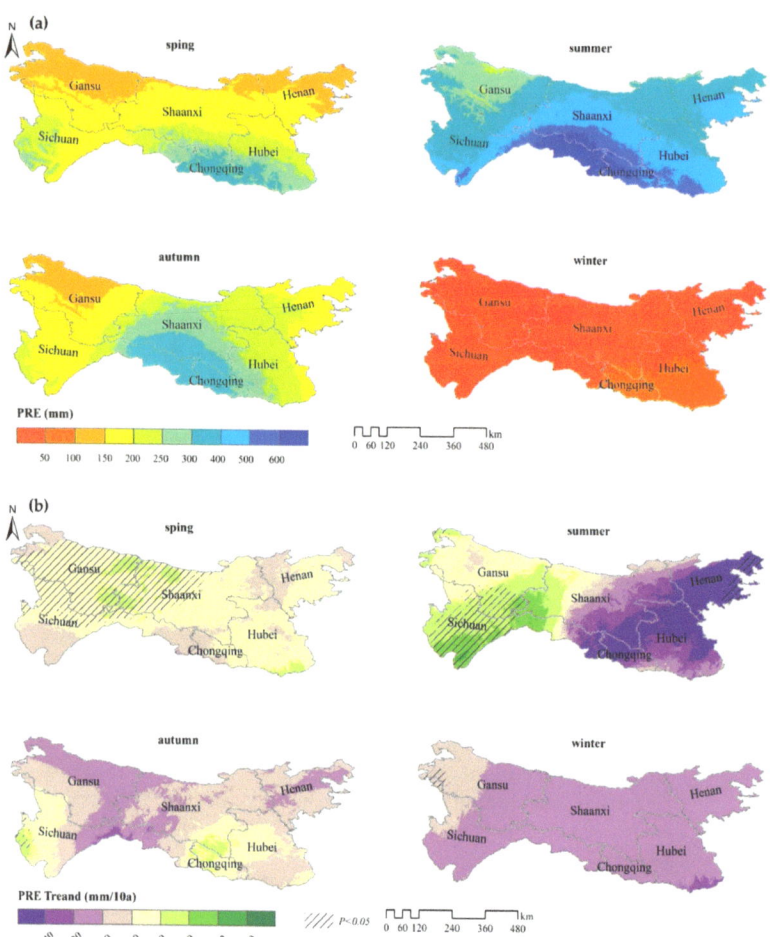

Figure 10. Change of seasonal average PRE spatial pattern in the QBM from 2000–2020. (a) Seasonal average; (b) Seasonal change trend rate.

Table 7. Seasonal average and change trend rate of PRE in the QBM.

		Average Value				Slope (°C/10a)			
		Spring	Summer	Autumn	Winter	Spring	Summer	Autumn	Winter
	QB	191.28	410.50	225.31	34.49	25.16	7.14	15.51	−3.33
	SC	215.17	446.09	232.55	27.61	22.72	53.23	14.76	−1.89
	SX	185.44	436.14	264.00	33.56	29.65	1.62	7.30	−4.89
PRE	HB	239.92	441.49	232.29	58.01	25.88	−36.50	19.54	−11.19
	HN	140.73	382.76	192.44	39.03	22.29	−63.85	2.75	−6.37
	GS	143.21	304.00	158.43	16.85	30.31	39.94	−1.32	−0.44
	CQ	311.38	552.02	331.33	60.67	19.32	−39.93	33.06	−7.85

Table 8. Seasonal average and change trend rate of *SSD* in the QBM.

		Average Value				Slope (°C/10a)			
		Spring	Summer	Autumn	Winter	Spring	Summer	Autumn	Winter
SSD	QB	515.66	520.55	375.43	402.28	−14.81	0.72	−44.85	5.96
	SC	465.32	461.86	364.47	413.70	−5.76	10.63	−32.48	12.22
	SX	519.31	540.01	347.59	371.90	−20.57	20.48	−40.60	24.90
	HB	481.35	520.23	368.53	337.14	−30.93	4.22	−51.62	8.97
	HN	583.11	530.99	418.72	417.47	−42.40	9.95	−51.85	23.49
	GS	568.03	549.34	414.26	498.82	−16.88	−36.82	−49.52	−2.27
	CQ	419.90	514.51	325.74	271.79	−19.72	13.81	−47.41	13.48

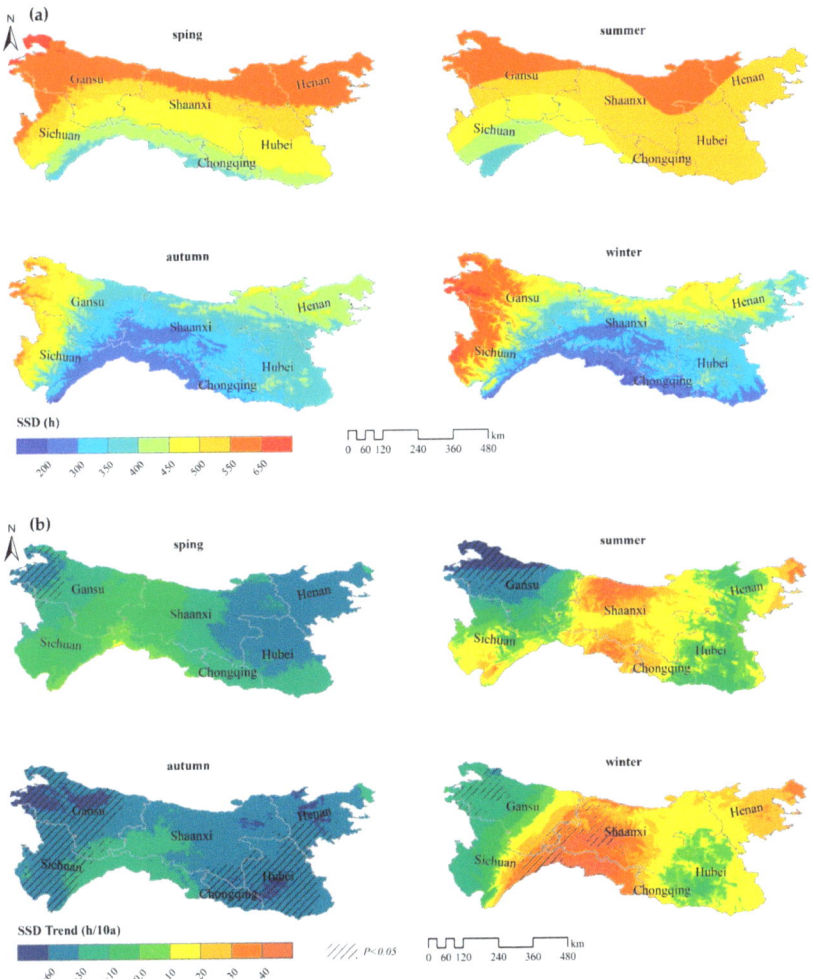

Figure 11. Change of seasonal average *SSD* spatial pattern in the QBM from 2000–2020. (**a**) Seasonal average; (**b**) Seasonal change trend rate.

The change trend rate of *SSD* in the four seasons showed clear spatial differences. The *SSD* in spring generally showed a decreasing trend, and the trend rate decreased from the center to both sides. A small increase was mainly in the south at the edge of Sichuan Basin (Figure 11b). In summer, decreases and increases in *SSD* were distributed interactively; and decrease were mainly on the east and west sides; and increases were in the central area, Qinling Mountains in Shaanxi, and Daba Mountain area, which exhibited the fastest growth rate. Similar to spring, *SSD* in autumn also decreased. The regions with the fastest decline rate were mainly in northern Gansu, central Henan, and the areas around Shennongjia in Hubei. In winter, the *SSD* in most areas increased and the growth rate in the center was the fastest. The areas with reduced *SSD* were concentrated in the west.

In Sichuan, Shaanxi, Hubei, Henan, and Chongqing, the rate of change of *SSD* in each season was consistent, showing a "N" pattern, with the maximum in winter and the minimum in autumn. The rate of change showed a "V" pattern, only in Gansu, with the maximum appearing in winter and the minimum in autumn.

3.3. Response of FVC Change to Climatic Factors

3.3.1. Response of FVC Change to a Single Climate Factor

The correlation coefficients between the annual, and seasonal average *FVC* and average *TEM*, *PRE*, and *SSD* were calculated (Table 9). Overall, from 2000–2020, the annual average *FVC* of the QBM was positively correlated with *TEM* and *PRE* but negatively correlated with *SSD*, with the correlation with *TEM* being stronger than those with *PRE* and *SSD*. From the perspective of each season, *FVC* was positively correlated with average *TEM*, with the strongest correlation in spring. The *FVC* was positively correlated with *PRE* in all seasons except winter. In contrast to *PRE*, only *SSD* in winter had a weak positive correlation with *FVC* and the other seasons had a negative correlation, with the maximum value appearing in autumn. The above results show that the vegetation coverage in the QBM is greatly affected by *TEM*, whereas the influence of *PRE* and *SSD* is relatively weak. However, in spring, the impact of *PRE* may be more obvious than that of the average *TEM*, whereas in autumn and winter, it is more affected by *SSD*.

Table 9. Correlation coefficient between *FVC* and climatic factors in the QBM.

		QB	SC	SX	HB	HN	GS	CQ
Year	TEM	0.26	0.22	0.24	0.29	0.25	0.32	0.41
	PRE	0.18	0.25	0.09	0.10	−0.05	0.44	0.09
	SSD	−0.20	−0.01	−0.05	−0.33	−0.36	−0.39	−0.23
Spring	TEM	0.22	0.25	0.21	0.10	0.07	0.36	0.37
	PRE	0.35	0.26	0.35	0.37	0.27	0.50	0.27
	SSD	−0.10	0.04	−0.03	−0.20	−0.34	−0.08	−0.10
Summer	TEM	0.04	0.03	0.08	0.06	0.14	−0.06	0.09
	PRE	0.05	0.06	−0.02	−0.03	−0.08	0.29	0.00
	SSD	−0.05	0.13	0.01	−0.08	−0.06	−0.26	−0.03
Autumn	TEM	0.19	0.18	0.22	0.17	0.05	0.25	0.22
	PRE	0.09	0.09	0.03	0.09	0.16	0.14	0.21
	SSD	−0.24	−0.14	−0.13	−0.30	−0.29	−0.38	−0.33
Winter	TEM	0.09	0.06	0.03	0.08	0.23	0.13	0.11
	PRE	−0.09	−0.11	−0.11	−0.01	0.00	−0.16	0.05
	SSD	0.15	0.21	0.20	0.05	0.07	0.13	0.26

Figure 12 shows the spatial distribution of the correlation between the annual average *FVC* and the annual average *TEM*, *PRE*, and *SSD* in the QBM from 2000–2020. Over the 21 years, there was a positive correlation between the annual average *FVC* and the annual average *TEM* in nearly 90% of the regions (Table 10), especially in the south of Gansu and northeast of Henan. A generally weak negative correlation was mostly observed in the southern region, Guangyuan; however, the negative correlation was generally weak (Figure 12a). This shows that the increase in *TEM* in most parts of the study area over the past 21 years was conducive to the growth and development of vegetation, thereby promoting an improvement in regional vegetation coverage.

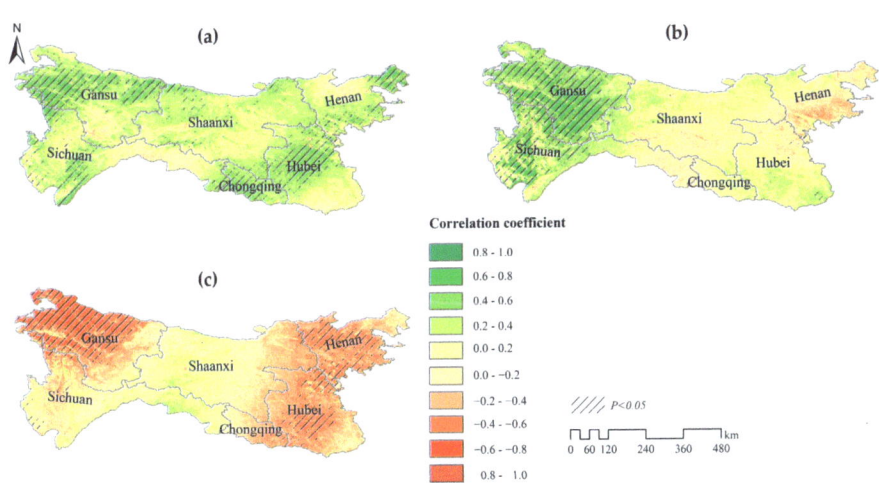

Figure 12. Spatial distribution of the correlation between annual *FVC* and climatic factors in the QBM from 2000–2020: (**a**) *TEM*; (**b**) *PRE*; (**c**) *SSD*.

Table 10. Area proportion of correlation level between *FVC* and climate factors /%.

	1	2	3	4	5
	Significant Positive Correlation	No significant Positive Correlation	Significant Negative Correlation	No significant Negative Correlation	No Correlation
TEM	21.10	68.95	0.20	9.02	0.73
PRE	18.55	55.69	0.27	24.75	0.73
SSD	0.26	25.32	20.57	53.13	0.73

Overall (Figure 12b), the annual average *FVC* was positively correlated with the annual *PRE*, mainly in the western region. The strongest positive correlation was in southern Gansu and western Sichuan. Approximately 26% of the regions have a negative correlation between the two, which was mainly distributed in the eastern region, particularly in Nanyang, Henan. This indicates that in the western region the increase in *PRE* over 21 years has improved regional vegetation coverage, whereas in the east, the reduction in *PRE* has increased vegetation coverage.

In the last 21 years, 73.70% of the region's annual average *FVC* had a significant negative correlation with the annual *SSD* (Figure 12c), which was concentrated in the northwest and east. In contrast, the proportion of positive correlation between them was mainly in the central region, but this correlation is weak. This shows that in the 21 years, the increase in annual *SSD* has played a weak role in promoting the improvement of regional vegetation coverage only in the central region. The reduction in annual *SSD* in other regions has greatly improved regional vegetation coverage.

For more than 80% of the regions, there was a positive correlation between the average *FVC* and the average *TEM* in the same season in spring and autumn (Table 11), mainly in the north and west (Figure 13). The negative correlation was mostly to the east of Ankang, and in autumn was mainly at the edge of the Sichuan Basin and around the mountains in western Henan. In summer and winter, the proportion of *FVC* positively correlated with the seasonal average *TEM* was in the range of 60%–70%, but this correlation was weak. In summer, the positive correlation area was concentrated east of Hanzhong, while the negative correlation area was in the south of Gansu. In winter, the positive correlation region was mainly in the northwest and northeast, while the negative correlation was in the central region. In spring, summer, and autumn, the increase in regional vegetation coverage was promoted by an increase in *TEM* in the concurrent season. A decrease in the average *TEM* in the concurrent season was more conducive to the improvement of vegetation coverage in only a few areas. In contrast, in winter, although the TEM in most areas dropped, vegetation coverage still improved.

Table 11. Area proportion of correlation level between FVC and TEM/%.

	1 Significant Positive Correlation	2 No significant Positive Correlation	3 Significant Negative Correlation	4 No significant Negative Correlation	5 No Correlation
Spring	17.75	64.88	0.19	15.80	1.38
Summer	1.52	59.48	1.54	37.24	0.22
Autumn	10.88	71.42	0.27	16.97	0.45
Winter	3.66	62.41	0.14	31.50	2.29

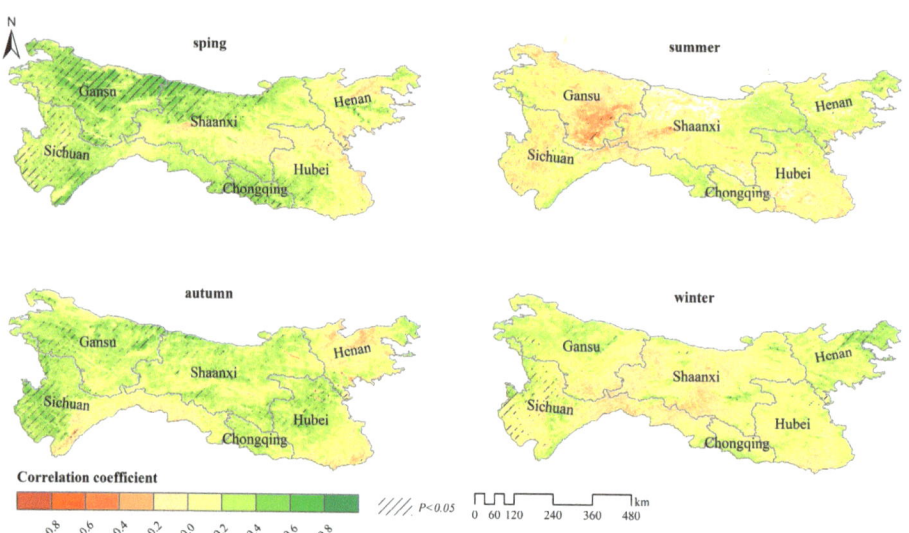

Figure 13. Spatial distribution of the correlation between seasonal FVC and TEM in the QBM from 2000–2020.

Figure 14 shows that from spring to winter, the proportion of positive correlation area between seasonal average FVC and PRE in the concurrent season decreased, indicating that the negative correlation between them gradually increased. In spring, regions with a negative correlation accounted for less than 6% (Table 12). In summer and autumn, the proportion of the area with a positive correlation between FVC and PRE was 55%–68%, but the correlation was slightly weak. In winter, the area of negative correlation was twice that of positive correlation, and it was strong in Tianshui, Baoji, and Mianyang. For the western region, regardless of the season, an increase in PRE promoted the improvement of regional vegetation coverage. In most areas, the winter PRE decreased, but the vegetation coverage in the current season continued to increase, indicating that the reduction in winter PRE may be beneficial to the growth of vegetation.

From the correlation between FVC and SSD in each season, it was found that the area with negative correlation was largest in autumn, followed by spring and summer, and the area with positive correlation was largest in winter (Table 13). In spring, the negative correlation was mostly in the east, particularly in Henan (Figure 15). The areas with strong negative correlation in summer were concentrated in northwest Gansu. In autumn, the negative correlation area was mainly in the east and west, and the negative correlation was strong. The proportion of the negative correlation area in winter was less than 30%. Although it was distributed in the west and east, it showed a weak negative correlation. Combined with the change trend of seasonal SSD, it was found that there was a decreasing trend in the negative correlation regions, indicating that for most regions, its shortening promotes regional vegetation coverage. However, for the central and Daba Mountain regions, the increase in *SSD* in winter was more beneficial to the increase in vegetation coverage.

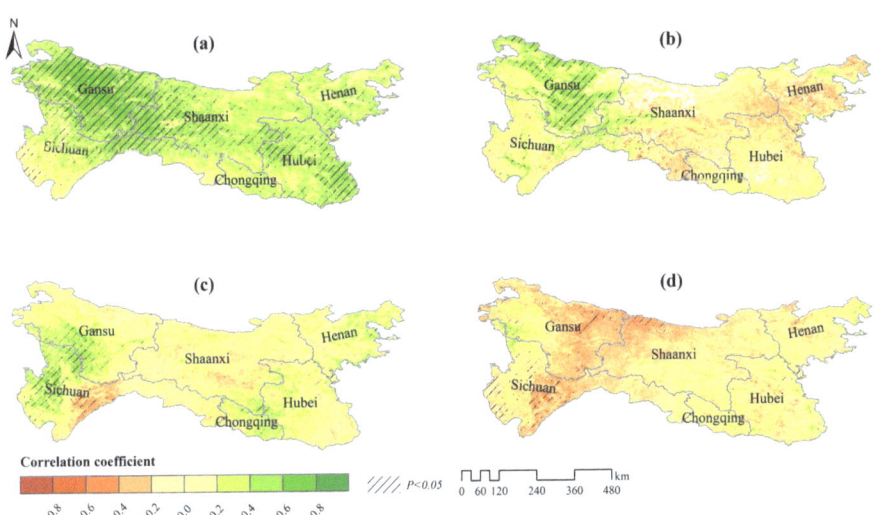

Figure 14. Spatial distribution of the correlation between seasonal *FVC* and *PRE* in the QBM from 2000–2020.

Table 12. Area proportion of correlation level between *FVC* and *PRE*/%.

	1	2	3	4	5
	Significant Positive Correlation	No significant Positive Correlation	Significant Negative Correlation	No significant Negative Correlation	No Correlation
Spring	34.84	59.76	0.05	3.96	1.38
Summer	8.84	47.02	2.31	41.61	0.22
Autumn	6.22	60.89	1.13	31.32	0.45
Winter	1.01	32.19	4.62	59.89	2.29

Table 13. Area proportion of correlation level between *FVC* and *SSD*/%.

	1	2	3	4	5
	Significant Positive Correlation	No significant Positive Correlation	Significant Negative Correlation	No significant Negative Correlation	No Correlation
Spring	0.45	34.51	7.11	56.54	1.38
Summer	2.70	41.98	7.93	47.15	0.23
Autumn	0.41	17.77	23.69	57.67	0.47
Winter	11.08	62.09	0.39	24.16	2.29

3.3.2. Main Climate Factors of *FVC* Change

From the above analysis, it can be seen that there are obvious differences in the response of *FVC* in different regions to changes in a single climate factor. Therefore, a multiple regression analysis method was adopted to calculate the relative importance of each factor to *FVC* change pixel-by-pixel. In the 21 years in the QBM, the relative importance of the annual average *TEM* (0.33) to *FVC* was greater than that of *PRE* (0.17) and *SSD* (0.20) (Table 14) However, there were clear regional differences in the spatial distribution of the dominant factors. The areas dominated by *TEM* are mainly in the central and eastern regions. The areas mainly controlled by *PRE* change are in western Sichuan, southern Gansu and western Shaanxi, whereas the northwest and east were most affected by *SSD* (Figure 16a).

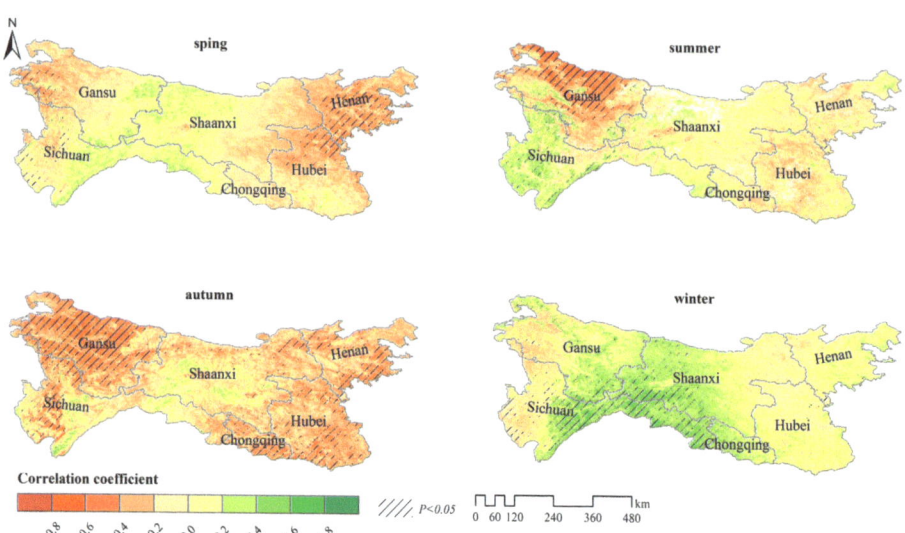

Figure 15. Spatial distribution of the correlation between seasonal FVC and SSD in the QBM from 2000–2020.

Table 14. Relative importance index of climate factors in the QBM.

		QB	SC	SX	HB	HN	GS	CQ
Year	TEM	0.33	0.21	0.30	0.45	0.37	0.36	0.49
	PRE	0.17	0.27	0.17	0.05	−0.01	0.26	0.18
	SSD	−0.20	0.07	−0.05	−0.46	−0.46	−0.31	−0.20
Spring	TEM	0.27	0.22	0.24	0.33	0.36	0.27	0.42
	PRE	0.32	0.27	0.38	0.37	0.16	0.40	0.22
	SSD	−0.09	0.06	0.03	−0.22	−0.44	−0.06	−0.16
Summer	TEM	0.10	0.00	0.11	0.18	0.29	0.03	0.18
	PRE	0.04	0.08	0.00	−0.06	−0.08	0.20	0.05
	SSD	−0.09	0.15	−0.05	−0.24	−0.28	−0.19	−0.10
Autumn	TEM	0.18	0.17	0.23	0.15	0.07	0.20	0.18
	PRE	−0.08	−0.04	−0.14	−0.05	0.02	−0.12	0.07
	SSD	−0.26	−0.13	−0.19	−0.29	−0.28	−0.41	−0.26
Winter	TEM	0.06	0.02	−0.01	0.07	0.26	0.06	0.11
	PRE	−0.09	−0.10	−0.12	−0.02	0.03	−0.14	−0.11
	SSD	0.11	0.22	0.19	0.04	−0.04	0.03	0.32

In combination with various regions, Sichuan has the largest relative importance and area proportion of the PRE (Figure 17a), mainly in western Sichuan. Although Shaanxi and Chongqing were more affected by TEM, Hanzhong was mainly controlled by PRE. In Hubei and Henan, the difference between the relative importance index of SSD and the average TEM was small, indicating that the regional annual FVC change was a result of their joint influence. However, in Shiyan and Zhengzhou, the change in FVC was mainly controlled by TEM. In contrast to other regions, the relative importance index of the three climate factors in Gansu was basically the same, which indicates that the annual FVC changes in this region were affected by the three factors together. The south was mainly controlled by PRE, the central and eastern regions are more affected by TEM, and the rest are more affected by SSD.

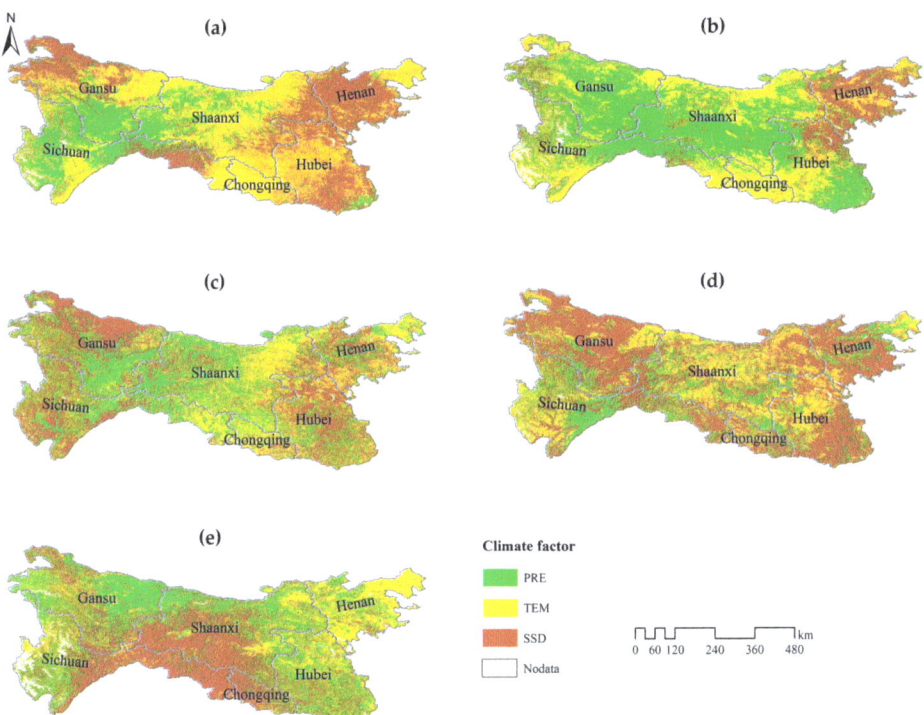

Figure 16. Main climate factors of average *FVC* in the QBM. (**a**) Year; (**b**) Spring; (**c**) Summer; (**d**) Autumn; (**e**) Winter.

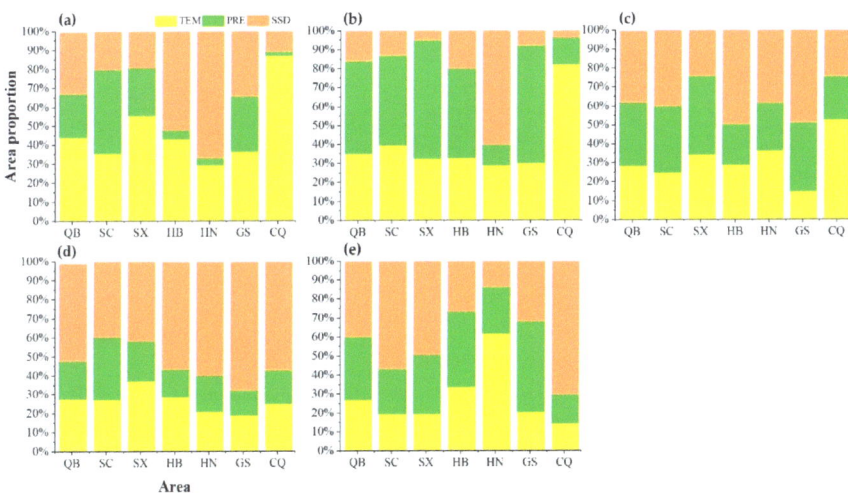

Figure 17. Proportion of the main control area of climate factors in the QBM. (**a**) Year; (**b**) Spring; (**c**) Summer; (**d**) Autumn; (**e**) Winter.

In the past 21 years, the climatic factors leading to *FVC* changes in the QBM have changed significantly during different seasons. In spring, the QBM was more affected by *TEM* and *PRE* (Figure 17b). The former mainly affects the high-altitude areas at the edge, while the latter mainly affects the middle and low-altitude areas in the central and west (Figure 16b). In summer, although the difference in their relative importance index was small, the area dominated by SSD was the largest proportionally (Figure 17c). In autumn, the relative importance of TEM and SSD was greater. The former was mainly in the east of Shaanxi, whereas the latter is distributed in the east and west (Figure 16d). It was most affected by SSD in winter, which was concentrated in the south of Qinling Mountains (Figure 16e).

In addition, the changes in dominant climate factors in different seasons in different regions were also obvious. In spring, Sichuan, Gansu, Shaanxi, and Hubei were affected by TEM and PRE. The importance index of SSD in Henan was the largest, and the TEM in Chongqing was the largest.

In summer, the relative importance of TEM and PRE weakened, while the influence of SSD generally increased. In Sichuan and Gansu, SSD and PRE had a relatively significant influence on FVC changes. Henan and Hubei were affected by SSD and TEM. Chongqing was mainly affected by TEM changes. Although the relative importance of TEM in Shaanxi was the largest, the area dominated by PRE was larger than that dominated by TEM. The change in FVC during the summer in this area may be the result of their joint influence.

In autumn, the relative importance of TEM and SSD on FVC changes in most regions increased, SSD more than TEM, but the impact of PRE remained weak. In Sichuan, the relative contribution rate of TEM was greater than that of SSD and PRE, but the areas dominated by the latter two were greater than those of TEM, indicating that the region is jointly influenced by the three factors. In Shaanxi, the relative contribution rate of TEM was the largest, but the area under the main control of SSD was the largest, indicating that the region was jointly affected by TEM and SSD. In Hubei, Henan, Gansu and Chongqing, the change of FVC in autumn in these regions was mainly affected by SSD.

In winter, the relative importance of TEM decreased, the importance of PRE in some areas increased, and SSD increased and decreased. In Sichuan and Shaanxi, FVC in winter was most influenced by SSD, mainly in the north of the Sichuan Basin and Hanshui Valley, whereas other areas were significantly affected by PRE. In Henan, Chongqing, and Gansu, the most important factors were TEM, SSD, and PRE. Although the relative importance of the winter TEM in Hubei was the greatest, the main area affected by PRE was greater. This shows that the FVC changes in winter in this region were the result of the joint action of TEM and PRE.

4. Discussion

Vegetation activity on a global scale is showing an upward trend [24]. Many studies have shown that in the past 20 years, China's vegetation has been restored, and vegetation coverage has shown a significant increase. However, the average vegetation coverage and its dynamic change trends show obvious spatial heterogeneity. In general, the vegetation coverage in southeast China is higher than that in northwest China. At the same time, the growth trends of the central, eastern, and southwest regions are significant (such as Shaanxi, Gansu, Chongqing, and surrounding regions). This may be related to changes in climate conditions and the adjustment of human activities, such as ecological restoration [25,26]. However, for regions such as the southeast of the Qinghai Tibet Plateau and the northwest Sichuan Plateau, as well as large urban agglomerations with developed economies, because of the terrain, hydrothermal conditions, urban development, and population expansion, the vegetation has showed a significant degradation trend [27–29]. Based on MODWAS NDVI data, this study analyzed the dynamic changes in FVC in the QBM from 2000–2020. The results showed that vegetation coverage in the QBM was relatively high and showed a significant increasing trend over this period. The regions with rapid growth rates are mainly in the east of Shaanxi, south of Gansu, central Sichuan, and Chongqing. There are a few scattered areas with degraded vegetation in Chengdu and Aba in Sichuan, Hanzhong in Shaanxi, Shiyan in Hubei, Luoyang in Henan, and other urban areas. This result is consistent with the above conclusions.

Changes of climate conditions are widely considered important abiotic factors in the spatial distribution and dynamic change of vegetation [30–32]. In this study, the annual average *TEM* was the main driving factor for changes in vegetation coverage in the QBM. The annual average *FVC* value increased with an increase in annual average *TEM*. There was a positive and negative correlation between annual *PRE* and annual *SSD*, respectively, but these were relatively weak. The area mainly controlled by *TEM* is in the southern foothills of Qinling and Daba Mountains, and the area mainly controlled by *PRE* is in the western high-altitude area. The areas dominated by *SSD* are in the eastern, northwestern, and northern edges of the Sichuan Basin. Previous studies have shown that before reaching a threshold, an increase in *TEM* promotes photosynthesis, prolonging the growth cycle of vegetation, and thus promoting vegetation growth [33,34]. Gao [35] confirmed the reliability of these results. His research results show that in southeast and southwest China, dynamic change in vegetation is most significantly controlled by *TEM*. These areas have abundant rainfall, short *SSD*, and relatively high soil moisture content; therefore, the relationship between vegetation activity and *TEM* was closer. However, for the northern and northwestern regions of China, changes in vegetation cover are most closely related to *PRE* and *SSD* [36–39]. Most of these areas are arid and semi-arid, with long sunshine duration and vegetation growth limited by available moisture. Therefore, an increase in *PRE* and decrease in *SSD* can offset the negative impact of *TEM* rise on vegetation growth, and increase regional vegetation coverage by promoting vegetation growth. However, the results of Kong's study in the Northern Hemisphere for nearly 30 years showed that temperature is a major factor in vegetation greening at high latitudes, especially spring and autumn temperature in North America and Siberia. Solar radiation corresponds well with vegetation trends in northern North America and eastern China [40]. The differences in these results may be related to the differences in the study period and region.

However, under different space-time scales, the main driving factors of changes in vegetation coverage are also different, which is often ignored. Relevant research has shown that dynamic change in vegetation in most regions of China gives the strongest response to spring *TEM*, whereas its positive and negative correlations with *PRE* vary from place to place [41,42]. The present study found that, in the QBM, the positive correlation between changes in vegetation coverage in spring and *PRE* was stronger, the relative impact of the three climate factors was smaller in summer, and the impact of *SSD* was the greatest in autumn and winter. Qi [18] also confirmed the reliability of the results in Qinling area. The results showed that the vegetation growth in the north and south of Qinling Mountains was significantly affected by changes in dry and wet conditions in spring and summer, and that the moisture in spring could promote vegetation growth.

Changes in vegetation coverage are the result of the comprehensive action of climatic and non-climatic factors. In the present study, only climatic factors were considered when focusing on the factors driving changes in vegetation coverage in the QBM. Many other environmental factors also affect the spatial heterogeneity of vegetation dynamic activities, including terrain, soil conditions, and CO_2 concentration. For example, Liu [43] showed that vegetation coverage in the karst region in southwest China is greatly affected by elevation and slope, and vegetation degradation mainly occurs on low-altitude gentle slopes. Shang [44] showed that different soil types could cause differences in vegetation types and attributes. Piao [45] showed that the increase in atmospheric CO_2 concentration and nitrogen deposition are the most likely reasons for the greening trend in China over the past three decades. Therefore, the impact of environmental factors on vegetation should be fully considered in future research.

In addition, on a short time scale, the impact of anthropogenic activities on vegetation growth is very important. In China, the improvement of vegetation coverage in most areas is largely due to the implementation of ecological projects, such as returning farmland to forests/grass, artificial afforestation and grass planting, comprehensive management of rocky desertification [46,47], and appropriate agricultural management measures [48]. However, various human development and utilization activities, such as urban expansion, massive transfer of construction land, illegal logging, and overgrazing have led to the destruction of surface vegetation to a certain extent [49]. This negative impact was particularly evident in large urban agglomerations [24]. Therefore, in subsequent research, it is necessary to quantitatively analyze and evaluate the driving role and mechanism of anthropogenic factors in regional change in vegetation coverage. Furthermore, the separation of the relationship between climatic factors and anthropogenic activity factors will also be the focus of future work.

5. Conclusions

In the present study, we revealed the dynamic change characteristics of vegetation coverage in the QBM of China from 2000–2020 and its response to major climate factors and analyzed the changes in the importance of three climate factors to these changes. The overall distribution of vegetation coverage in the QBM shows a spatial pattern of "low in the east and west and, high in the central." In the past 21 years, regional vegetation coverage has continuously improved, with a faster rising trend in winter and spring. *TEM* showed a spatial distribution pattern that was low in the northwest and high in the southeast, with an obvious warming trend, and warming faster in summer and autumn. The spatial distribution of *PRE* gradually decreased from southeast to northwest. The fluctuation increased in 21 years, with the fastest growth rate occurring in spring. *SSD* generally showed the distribution characteristics of low in the south and high in the north, with a decreasing trend in fluctuation, and the largest decreasing rate in autumn.

The results showed that in the past 21 years, in most areas of the QBM, the change in vegetation coverage was positively correlated with *TEM*, negatively correlated with *SSD*, and both positively and negatively correlated with PRE. On an annual scale, the area mainly affected by *TEM* accounts for approximately 44% of the study area and was mainly in the central and eastern regions. The *SSD* was the main controlling factor affecting the vegetation coverage change in the northwest and east regions. Water condition was the main factor affecting changes in vegetation coverage in western Sichuan, southern Gansu and Hanzhong in Shaanxi. On a seasonal scale, the area dominated by *PRE* in spring was larger. In summer, the relative importance of *TEM* and *PRE* began to weaken, but the area dominated by *SSD* expanded significantly. In autumn, the influence of *TEM* and *SSD* increased, and *SSD* was the main factor affecting vegetation coverage in most areas. Although the relative importance of the three factors was greatly reduced in winter, *SSD* remained the main controlling factor for the change in vegetation coverage in most regions. The research results are helpful in understanding the impact of climate factors on vegetation coverage change in the north-south transitional region of China to better carry out vegetation restoration and protection against the background of global climate change.

Author Contributions: Conceptualization, H.R.; methodology, H.R., C.C. and L.Z. (Lijuan Zhang); software, H.R., Y.L. and L.W.; validation, H.R. and W.Z.; formal analysis, H.R.; investigation, H.R.; resources, H.R.; data curation, H.R.; writing—original draft preparation, H.R.; writing—review and editing, W.Z.; supervision, W.Z.; funding acquisition, L.Z. (Lianqi Zhu). All authors have read and agreed to the published version of the manuscript.

Funding: This research was funded by the National Key Research and Development Program of China, grant number 2021YFE0106700; the National Science and Technology Basic Resource Investigation Program of China, grant number 2017FY100902.

Data Availability Statement: The data presented in this study are available on request from the corresponding author.

Acknowledgments: We thank the School of Geography and Environment, Henan University for providing support.

Conflicts of Interest: The authors declare no conflict of interest.

References

1. Sitch, S.; Smith, B.; Prentice, I.C.; Arneth, A.; Bondeau, A.; Cramer, W.; Kaplan, J.O.; Levis, S.; Lucht, W.; Sykes, M.T.; et al. Evaluation of ecosystem dynamics, plant geography and terrestrial carbon cycling in the LPJ dynamic global vegetation model. *Glob. Chang. Biol.* **2003**, *9*, 161–185. [CrossRef]
2. Sun, R.; Chen, S.; Su, H. Climate Dynamics of the Spatiotemporal Changes of Vegetation NDVI in Northern China from 1982 to 2015. *Remote Sens.* **2021**, *13*, 187. [CrossRef]
3. Myneni, R.B.; Keeling, C.D.; Tucker, C.J.; Asrar, G.; Nemani, R.R. Increased plant growth in the northern high latitudes from 1981 to 1991. *Nature* **1997**, *386*, 698–702. [CrossRef]
4. Tucker, C.J.; Slayback, D.; Pinzon, J.E.; Los, S.; Myneni, R.; Taylor, M.G. Higher northern latitude normalized difference vegetation index and growing season trends from 1982 to 1999. *Int. J. Biometeorol.* **2001**, *45*, 184–190. [CrossRef] [PubMed]
5. Zhang, X. Main Models of Variations of Autumn Vegetation Greenness in the Mid-latitude of North Hemisphere in 1982–2011. *Sci. Geogr. Sin.* **2014**, *34*, 1226–1232. [CrossRef]
6. Myers-Smith, I.H.; Kerby, J.T.; Phoenix, G.K.; Bjerke, J.W.; Epstein, H.E.; Assmann, J.J.; John, C.; Andreu-Hayles, L.; Angers-Blondin, S.; Beck, P.S.A.; et al. Complexity revealed in the greening of the Arctic. *Nat. Clim. Chang.* **2020**, *10*, 106–117. [CrossRef]
7. Jiao, K.; Gao, J.; Wu, S. Climatic determinants impacting the distribution of greenness in China: Regional differentiation and spatial variability. *Int. J. Biometeorol.* **2019**, *63*, 523–533. [CrossRef]
8. Piao, S.; Nan, H.; Huntingford, C.; Ciais, P.; Friedlingstein, P.; Sitch, S.; Peng, S.; Ahlström, A.; Canadell, J.G.; Cong, N.; et al. Evidence for a weakening relationship between interannual temperature variability and northern vegetation activity. *Nat. Commun.* **2014**, *5*, 5018. [CrossRef]
9. Liu, Y.; Li, Y.; Li, S.; Motesharrei, S. Spatial and Temporal Patterns of Global NDVI Trends: Correlations with Climate and Human Factors. *Remote Sens.* **2015**, *7*, 13233–13250. [CrossRef]
10. Zuo, Y.; Li, Y.; He, K.; Wen, Y. Temporal and spatial variation characteristics of vegetation coverage and quantitative analysis of its potential driving forces in the Qilian Mountains, China, 2000–2020. *Ecol. Indic.* **2022**, *143*, 109429. [CrossRef]
11. Wang, S.-H.; Sun, W.; Li, S.-W.; Shen, Z.-X.; Fu, G. Interannual Variation of the Growing Season Maximum Normalized Difference Vegetation Index, MNDVI, and Its Relationship with Climatic Factors on the Tibetan Plateau. *Pol. J. Ecol.* **2015**, *63*, 424–439. [CrossRef]
12. Cui, X.; Bai, H.; Shang, X. The vegetation dynamic in Qinling area based on MODWAS NDVI. *J. Northwest Univ.* **2012**, *42*, 1021–1026. [CrossRef]
13. Guan, L.; Wang, H.; Wang, Y. Dynamic Change of Vegetation Coverage in Sichuan Region Based on GIMMS AVHRR NDVI. *Bullentin Sci. Technol.* **2016**, *32*, 31–36+41. [CrossRef]
14. Zhu, X.; Liu, K.; Li, J.; Zhu, J. Analyswas on Vegetation-Environment Gradient Correlation in Qinling Mountain Based on GWAS. *Res. Soil Water Conserv.* **2009**, *16*, 169–175.
15. Zhang, S.; Bai, H.; Gao, X.; He, Y.; Ren, Y. Spatial-temporal Changes of Vegetation Index and Its Responses to Regional Temperature in Taibai Mountain. *J. Nat. Resour.* **2011**, *26*, 1377–1386. [CrossRef]
16. Ren, Y.; Zhang, Z.; Hou, Q.; He, Y.; Yuan, B. Response of Vegetation Cover Changes to Climate Change in Daba Maintains. *Bullentin Sci. Technol.* **2012**, *2*, 56–59. [CrossRef]
17. Chen, X.; Jiang, H. Climate response of NDVI index on Qinling Mountains in 25 years. *Bull. Surv. Mapp.* **2019**, *3*, 103–107. [CrossRef]
18. Qi, G.; Bai, H.; Zhao, T.; Meng, Q.; Zhang, S. Sensitivity and areal differentiation of vegetation responses to hydrothermal dynamics on the southern and northern slopes of the Qinling Mountains in Shaanxi province. *Acta Geogr. Sin.* **2021**, *76*, 44–56. [CrossRef]
19. Zhang, B. Ten major scientific issues concerning the study of China's north-south transitional zone. *Prog. Geogr.* **2019**, *38*, 305–311. [CrossRef]
20. Long, S.; Guo, Z.; Xu, L.; Zhou, H.; Fang, W.; Xu, Y. Spatiotemporal Variations of Fractional Vegetation Coverage in China based on Google Earth Engine. *Remote Sens. Technol. Appl.* **2020**, *35*, 326–334. [CrossRef]
21. Zhang, X.; Zhu, W.; Cui, Y.; Zhang, J.J.; Zhu, L.Q. The response of forest dynamics to hydro-thermal change in Funiu Mountain. *Geogr. Res.* **2016**, *35*, 1029–1040. [CrossRef]
22. Wang, X.; Hou, X. Variation of Normalized Difference Vegetation Index and its response to extreme climate in coastal China during 1982–2014. *Geogr. Res.* **2019**, *38*, 807–821. [CrossRef]
23. Wang, J.; Wang, K.; Zhang, M.; Zhang, C. Impacts of climate change and human activities on vegetation cover in hilly southern China. *Ecol. Eng.* **2015**, *81*, 451–461. [CrossRef]
24. Eastman, J.R.; Sangermano, F.; Machado, E.A.; Rogan, J.; Anyamba, A. Global Trends in Seasonality of Normalized Difference Vegetation Index (NDVI), 1982–2011. *Remote. Sens.* **2013**, *5*, 4799–4818. [CrossRef]
25. Jin, K.; Wang, F.; Han, Q.; Shi, S.; Ding, W. Contribution of climatic change and human activities to vegetation NDVI change over China during 1982-2015. *Acta Geogr. Sin.* **2020**, *75*, 961–974. [CrossRef]
26. Luo, S.; Liu, Y.; Long, H. Nonlinear trends and spatial pattern analysis of vegetation cover change in China from 1982 to 2018. *Acta Ecol. Sin.* **2022**, *42*, 8331–8342. [CrossRef]

27. Liu, X.; Zhu, X.; Pan, Y.; Li, Y.; Zhao, A. Spatiotemporal changes in vegetation coverage in China during 1982–2012. *Acta Ecol. Sin.* **2015**, *35*, 5331–5342. [CrossRef]
28. Peng, W.; Zhang, D.; Luo, Y.; Tao, S.; Xu, X. Influence of natural factors on vegetation NDVI using geographical detection in Sichuan Province. *Acta Geogr. Sin.* **2019**, *74*, 1758–1776. [CrossRef]
29. Li, H.; Zhang, C.G.; Wang, S.Z.; Ma, W.D.; Liu, F.G.; Chen, Q.; Zhou, Q.; Xia, X.S.; Niu, B.C. Response of vegetation dynamics to hydrothermal conditions on the Qinghai-Tibet Plateau in the last 40 years. *Acta Ecol. Sin.* **2022**, *42*, 4770–4783. [CrossRef]
30. Chen, X.; Wang, H. Spatial and Temporal Variations of Vegetation Belts and Vegetation Cover Degrees in Inner Mongolia from 1982 to 2003. *Acta Geogr. Sin.* **2009**, *64*, 84–94. [CrossRef]
31. Zhu, Z.; Piao, S.; Myneni, R.B.; Huang, M.; Zeng, Z.; Canadell, J.G.; Ciais, P.; Sitch, S.; Friedlingstein, P.; Arneth, A.; et al. Greening of the Earth and its drivers. *Nat. Clim. Chang.* **2016**, *6*, 791–795. [CrossRef]
32. Prăvălie, R.; Sîrodoev, I.; Nita, I.-A.; Patriche, C.; Dumitraşcu, M.; Roşca, B.; Tişcovschi, A.; Bandoc, G.; Săvulescu, I.; Mănoiu, V.; et al. NDVI-based ecological dynamics of forest vegetation and its relationship to climate change in Romania during 1987–2018. *Ecol. Indic.* **2022**, *136*, 108629. [CrossRef]
33. Li, J.; Liu, H.; Li, C.; Li, L. Changes of Green-up Day of Vegetation Growing Season Based on GIMMS 3g NDVI in Northern China in Recent 30 Years. *Sci. Geogr. Sin.* **2017**, *37*, 620–629. [CrossRef]
34. He, Y.; Liu, X.; Wang, H. Variation characteristics of vegetation cover in the latest 10 years over China based on EVI. *J. Meteorol. Sci.* **2017**, *37*, 51–59. [CrossRef]
35. Gao, J.; Jiao, K.; Wu, S. Revealing the climatic impacts on spatial heterogeneity of NDVI in China during 1982–2013. *Acta Geogr. Sin.* **2019**, *74*, 534–543. [CrossRef]
36. Lu, C.; Hou, M.; Liu, Z.; Li, H.; Lu, C. Variation Characterwastic of NDVI and its Response to Climate Change in the Middle and Upper Reaches of Yellow River Basin, China. *IEEE J. Sel. Top. Appl. Earth Obs. Remote Sens.* **2021**, *14*, 8484–8496. [CrossRef]
37. Wu, Z.; Bi, J.; Gao, Y. Drivers and Environmental Impacts of Vegetation Greening in a Semi-Arid Region of Northwest China since 2000. *Remote Sens.* **2021**, *13*, 4246. [CrossRef]
38. Zou, Y.; Chen, W.; Li, S.; Wang, T.; Yu, L.; Xu, M.; Singh, R.P.; Liu, C.-Q. Spatio-Temporal Changes in Vegetation in the Last Two Decades (2001–2020) in the Beijing–Tianjin–Hebei Region. *Remote Sens.* **2022**, *14*, 3958. [CrossRef]
39. Cao, W.; Xu, H.; Zhang, Z. Vegetation Growth Dynamic and Sensitivity to Changing Climate in a Watershed in Northern China. *Remote Sens.* **2022**, *14*, 4198. [CrossRef]
40. Kong, D.; Zhang, Q.; Singh, V.P.; Shi, P. Seasonal vegetation response to climate change in the Northern Hemisphere (1982–2013). *Glob. Planet. Chang.* **2016**, *148*, 1–8. [CrossRef]
41. Zhang, X.; Dai, J.; Ge, Q. Spatial Differences of Changes in Spring Vegetation activities across Eastern China during 1982–2006. *Acta Geogr. Sin.* **2012**, *67*, 53–61. [CrossRef]
42. Wang, Y.; Wang, Y.; Wang, J.; Li, W. Transient and Lagged Response of Seasonal Vegetation Changes to Climate in China during the Past Two Decades. *Geogr. Geo-Inf. Sci.* **2020**, *4*, 33–40. [CrossRef]
43. Liu, L.; Zhan, C.; Hu, S.; Dong, Y. Vegetation change and its topographic effects in the karst mountainous areas of Guizhou and Guangxi. *Geogr. Res.* **2018**, *37*, 2433–2446. [CrossRef]
44. Shang, B.; Zheng, B.; Zhou, Z.; Wang, Z.; Wang, L. Vegetation Growth and Its Influencing Factors under Different Soil Conditions in Mayi Lake Area of Kelamayi City. *J. Northeast. For. Univ.* **2021**, *1*, 44–49. [CrossRef]
45. Piao, S.; Yin, G.; Tan, J.; Cheng, L.; Huang, M.; Li, Y.; Liu, R.; Mao, J.; Myneni, R.B.; Peng, S.; et al. Detection and attribution of vegetation greening trend in China over the last 30 years. *Glob. Chang. Biol.* **2015**, *21*, 1601–1609. [CrossRef] [PubMed]
46. LV, Y.; Zhang, L.; Yan, H.; Ren, X.; Wang, J.; Niu, Z.; Gu, X.; He, H. Spatial and temporal patterns of changing vegetation and the influence of environmental factors in the karst region of Southwest China. *Acta Ecol. Sin.* **2018**, *38*, 8774–8786. [CrossRef]
47. Zhao, Z.; Han, R.; Guan, X.; Xiao, W.; Li, J. Change of vegetation coverage and the driving factor in the Beijing-Tianjin-Hebei region from 2000 to 2019. *Acta Ecol. Sin.* **2022**, *42*, 8860–8868. [CrossRef]
48. Zhang, L.; Liang, C.; Ma, H.; Chen, X.; Cai, J.; Guo, R.; Chen, T. The potential contribution of land managements to vegetation greening: A case study of the northeast agricultural region in China. *Acta Ecol. Sin.* **2022**, *42*, 720–731. [CrossRef]
49. Sun, Y.; Yi, L.; Yin, S. Vegetation Cover Change in Dongting Lake Basin and Its Coordination Governance. *Econ. Geogr.* **2022**, *4*, 190–201. [CrossRef]

Disclaimer/Publisher's Note: The statements, opinions and data contained in all publications are solely those of the individual author(s) and contributor(s) and not of MDPI and/or the editor(s). MDPI and/or the editor(s) disclaim responsibility for any injury to people or property resulting from any ideas, methods, instructions or products referred to in the content.

Article

Spatial and Temporal Variations of Vegetation Phenology and Its Response to Land Surface Temperature in the Yangtze River Delta Urban Agglomeration

Yi Yang [1], Lei Yao [1,*], Xuecheng Fu [1,2], Ruihua Shen [1], Xu Wang [1] and Yingying Liu [1]

1 College of Geography and Environment, Shandong Normal University, Jinan 250358, China; yang473087421@163.com (Y.Y.); ferwinjohn@163.com (X.F.); srh2455089583@163.com (R.S.); wangx_000111@163.com (X.W.); 201814010446@stu.sdnu.edu.cn (Y.L.)
2 School of Architecture and Urban Planning, Chongqing University, Chongqing 400030, China
* Correspondence: alex_yaolei@126.com

Abstract: In the Yangtze River Delta urban agglomeration, which is the region with the highest urbanization intensity in China, the development of cities leads to changes in land surface temperature (LST), while vegetation phenology varies with LST. To investigate the spatial and temporal changes in vegetation phenology and its response to LST in the study area, this study reconstructed the time series of the enhanced vegetation index (EVI) based on the MODIS EVI product and extracted the vegetation phenology indicators in the study area from 2002 to 2020, including the start of the growing season (SOS), the end of the growing season (EOS), and the growing season length (GSL), and analyzed the temporal–spatial patterns of vegetation phenology and LST in the study area, as well as the correlation between them. The results show that (1) SOS was advanced, EOS was postponed, and GSL was extended in the study area from 2002 to 2020, and there were obvious differences in the vegetation phenology indicators under different land covers and cities; (2) LST was higher in the southeast than in the northwest of the study area from 2002 to 2020, with an increasing trend; and (3) there are differences in the response of vegetation phenology to LST across land covers and cities, and SOS responds differently to LST at different times of the year. EOS shows a significant postponement trend with the annual mean LST increase. Overall, we found differences in vegetation phenology and its response to LST under different land covers and cities, which is important for scholars to understand the response of vegetation phenology to urbanization.

Keywords: vegetation phenology; land surface temperature; land covers; temporal–spatial pattern; partial correlation

Citation: Yang, Y.; Yao, L.; Fu, X.; Shen, R.; Wang, X.; Liu, Y. Spatial and Temporal Variations of Vegetation Phenology and Its Response to Land Surface Temperature in the Yangtze River Delta Urban Agglomeration. *Forests* **2024**, *15*, 1363. https://doi.org/10.3390/f15081363

Academic Editor: Guojie Wang

Received: 16 June 2024
Revised: 14 July 2024
Accepted: 2 August 2024
Published: 4 August 2024

Copyright: © 2024 by the authors. Licensee MDPI, Basel, Switzerland. This article is an open access article distributed under the terms and conditions of the Creative Commons Attribution (CC BY) license (https://creativecommons.org/licenses/by/4.0/).

1. Introduction

The rapid development of human society and economy has led to accelerated urbanization and continuous changes in the global climate [1,2]. As a result, the environment of urban areas has been greatly affected, including urban heat stress, urban pollution, and urban ecological changes [3]. One of these effects is the increase in land surface temperature (LST) [4]. With the changes in LST, corresponding changes in vegetation phenology in these areas are induced, which in turn affects the energy, carbon, and water cycles; vegetation dynamics; and public health, among others [5–9].

Vegetation phenology is the seasonal timing of plant growth and reproduction such as budding, leaf development, flowering, fruiting, yellowing, and so on, and is influenced by a combination of environmental factors such as the climate, hydrology, soil, and human activities [10–13]. Vegetation phenology is highly sensitive to environmental change, and it can also regulate climate by altering the exchange of energy, water, and carbon between the land surface and the atmosphere [14–16]. With the emergence of global warming, a large number of scholars have found that vegetation phenology changes with temperature, and the main trend is that as the temperature increases, the start of the growing season (SOS) is advanced, the end of the growing season (EOS) is delayed, and the growing season length (GSL) is extended [17–20].

The changes in the underlying surface in urban construction, especially the increase in impervious surfaces and buildings made of asphalt, cement, and other materials, lead to an increase in the temperature in these areas [21]. Scholars have conducted a lot of research on the impact of urbanization on vegetation phenology, which mainly focuses on the changes in LST and vegetation phenology in cities and the surrounding areas due to the intensity of urbanization, urban–rural differences, and so on. Some scholars found that the effect of urban climate on vegetation phenology decayed exponentially with distance from urban areas [22]; others found that the effect of urbanization on phenology is related to LST [23]. Many of these studies have shown that the changes in LST are closely related to the changes in vegetation phenology and that vegetation phenology responds differently to LST in different cities [24]—SOS and EOS tend to change as the LST rises [5,23–25]. At the same time, the temperature sensitivity of vegetation phenology tends to decrease with increasing LST, such as when LST increases; changes in vegetation phenology in areas with higher LST tend to be less pronounced than changes in areas with lower LST [26]. In addition, due to inter-city heterogeneity, there are some differences in the vegetation phenology of different cities [27]. However, there may be similar differences in the response of vegetation phenology to environmental factors under different land covers [28]. Previous studies have estimated that vegetation phenology in China can be affected by cities over an area of up to 20 km [29] and that these areas may encompass a wide range of different land covers. Meanwhile, previous studies on urban vegetation phenology have focused on the urban–rural gradient, whereas urban agglomerations, as areas with a large concentration of population and economy, are bound to have an impact on vegetation phenology due to the strong human activities within them and the expansion of the city limits, which cannot be ignored. Therefore, it is necessary to conduct an analysis from the perspective of different land covers and cities within the urban agglomerations. By studying LST and vegetation phenology under different land covers and cities, we can reveal the changing pattern and response relationship between vegetation phenology and LST in different land covers and cities and help residents better cope with urbanization and environmental changes.

In summary, vegetation phenology often responds to changes in LST, and changes in LST and vegetation phenology, as well as the response of vegetation phenology to LST, may vary under different land covers and different cities. We aimed to analyze the changes in LST, vegetation phenology, and the response of vegetation phenology to LST for different land covers and different cities in the Yangtze River Delta urban agglomeration from 2002 to 2020. We extracted vegetation phenology indicators (SOS, EOS, and GSL) for the Yangtze River Delta urban agglomeration from enhanced vegetation index (EVI) data. We combined phenological indicators with LST data to explore (1) the spatial and temporal changes in vegetation phenology and LST of different land covers and different cities in the Yangtze River Delta urban agglomeration during 2002–2020 and (2) the response of vegetation phenology to LST under different land covers and different cities.

2. Materials and Methods

2.1. Study Area

The Yangtze River Delta Urban Agglomeration (29°20′–32°34′ N, 115°46′–123°25′ E), located in the lower reaches of the Yangtze River Basin, on the west coast of the Pacific Ocean, has a well-developed regional economy. The area includes 26 cities—Shanghai; Nanjing, Wuxi, Changzhou, Suzhou, Nantong, Yancheng, Yangzhou, Zhenjiang, and Taizhou in Jiangsu Province; Hangzhou, Ningbo, Jiaxing, Huzhou, Shaoxing, Jinhua, Zhoushan, and Taizhou in Zhejiang Province; and Hefei, Wuhu, Ma'anshan, Tongling, Anqing, Chuzhou, Chizhou, and Xuancheng in Anhui Province. The climate is predominantly subtropical monsoon, warm, and humid all year round; the average annual temperature is 15–17 °C, and the average annual precipitation is about 1000–1800 mm. The main crops are rice, winter wheat, and maize. Forest vegetation types are mainly mixed evergreen deciduous broad-leaf forests, green broad-leaved forests, and temperate deciduous broad-leaved forests. The land cover in the northern part of the study area is dominated by cropland and the southern part is dominated by forests, while large areas of impervious surfaces still exist in cities such as Shanghai, Nanjing, Suzhou, etc. The land cover in the entire study area is mainly composed of three types: cropland (46.38%), forests (26.70%), and impervious surfaces (6.31%), and the other types (water is not taken into account) are very few and can be ignored (Figure 1).

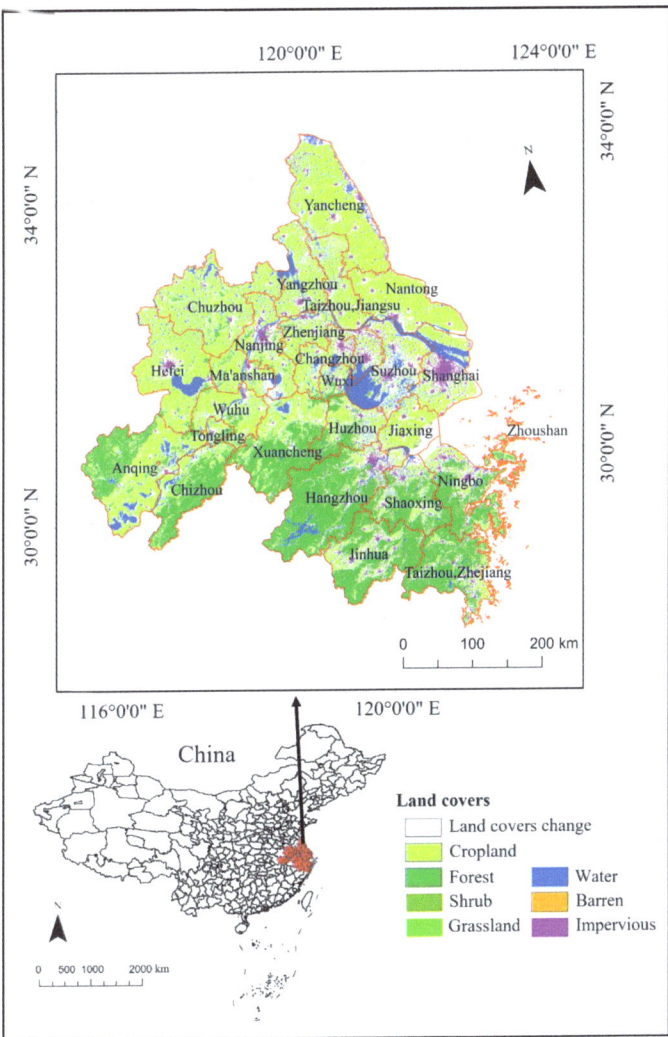

Figure 1. Land cover conditions in the Yangtze River Delta urban agglomeration in 2002–2020. "Land covers change" for areas where there has been a change in land covers from 2002 to 2020. Others for no change in land cover from 2002 to 2020.

2.2. Data

2.2.1. EVI Data

Enhanced vegetation index data were obtained from MOD13Q1 vegetation index data provided by the National Aeronautics and Space Administration (NASA), which contains both normalized difference vegetation index (NDVI) and EVI data, and we extracted phenology indicators from the EVI data. Compared to NDVI data, EVI data not only removes the influence of the vegetation canopy background but also reduces the influence of atmospheric and soil noise and mitigates the oversaturation problem of NDVI [30]; it is more sensitive to changes in dense vegetation [31]. The data have a temporal resolution of 16d and a spatial resolution of 250 m. We used the data from 2001 to 2021, in which the Yangtze River Delta urban agglomeration covered four image strips, thus containing a total of 1932 images for 21 years (the extraction of vegetation phenology indicators requires the extraction of vegetation phenology parameters for the intermediate year using at least three consecutive years of vegetation index data; for example, extracting vegetation phenology indicators for 2002 requires the use of vegetation index data for 2001–2003.

2.2.2. LST Data

We considered the following two factors: (1) LST is closely related to both urbanization and vegetation phenology [8,32,33]; (2) urban ecosystems are affected by both global warming and urbanization compared to natural ecosystems, and in previous studies of vegetation phenology, atmospheric temperature was more often used to reflect changes in background climate due to global warming, whereas LST was more often used to reflect changes in temperature due to urbanization because of the significant impact of changes in the subsurface caused by urbanization [34]. Therefore, we used LST data instead of atmospheric temperature data. The LST data used comprised a daily 1 km all-weather land surface temperature dataset for China's landmass and its surrounding areas (TRIMS LST; 2000–2022) [35–38]. The dataset was prepared following an enhanced satellite thermal infrared remote sensing–reanalysis data integration method. The data are an all-weather high-quality LST dataset with good precision and quality, having a high agreement with the daily 1 km Terra/Aqua MODIS LST product, which is now widely used in academia every day. The difference between clear-sky and non-clear-sky conditions is not significant. The mean bias deviation (MBD) of the dataset is 0.09 K and 0.03 K, and the standard deviation of bias (STD) is 1.45 K and 1.17 K for daytime and nighttime when using MODIS LST as a reference. The spatial resolution is 1 km and seamless, and the temporal resolution is 4 times a day throughout 2000–2021—of which we use the data from 2001 to 2020—and the data of each year are divided into four periods: the winter (including December of the previous year and January and February of the current year), March, April, and the annual mean because previous studies have shown that the SOS response is more pronounced in winter and spring temperatures [39,40]; additionally, considering that SOS in the study area occurs mainly before May, the response of SOS to May temperatures is negligible, whereas EOS occurs later and may be affected by year-round temperatures [27]. Using LST data from different time periods, we analyze the response of SOS to winter, March, and April LSTs and the response of EOS to the annual average LST.

2.2.3. Precipitation Data

We used precipitation data—a new daily gridded precipitation dataset for the Chinese mainland based on gauge observations [41–43]. The dataset is based on daily precipitation observations at the station and is prepared by applying monthly precipitation constraints and topographic corrections based on the idea of precipitation background field and precipitation ratio field construction. The dataset precision and quality are good and in good agreement with currently used precipitation datasets (CGDPA, CN05.1, and CMA V2.0). The spatial variability of precipitation can be better characterized. The median correlation coefficient between the daily valued time series of the dataset and the daily valued precipitation observations at the high-density sites was 0.78, the median root-mean-square error (RMSE) was 8.8 mm/d, and the median Kling–Gupta efficiency (KGE) value was 0.69. The spatial resolutions are 0.1°, 0.25°, and 0.5°, and the temporal resolution is 1 time per day, of which we used the 2001–2020 data with a spatial resolution of 0.1°.

Similar to the LST data, we also divided the precipitation data for each year into four time periods—winter, March, April, and the annual mean.

2.2.4. Land Cover Data

We used land cover data—the 30 m annual land cover datasets and dynamics in China from 1985 to 2022 [44]. The data are based on satellite image data, China land use/cover datasets, third-party validation data, and existing year-by-year land use/cover data, and are obtained through a series of processes, such as the generation of training and test samples, the construction of yearly input features, the checking of classifications and spatial and temporal consistency, the assessment of accuracy, and the comparison of products. The overall precision is 76.45% < overall accuracy (OA) < 82.51%, with an average OA of 79.30 ± 1.99%. The data are better than MCD12Q1 and ESACCI_LC in terms of overall precision for all years. The data span the period of 1985–2021 with a spatial resolution of 30 m and a temporal resolution of 1 year. We used data from 2002 to 2020 in it.

2.3. Methodology

We explored the response of vegetation phenology to LST using the proposed framework in Figure 2. Firstly, we collected and preprocessed datasets from different years, including EVI data, land cover data, and LST data; secondly, we extracted the vegetation phenology indicators SOS, EOS, and GSL from the EVI data; thirdly, we used these datasets to analyze the spatial and temporal patterns of vegetation phenology and LST and the correlation between vegetation phenology and LST in different cities and different land covers.

2.3.1. Reconstruction of Vegetation Index Time Series

Although the MOD13Q1 data have been subjected to noise rejection, data acquired through remote sensing are inevitably affected by the sensors themselves, as well as by atmospheric aerosols and clouds, resulting in a reduction in the precision of the data. Therefore, we need to reconstruct the vegetation index time series as well as smoothing. Commonly used methods for vegetation index time series reconstruction include the adaptive Savitzky–Golay filtering (S-G) [45], the fits to asymmetric Gaussians (A-G) [46], the double Logistic function [47], and so on.

It was found that the A-G fitting algorithms not only generate more consistently reconstructed NDVI time series to the original NDVI temporal curve, but also perform extremely similarly to keep the fidelity of high-quality NDVI samples, and their fitting NDVI series are better [48]. Therefore, we used the A-G method to reconstruct the vegetation index time series. Firstly, we extracted the values of all remotely sensed images for a given year at the same location. Then, we fitted the values at that location using the A-G method to obtain a continuous time series curve for one year at that location.

The method fits the vegetation data with a local model function based on the interval between the maximum and minimum values in the time series. The local model function is as follows:

$$f(t) = f(t; c, m) = a + bg(t; m) \quad (1)$$

where the base level and the amplitude are determined by the linear parameters $c = (a, b)$ and the shape of the basis function $g(t, m)$ is determined by the nonlinear parameters $m = (m_1, m_2, \ldots, m_n)$. The basis function of the asymmetric Gaussian function is:

$$g(t; m_1, m_2, \ldots, m_5) = \begin{cases} \exp\left[-\left(\frac{t-m_1}{m_2}\right)^{m_3}\right] & t > m_1 \\ \exp\left[-\left(\frac{t-m_1}{m_4}\right)^{m_5}\right] & t < m_1 \end{cases} \quad (2)$$

where m_1 is used to determine the position of the maximum or minimum for the independent time variable t; m_2 and m_3 determine the width and flatness (kurtosis) of the right function half; and m_4 and m_5 determine the width and flatness of the left half. The time series in the whole time interval $[t_L, t_R]$ is modelled using the global function $F(t)$, $F(t)$ is

$$F(t) = \begin{cases} \alpha(t)f_L(t) + [1 - \alpha(t)]f_C(t) & t_L < t < t_C \\ \beta(t)f_C(t) + [1 - \beta(t)]f_R(t) & t_C < t < t_R \end{cases} \quad (3)$$

where $\alpha(t)$ and $\beta(t)$ are cut-off functions that in small intervals around $(t_L + t_C)/2$ and $(t_C + t_R)/2$, respectively, smoothly drop from 1 to 0. $f_L(t)$, $f_C(t)$, and $f_R(t)$ are local model functions fitted to the left minimum, the central maximum, and the right minimum, respectively, and are located in the interval $[t_L, t_R]$. By merging local functions into global functions, it is possible to ensure that the fitting function follows the time series while increasing flexibility [46].

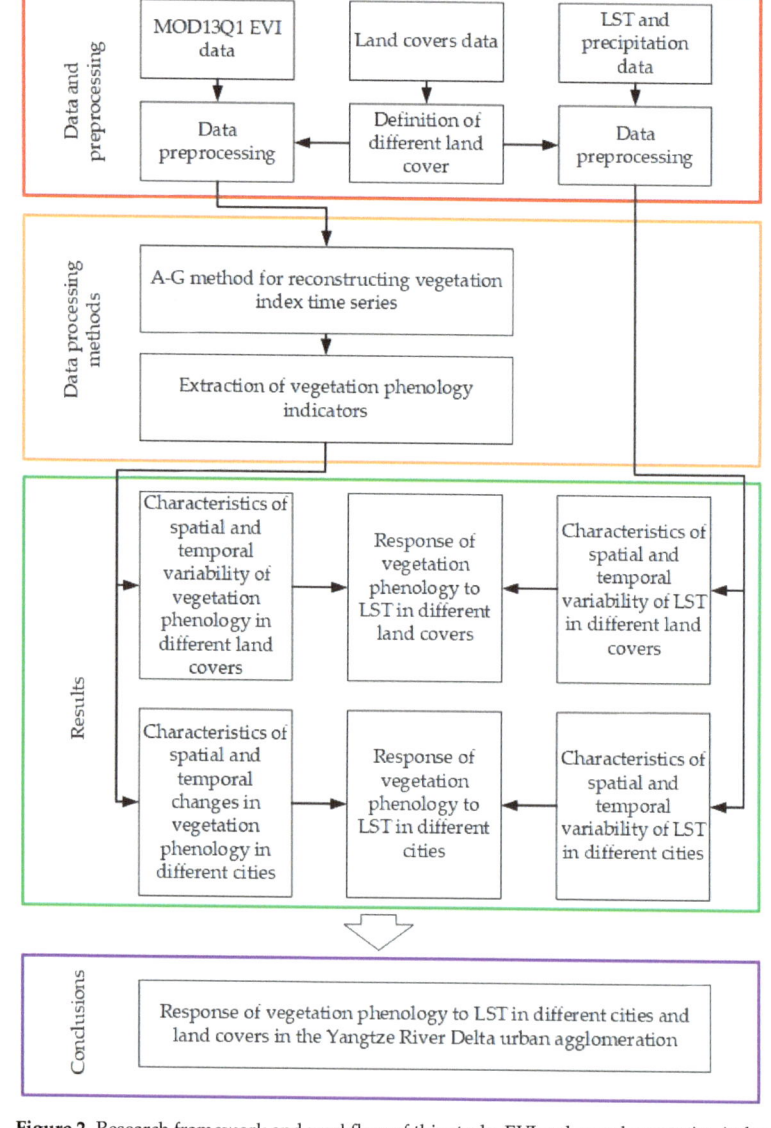

Figure 2. Research framework and workflow of this study. EVI: enhanced vegetation index. LST: land surface temperature. A-G: the fits to asymmetric Gaussians.

2.3.2. Extraction of Vegetation Phenology Indicators

It was found that vegetation phenology indicators derived from MODIS vegetation index data were in good agreement with field observations in terms of absolute errors and time trends [49,50]. The vegetation phenology indicators we used were SOS, EOS, and GSL. Common methods used in previous studies to extract these indicators include thresholding, curve fitting, maximum rate of change, and so on.

Among them, the dynamic threshold method within the threshold method has better adaptability, and considering that the EVI in the city is lower compared with that in the countryside, the use of the fixed threshold method will lead to errors in the obtained phenological indicators, and previous studies have shown that the vegetation phenology extracted using the dynamic threshold method is consistent with the time recorded at the observatory [51]. Therefore, we adopted the dynamic threshold method to extract vegetation phenology parameters; that is, the time when the fitted curves on the left and right sides reached a certain proportion of the seasonal amplitude was taken as SOS and EOS, respectively, and the difference between EOS and SOS was taken as GSL. When scholars initially proposed the dynamic threshold method, they suggested that the SOS and EOS thresholds be set at about 20% of the annual amplitude of the NDVI [46]; in other studies, scholars adopted 20% as the thresholds for SOS and EOS based on previous experience when studying the vegetation phenology of 32 major Chinese cities [23]. When studying the climatic changes in a typical vegetation sample area of the northern Tibetan Plateau, the SOS threshold was set at 10% and the EOS threshold at 20% based on previous experience and a large number of experiments [52]. In the study of vegetation climate change in Zhejiang Province, based on the characteristics of high vegetation cover and incomplete symmetry of vegetation growth curves in Zhejiang Province, the SOS threshold was set at 30% and the EOS threshold at 15% on the basis of a large number of experiments [51]. In conclusion, it is more flexible in using the dynamic threshold method to extract vegetation phenology parameters. Considering that Zhejiang Province is located in the southern part of the study area, with large forested areas and high vegetation coverage, this paper adopts the thresholds set by scholars in extracting the vegetation phenology in Zhejiang Province, with 30% as the SOS threshold and 15% as the EOS threshold for extracting the phenology indicators. Since the vegetation phenology indicators extracted from remotely sensed data may go beyond the range of vegetation physiological periods and lead to errors, to eliminate the uncertainty caused by these errors, the SOS and EOS were limited to 50–180 days and 240–330 days, respectively.

2.3.3. Methods of Statistical Analyses

1. Partial correlation analysis

Since temperature and precipitation are two important influences on vegetation phenology, and the effect of precipitation cannot be ignored in the analysis of vegetation phenology, we analyzed the correlation between vegetation phenology and LST after excluding the precipitation factor by partial correlation analysis, with the following formula:

$$c_{yx \times z} = \frac{c_{yx} - c_{yz}c_{xz}}{\sqrt{\left(1 - c_{yz}^2\right)\left(1 - c_{xz}^2\right)}} \tag{4}$$

where $c_{yx \times z}$ is the partial correlation coefficient between y and x after the control variable z, c_{yx} is the correlation coefficient between y and x, y is the vegetation phenology indicators, x is LST, and z is precipitation.

2. Geographical weighted regression analysis (GWR)

3. Considering the spatial heterogeneity in the response of vegetation phenology to LST, we used GWR to analyze the spatial differences in the response of vegetation phenology to LST, with the following formula:

$$y_i = \beta_0(\mu_i, v_i) + \sum_{k=1}^{p} \beta_k(\mu_i, v_i)x_{ik} + \varepsilon_i (i = 1, 2, 3, \cdots, n) \quad (5)$$

where y_i represents the vegetation phenology indicators of raster i, β_0 is the intercept, (μ_i, v_i) is the coordinate of raster i, k is the number of independent variables, β_k is the regression coefficient of the kth independent variable, x_{ik} is the kth independent variable of raster i, and ε_i is the random error.

3. Results

3.1. Spatial and Temporal Patterns of Vegetation Phenology

We extracted the main vegetation phenology indicators of the Yangtze River Delta urban agglomeration from 2002 to 2020, that is, SOS, EOS, and GSL, by fitting the EVI time series curves using the A-G method with a threshold of 30% for SOS and 15% for EOS. The results showed that the mean value of SOS in the study area from 2002 to 2020 was 7.64 days, the mean value of EOS was 294.47 days, and the GSL mean value was 206.77 days.

3.1.1. Spatial and Temporal Characteristics of Vegetation Phenology in the Study Area

Figures 3–5 show the summary statistics for vegetation phenology in the Yangtze River Delta urban agglomeration from 2002 to 2020. SOS is earliest in the area of concentrated impervious surfaces and forests in the southeast, followed by cropland in the north of Xuancheng and Huzhou and forests in the southwest of the study area, and latest in the east and west along the river basins and cropland distributed around impervious surfaces; SOS is significantly earlier than the surrounding areas in areas where impervious surfaces are concentrated (Figure 3). EOS is earliest in the northwestern and western river basins of the study area, followed by the northern cropland, and latest in the south-central and especially in the southern forest areas, which are similar to SOS, and the phenomenon of EOS is seen to occur later in the areas of impervious surface concentrations than in the surrounding areas (Figure 4). The GSL is the shortest in the northwest and in the east and west along the river basin it is shorter and longer in areas of concentrated impervious surfaces and in southern forested areas; similarly, the GSL shows longer characteristics in the concentrated areas of impervious surfaces than in the surrounding areas (Figure 5). Vegetation phenology in the concentrated area of impervious surfaces showed obvious differences compared to the surrounding areas; meanwhile, SOS was significantly delayed in the cropland surrounding the concentrated area of impervious surfaces compared to the cropland in other areas, and they became the two extremes of the earliest and the latest SOS. The reason for this phenomenon may be due to the fact that areas of concentrated impervious surfaces tend to have higher LSTs compared to other areas, and the increased LSTs lead to significant changes in vegetation phenology.

Figure 6 shows the trends in vegetation phenology from 2002 to 2020—SOS in the study area showed an overall advance trend of −0.58 days per year; EOS showed an overall delayed trend, which was smaller than SOS, with a change of only 0.08 days per year; and GSL showed an overall lengthening trend, with a change of 0.46 days per year. The three indicators of vegetation phenology showed a certain degree of fluctuation, with SOS showing local maximums in 2011 and 2017, while EOS and GSL showed local minimums in 2011 and 2016.

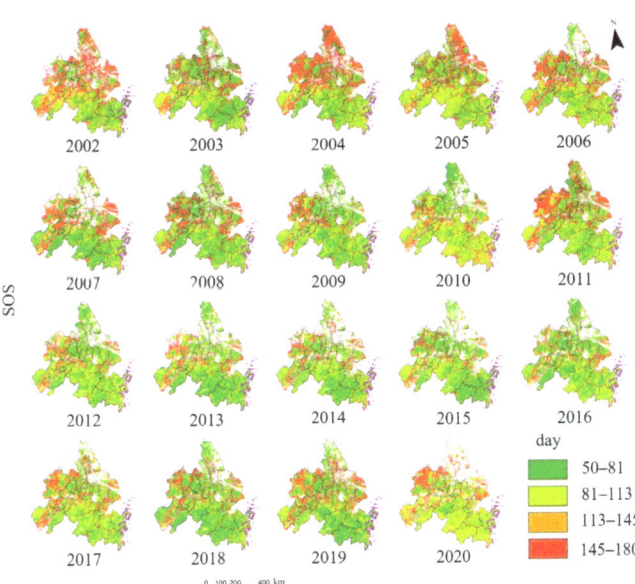

Figure 3. Spatial and temporal deviations in SOS in the Yangtze River Delta urban agglomeration in 2002–2020. SOS: the start of the growing season.

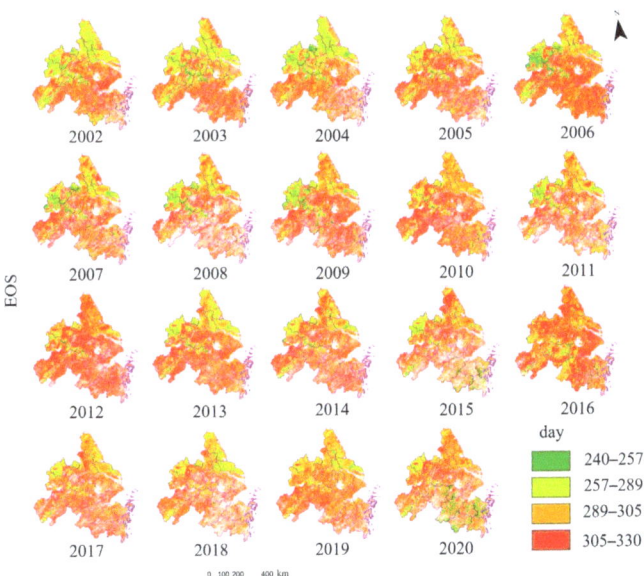

Figure 4. Spatial and temporal deviations in EOS in the Yangtze River Delta urban agglomeration in 2002–2020. EOS: the end of the growing season.

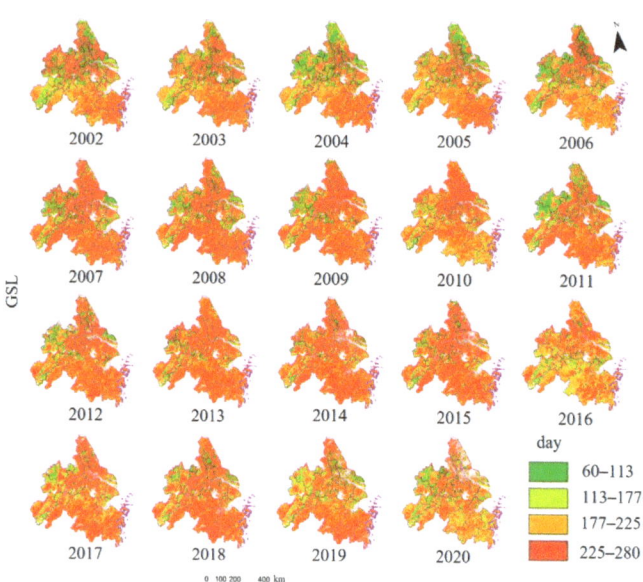

Figure 5. Spatial and temporal deviations in GSL in the Yangtze River Delta urban agglomeration in 2002–2020. GSL: the growing season length.

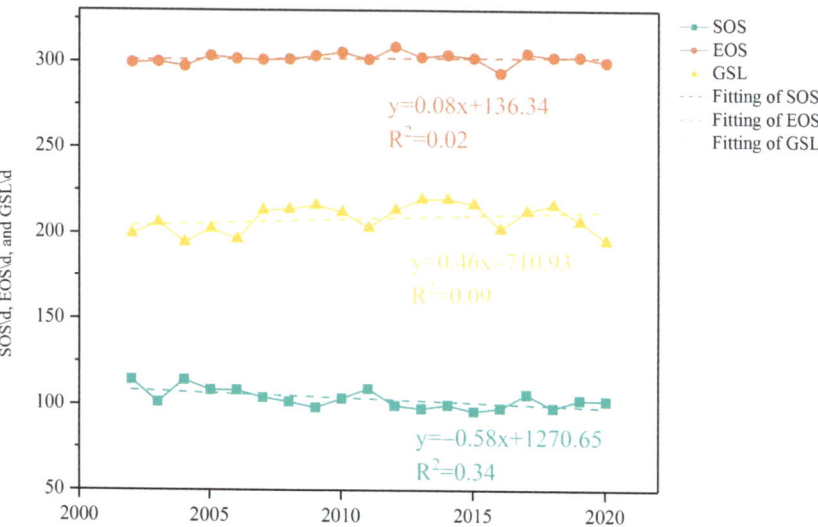

Figure 6. Time series of vegetation phenology in the Yangtze River Delta urban agglomeration in 2002–2020.

3.1.2. Differences in Vegetation Phenology between Land Covers

The results of the analysis of spatial and temporal patterns of vegetation phenology in the study area showed that there were some differences in vegetation phenology under different surface covers; we, therefore, analyzed the vegetation phenology under different land covers in the study area.

The results obtained from the preliminary analysis of vegetation phenology between different land covers are shown in Figure 7. There are significant differences in vegetation phenology under different land covers (the results of the analysis of variance (ANOVA) showed that $p < 0.05$ at the 0.05 level of significance, which indicates that there is a significant difference between the different land covers); in terms of the length of vegetation phenology indicators as well as earliness and lateness, the SOS in the study area is earlier in forests and impervious surfaces, at 92.27 and 92.56 days, respectively, and later in cropland, at 112.00 days; EOS was the latest in forests at 312.62 days and the earliest in cropland at 300.12 days; GSLs were longest in forests at 225.13 days, while GSL was the shortest in cropland at 190.61 days. In terms of the fluctuation of vegetation phenology indicators, forests vegetation phenology indicators had the least fluctuation with standard deviations in SOS, EOS, and GSL of 8.74, 8.27, and 13.90 days, respectively, and the greatest fluctuation was observed in cropland with standard deviations in SOS, EOS, and GSL of 25.17, 12.59, and 34.38 days, respectively. This may be due to differences in LST under different land covers and differences in vegetation types.

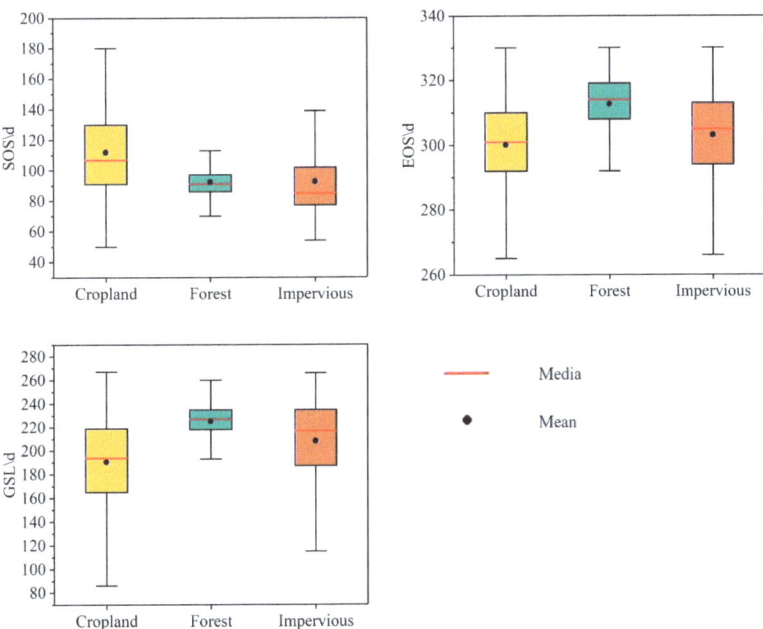

Figure 7. Differences in vegetation phenology under different land covers in the Yangtze River Delta urban agglomeration (2002–2020 average).

Figure 8 shows the changes in vegetation phenology under different land covers in the study area from 2002 to 2020. In terms of time, SOS under different land covers showed an overall trend of advancement, with the most pronounced trend in cropland at -1.00 days per year; EOS showed a significantly delayed trend in cropland, at 0.37 days per year, and forest EOS showed a significant advancement trend at -0.50 days per year, while EOS did not change significantly on impervious surfaces; GSL showed a lengthening trend on both cropland and impervious surfaces, which was most pronounced in agricultural areas at 0.96 days per year, and a shortening trend in forest GSL.

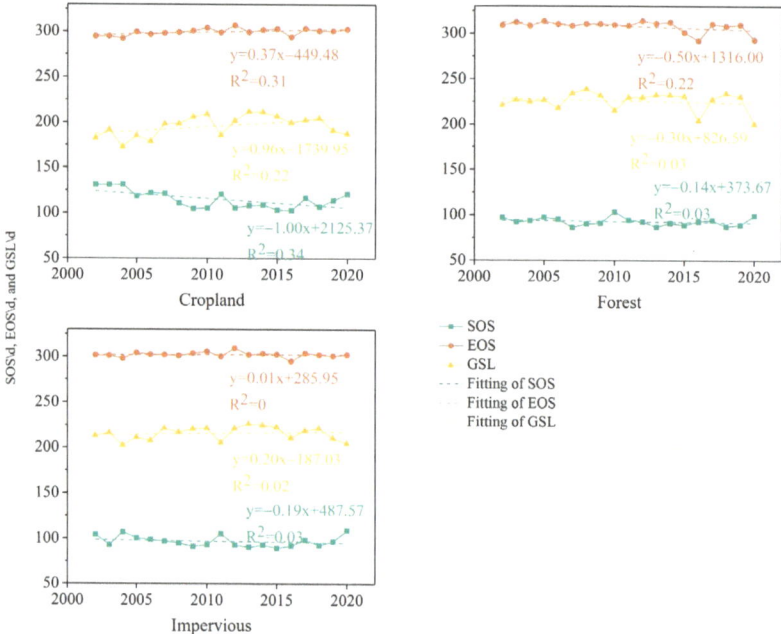

Figure 8. Time series of vegetation phenology for different land covers in the Yangtze River Delta urban agglomerations in 2002–2020.

3.1.3. Differences in Vegetation Phenology by City

Figure 9 shows the vegetation phenology indicators of the cities in the study area from 2002 to 2020; it was found that in terms of spatial distribution, the SOS was earlier in several cities located in the southern part of the study area, and later in several cities located along the river in the west and east of Tai Lake. Several cities in the eastern and western parts of the river basin have earlier EOSs and cities in the south have later EOSs; several cities located in the west have shorter GSLs and several cities in the southeast and north have longer GSLs. This may be due to LST and land cover differences. In general, several cities located in the west with shorter GSLs also show a trend in later SOSs and earlier EOSs.

In terms of temporal changes, the SOS of 22 cities showed an advanced trend, and only 4 cities showed a postponed trend, and their postponed trend was insignificant (less than 0.30 days per year); this may be due to elevated LST in winter and spring. The EOS of 15 cities showed a postponed trend, and the EOS of 11 cities showed an advanced trend, of which the cities with a postponed trend in EOS were mainly located in the northern part of the study area, and the cities with the advanced trend are mainly located in the southern part of the study area; this may be related to the predominant land covers of the city, as the northern cities have more cropland while the southern cities have more forests. Finally, 20 cities showed a lengthened trend in GSL and only 6 cities showed a shortened trend in GSL, which are mainly located in the southeastern part of the study area; this may indicate that the environment in the southeastern part of the study area has gradually begun to negatively affect the growth of native vegetation.

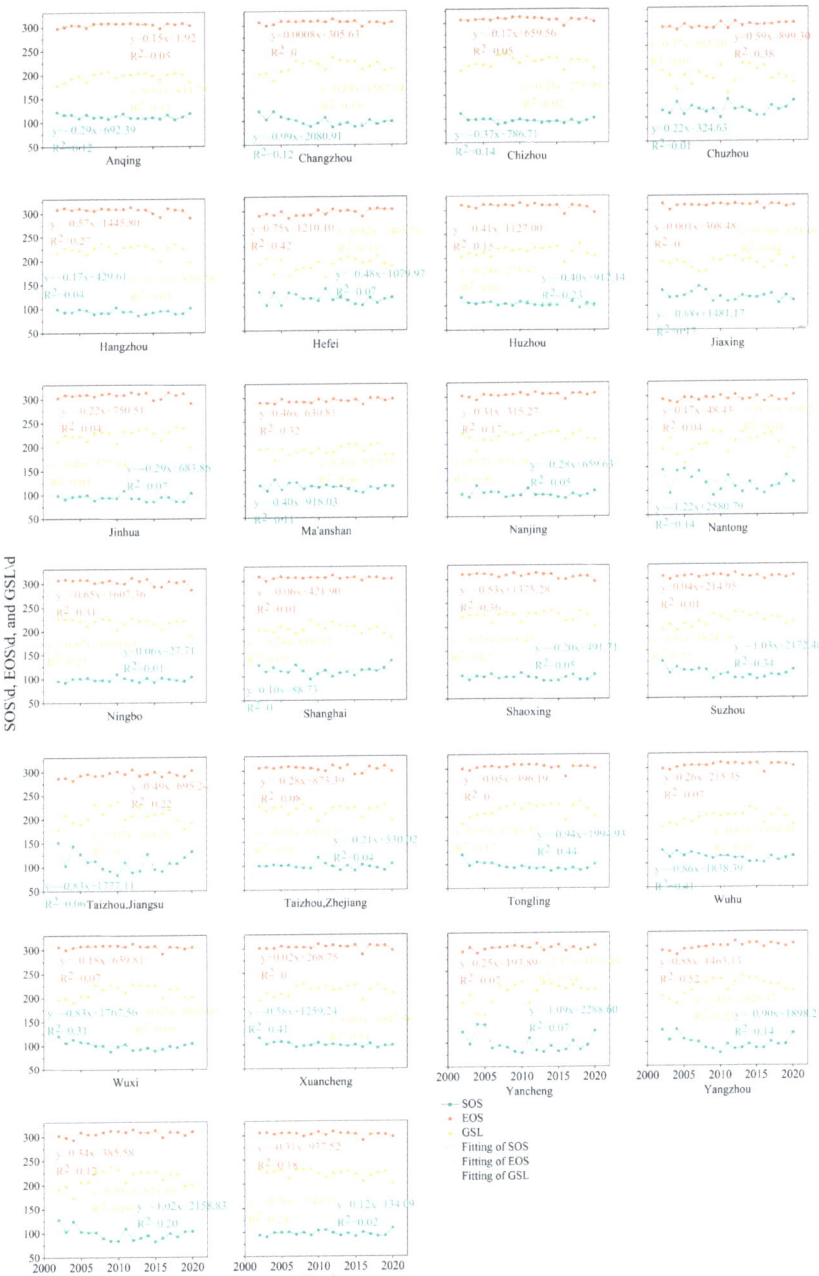

Figure 9. Time series of vegetation phenology in different cities of the Yangtze River Delta urban agglomeration in 2002–2020.

3.2. Spatial and Temporal Patterns of LST

3.2.1. Spatial and Temporal Characteristics of LST in the Study Area

As shown in Figures 10–13, the overall LST in the study area shows a gradual decrease from southeast to northwest, with the lowest LST in the north and the highest LST in the southeast, and this distribution may be related to the latitude as well as the monsoon climate of the study area. Meanwhile, the LST in the concentrated area of impervious surfaces exhibits significantly higher characteristics than the surrounding area due to the alteration in the subsurface.

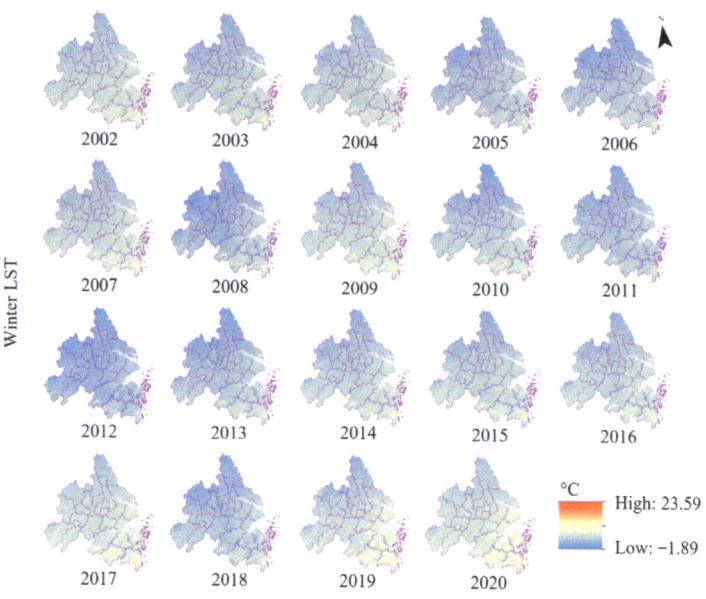

Figure 10. Spatial and temporal deviations in winter LST in the Yangtze River Delta urban agglomeration in 2002–2020.

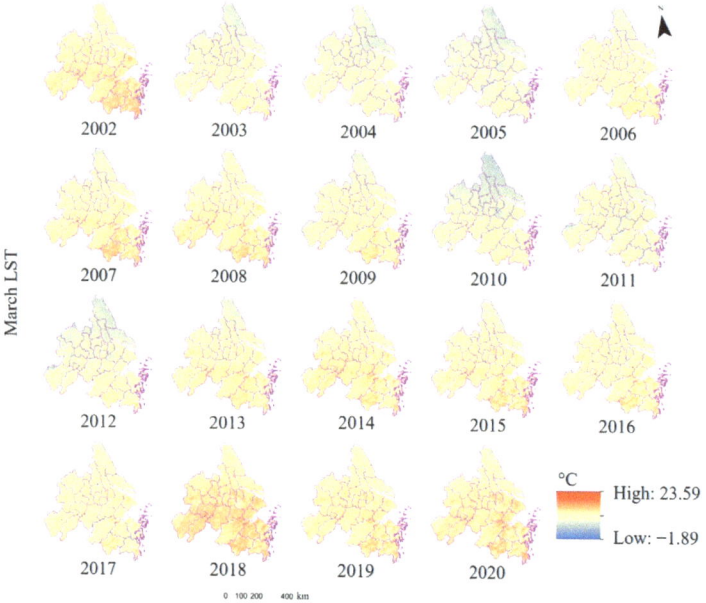

Figure 11. Spatial and temporal deviations in March LST in the Yangtze River Delta urban agglomeration in 2002–2020.

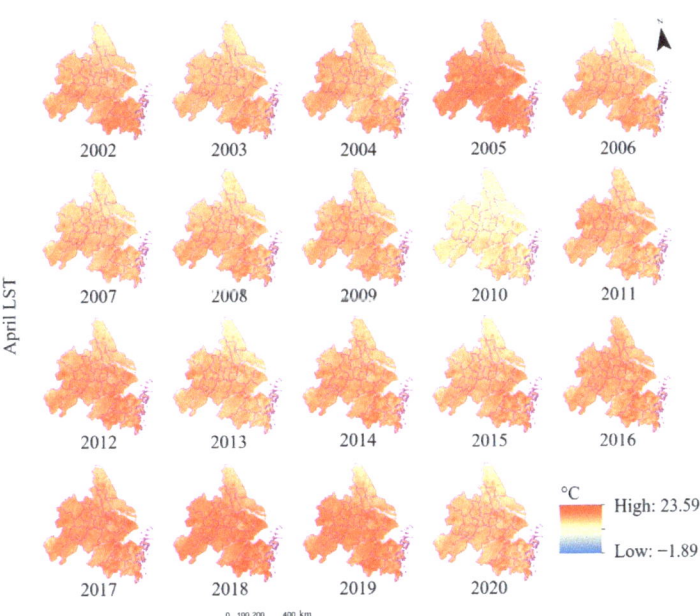

Figure 12. Spatial and temporal deviations in April LST in the Yangtze River Delta urban agglomeration in 2002–2020.

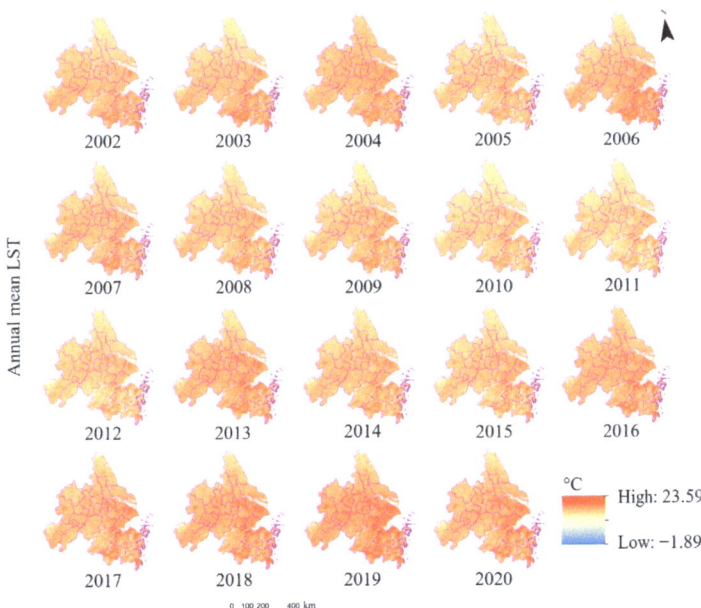

Figure 13. Spatial and temporal deviations in annual mean LST in the Yangtze River Delta urban agglomeration in 2002–2020.

Figure 14 shows the changes in LST in the study area from 2002 to 2020, and it was found that the winter LST, March LST, April LST, and annual mean LST in the study area generally showed an increased trend, with March LST showing the most pronounced increased trend, reaching 0.1 °C per year. However, LST showed a decreased trend before

2010, and an increased trend after 2010. This trend in LST was consistent with the changes in vegetation phenology, suggesting that the changes in vegetation phenology may be related to the changes in LST.

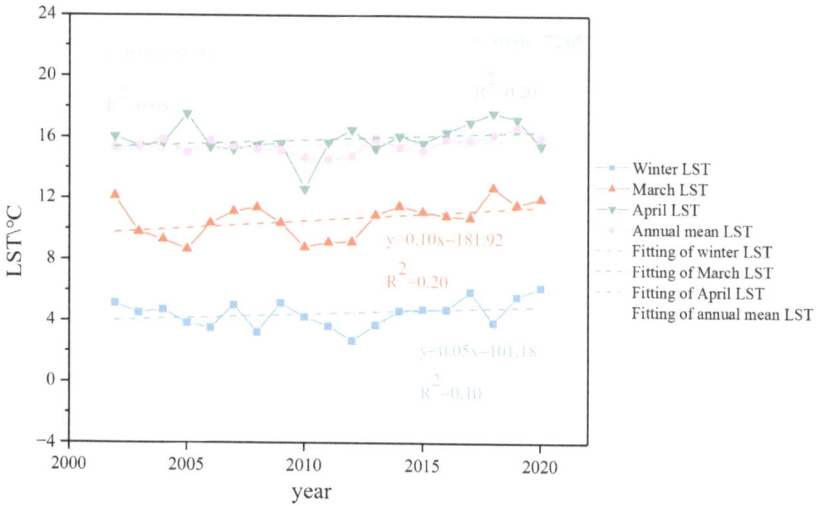

Figure 14. Time series of LST in the Yangtze River Delta urban agglomeration in 2002–2020.

3.2.2. Differences in LST between Land Covers

Figure 15 compares differences in LST under different land covers by counting the mean annual, winter, March, and April LST from 2002 to 2020 under different land covers (the results of the ANOVA showed that $p < 0.05$ at the 0.05 level of significance, which indicates that there is a significant difference between the different land covers). The LST of the cropland was lowest in winter, March, and April at 4.09 °C, 10.36 °C, and 15.71 °C, respectively, with some fluctuations. Forest LST was highest in winter and March at 5.00 °C and 11.03 °C, respectively, and the annual mean LST was the lowest at 15.28 °C; it had the least fluctuating LST values, with standard deviations of 0.96 °C, 0.85 °C, 0.87 °C, and 0.92 °C for winter, March, April, and annual mean LST, respectively. Impervious surface LST was highest in April and the annual mean at 16.47 °C and 16.15 °C, respectively, which had the greatest fluctuation in LST, with the standard deviations in the annual mean, winter, March, and April LST reaching 1.46 °C, 1.37 °C, 1.52 °C, and 1.32 °C, respectively. It is possible that the special subsurface of impervious surfaces, which elevates LST more significantly than forests and agricultural fields as solar radiation is enhanced, led to the highest April and annual average LSTs on impervious surfaces.

Figure 16 shows the interannual trend in LST under different land covers, which showed an increased trend under different land covers, similar to the results for the whole study area, with the most pronounced increased trend in LST in March under the three land covers. In terms of different land covers, impervious surface LST increased at the fastest rate, with winter, March, April, and annual mean LST increasing at 0.07 °C per year, 0.11 °C per year, 0.06 °C per year, and 0.08 °C per year, respectively, while forest LST increased at the slowest rate, with winter, March, April, and annual mean LST increasing at only 0.04 °C per year, 0.07 °C per year, 0.02 °C per year, and 0.01 °C per year. It is possible that differences in the rate of increase in LST between forests and impervious surfaces were due to differences in their subsurface.

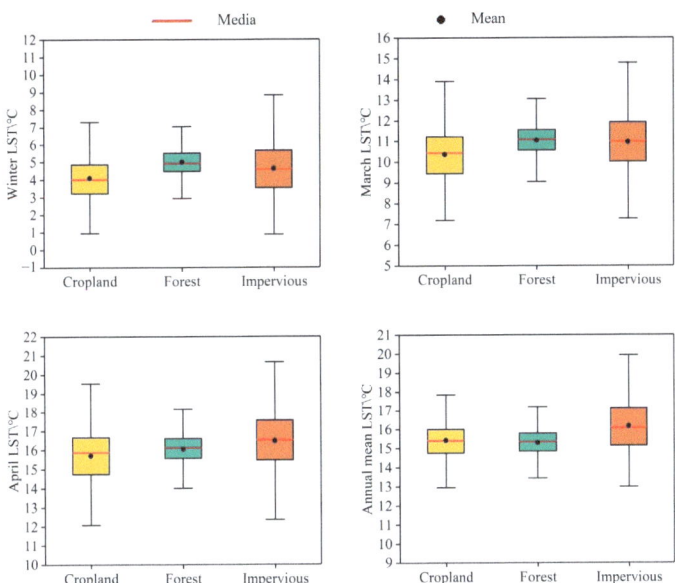

Figure 15. Differences in LST under different land covers in the Yangtze River Delta urban agglomeration (2002–2020 average).

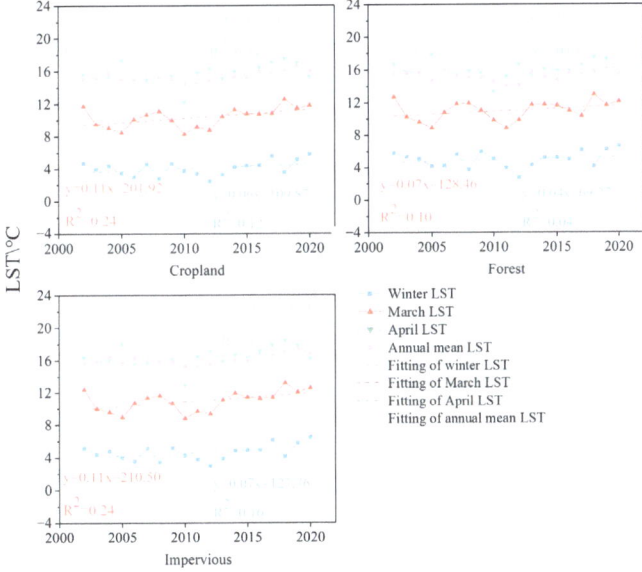

Figure 16. Time series of LST for different land covers in the Yangtze River Delta urban agglomerations in 2002–2020.

3.2.3. Differences in LST by City

Figure 17 shows the summary statistics for the LST status of each city in the study area from 2002 to 2020. In terms of spatial distribution, the cities with lower LST are mainly located in the northern part of the study area, and the cities with higher LST are mainly located in the eastern part of the study area, similar to the distribution of LST in the study area, which may be related to the latitude of the city and its exposure to the monsoon climate. What is interesting in this data is that Nantong, which is also located in the eastern part of the study area, is the northernmost city adjacent to Shanghai, but Nantong has a lower LST and Shanghai has a higher LST, which may be related to the fact that Shanghai has more impervious surfaces and strong human activities as a result of its higher urbanization intensity.

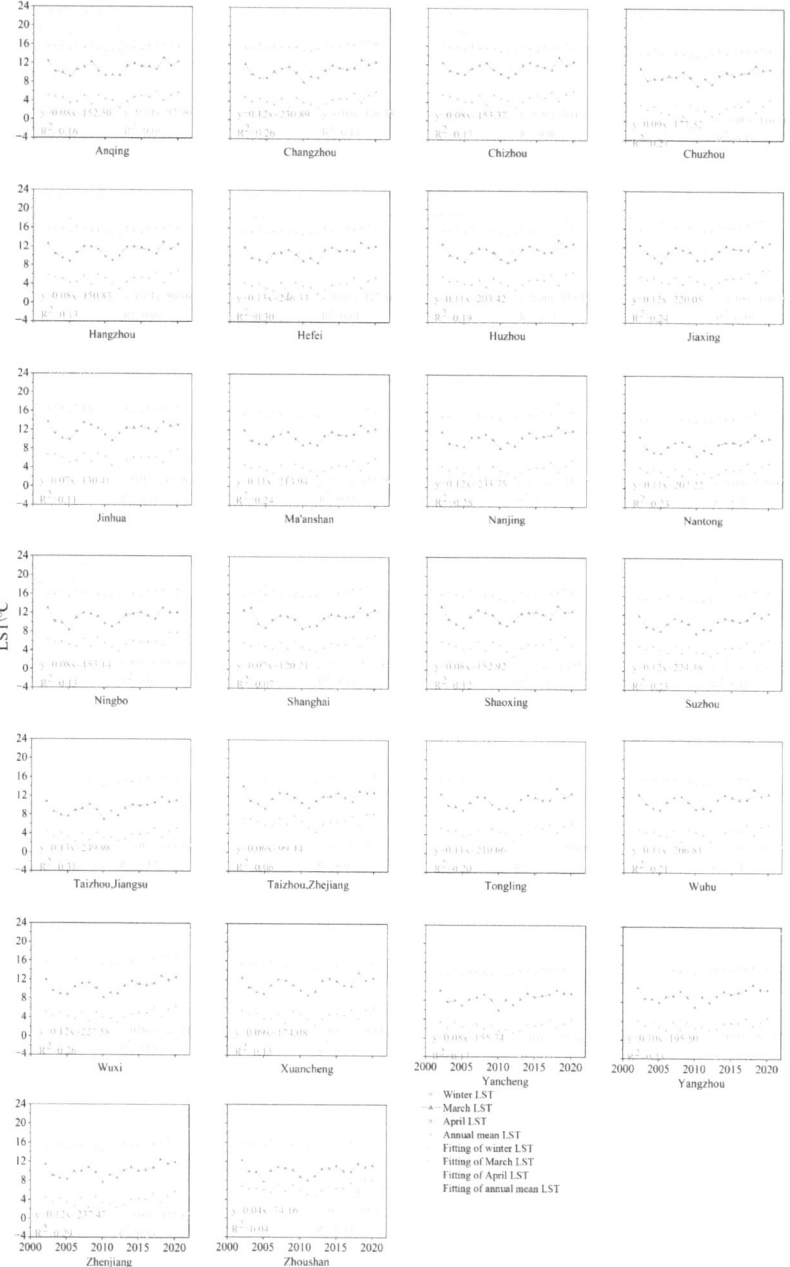

Figure 17. Time series of LST in different cities of the Yangtze River Delta urban agglomeration in 2002–2020.

In terms of temporal changes, LST showed an increased trend in all cities from 2002 to 2020, and the cities with a more pronounced LST increased trend were mainly located in the central part of the study area, while the cities in the southern part of the study area had a slower LST increased trend; this general upward trend in LST may be due to a combination of global warming and urbanization in the study area. The variation in LST in different periods was similar to that of the whole study area, with a more pronounced trend in elevated LST in March than in winter, April, and the annual mean LST.

3.3. Response of Vegetation Phenology to LST

Based on the spatial and temporal patterns of vegetation phenology and LST in the study area, we determined that there may be a certain correlation between changes in vegetation phenology and LST. To validate our analysis, we decided to use correlation coefficients to analyze the response of vegetation phenology to LST while considering that vegetation phenology responds to LST as well as precipitation; we calculated the partial correlation coefficients between vegetation phenology and LST after excluding the precipitation factor.

3.3.1. Partial Correlation Analysis between Vegetation Phenology and LST

Figure 18 shows the partial correlation between vegetation phenology and LST from 2002 to 2020. The partial correlation coefficients between SOS and LST in the study area showed a numerical trend of increasing and then decreasing, with the negative partial correlation gradually changing to a positive partial correlation from 2002, reaching the maximum value in 2010, and then changing to a negative partial correlation in 2020. This suggests that SOS is delayed with increasing LST around 2010, while SOS is advanced with increasing LST around 2002 and 2020. The overall partial correlation is very weak, and the absolute values of the partial coefficients between SOS and winter, March, and April LSTs are more than 0.1 in only 5, 4, and 4 years, respectively, among the 19 years. This suggests that the change in vegetation phenology with elevated LST is not very strong and that there may be a response to changes in other factors as well. From the perspective of different periods, SOS showed the most years of negatively partial correlation with winter LST and the least years of negatively partial correlation with April LST. EOS and annual LST in the study area mainly showed a significant positive partial correlation; only 2019 showed a very weak negative partial correlation, and the strength of its partial correlation was higher than that of SOS and LST, with partial correlation coefficients exceeding 0.1 in 12 years. This suggests that EOS exhibits a delayed trend with a higher average annual LST.

Figures 19–22 show the geographically weighted regression results in the study area from 2002 to 2020. There were more areas with negative regression coefficients of SOS and LST in the study area, in which the areas with smaller regression coefficients were mainly located in the northwestern part of the study area and the eastern part of Tai Lake, and those with positive regression coefficients were mainly located in the northern, southwestern, and southeastern parts of the study area. There were more areas with positive regression coefficients of the overall EOS and the annual average LST in the study area, and the regression coefficients were the largest in the northwestern part of the study area versus those located in the southwestern and central south side of the study area.

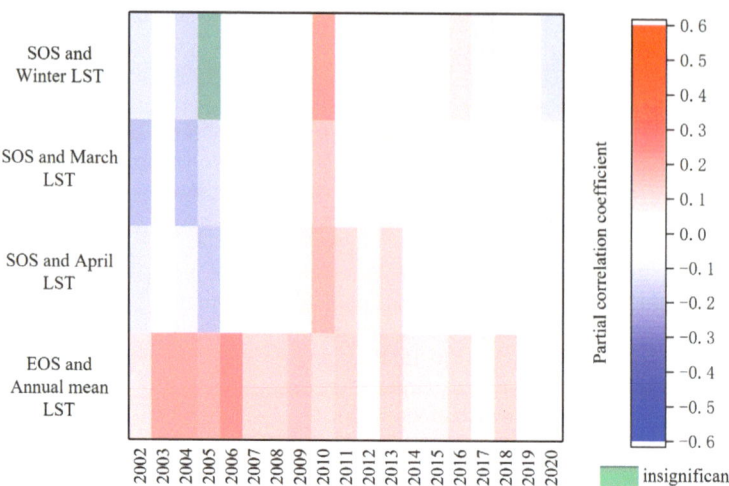

Figure 18. Partial correlation coefficients between vegetation phenology and LST in the Yangtze River Delta urban agglomeration in 2002–2020.

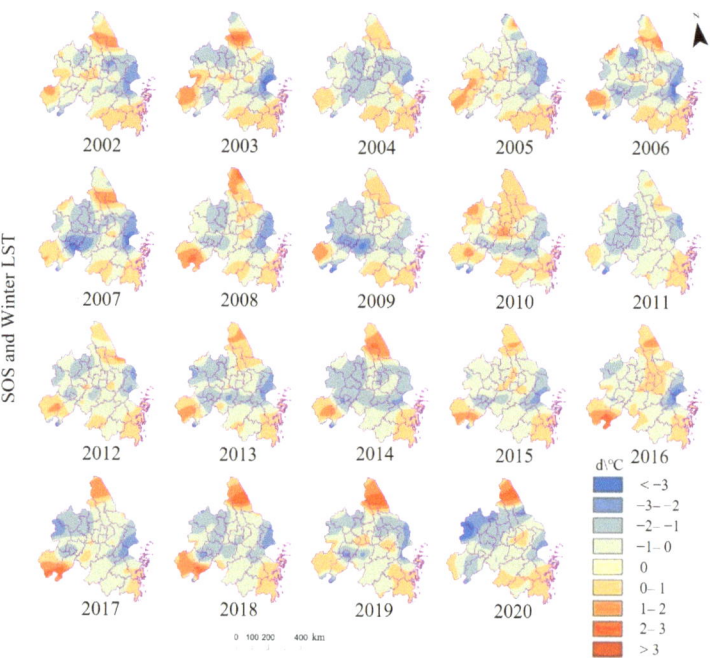

Figure 19. Regression coefficients of SOS and winter LST in the Yangtze River Delta urban agglomeration in 2002–2020.

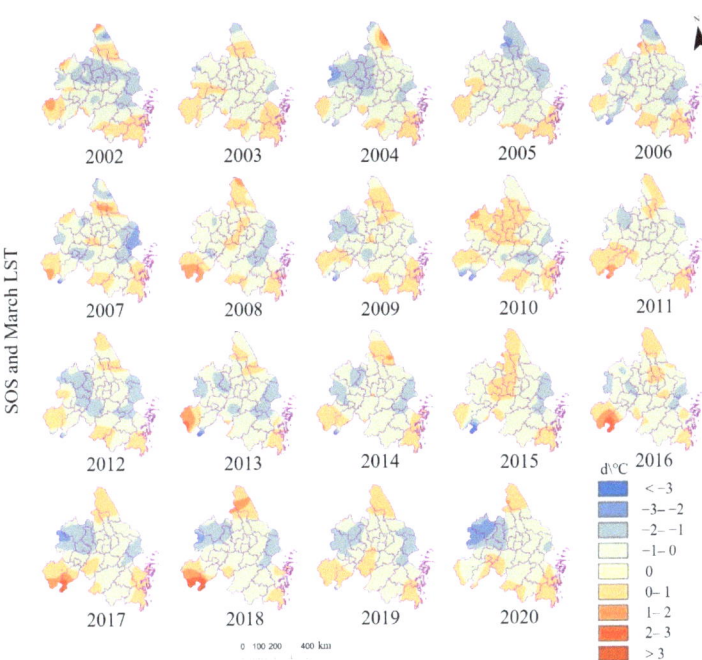

Figure 20. Regression coefficients of SOS and March LST in the Yangtze River Delta urban agglomeration in 2002–2020.

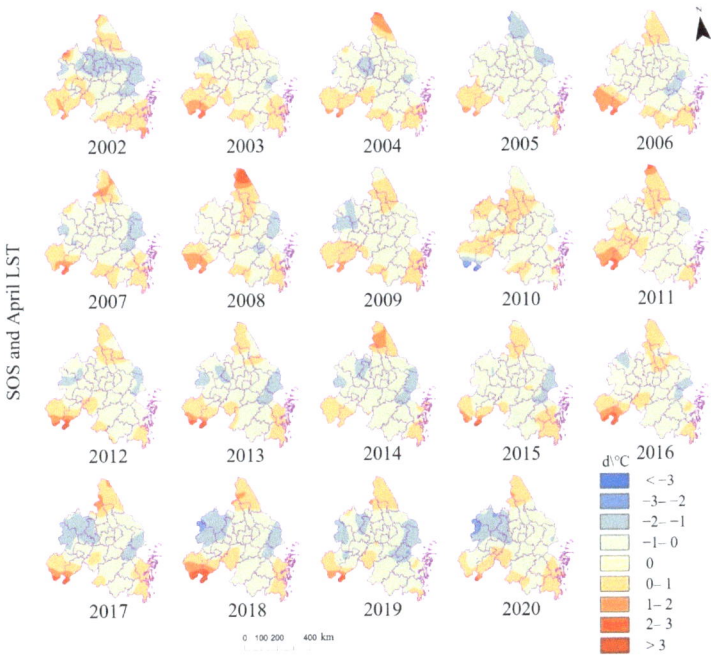

Figure 21. Regression coefficients of SOS and April LST in the Yangtze River Delta urban agglomeration in 2002–2020.

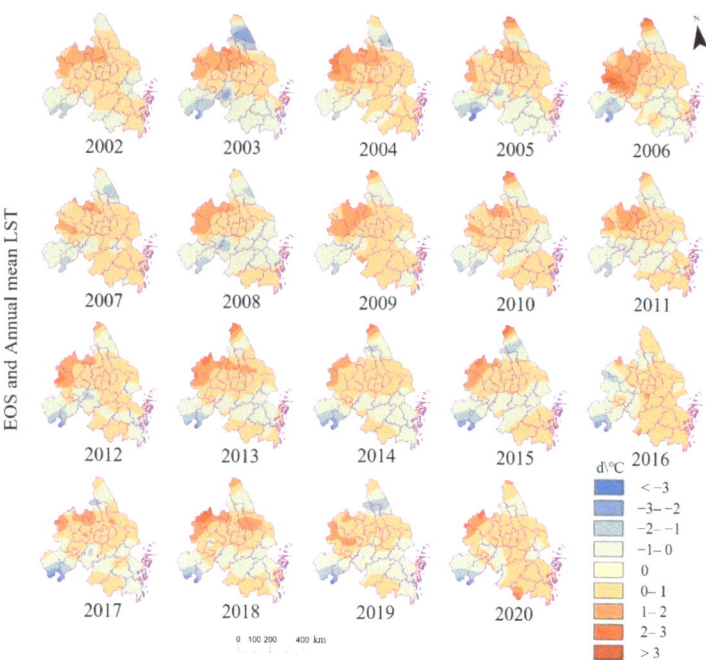

Figure 22. Regression coefficients of EOS and annual mean LST in the Yangtze River Delta urban agglomeration in 2002–2020.

3.3.2. Partial Correlation Analysis between Vegetation Phenology and LST for Different Land Covers

Figure 23 shows the partial correlation coefficients between vegetation phenology and LST from 2002 to 2020 for different land covers in the study area. In terms of the direction of the partial correlation coefficients, forest and impervious surface SOSs showed a significant negative partial correlation with LST, while cropland SOS mainly showed a significant positive partial correlation with LST. This suggests that forest and impervious surface SOSs exhibit an advancing trend with increasing LST, while cropland fields exhibit a delaying trend. This may be a result of differences in vegetation types under different land covers. EOS under different land covers showed a significant positive partial correlation with the average LST throughout the year. This suggests that EOS exhibits a postponement trend with increasing LST. In terms of the strength of the partial correlation, the partial correlation between vegetation phenology and LST was strongest on impervious surfaces and weakest on cropland and forest. This indicates that impervious surface vegetation phenology responds more strongly to changes in LST than forests and cropland.

Cropland SOS was most strongly and partially correlated with April LST at different times of the year; forest and impervious surfaces SOSs are strongly and partially correlated with LST in March and April. In terms of interannual trends, the partial correlation coefficients between SOS and cropland LST increased and then decreased, gradually changing from a negative partial correlation in 2002 to a positive partial correlation, reaching a maximum in 2010, and then gradually changing to a negative partial correlation in 2020. The change in the coefficient of partial correlation between SOS and forest LST fluctuates, and the strength of the negative partial correlation between SOS and impervious surfaces LST shows a tendency to decrease and then increase. The partial correlation coefficients between EOS and LST show a tendency to increase and then decrease on impervious surfaces, reaching the highest in 2005, while they fluctuate on cropland and forests. Differences in the response of vegetation phenology to LST under different land covers may result from differences in vegetation types.

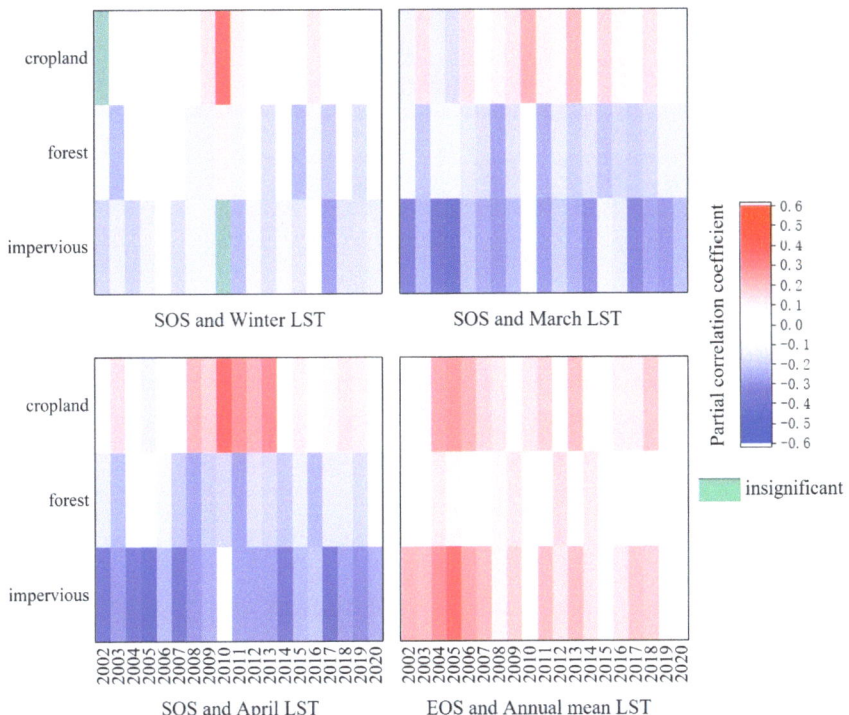

Figure 23. Partial correlation coefficients between vegetation phenology and LST of different land covers in the Yangtze River Delta urban agglomerations in 2002–2020.

3.3.3. Partial Correlation Analysis between Vegetation Phenology and LST for Different Cities

Figure 24 shows the partial correlation coefficients between vegetation phenology and LST from 2002 to 2020 for different cities in the study area, in terms of the direction of the partial correlation coefficients; SOS and LST in the cities of the study area showed a negative partial correlation in a larger number of cities. This suggests that SOS advances with increasing LST in most cities. The cities with a strong positive correlation between SOS and LST are mainly located in the northern and western riverine areas of the study area, and the land cover there is mainly cropland while the cities with a strong negative correlation between SOS and LST are mainly located in the central and eastern parts of the study area, and most of these cities have a large number of concentrated impervious surfaces. This suggests that inter-city differences in vegetation phenology in response to LST may be related to city land cover. The EOS and annual mean LST of each city mainly show positive partial correlations. This suggests that EOS exhibits a postponement trend with increasing LST.

Analyzing the strength of the LST partial correlation between SOS and different times of the year, SOS is more strongly correlated with the April LST partial correlation. In terms of temporal changes, the partial correlation coefficients between vegetation phenology and LST varied from year to year but generally showed a tendency to weaken and then strengthen. The strongest partial correlation between vegetation phenology and LST occurred mainly around 2002, 2007, and 2017, while the weaker partial correlation was mainly around 2010. This suggests that SOS responded more strongly to changes in LST in April, and the response of vegetation phenology to LST generally showed a trend of weakening and then strengthening and was significantly stronger in some years than in others.

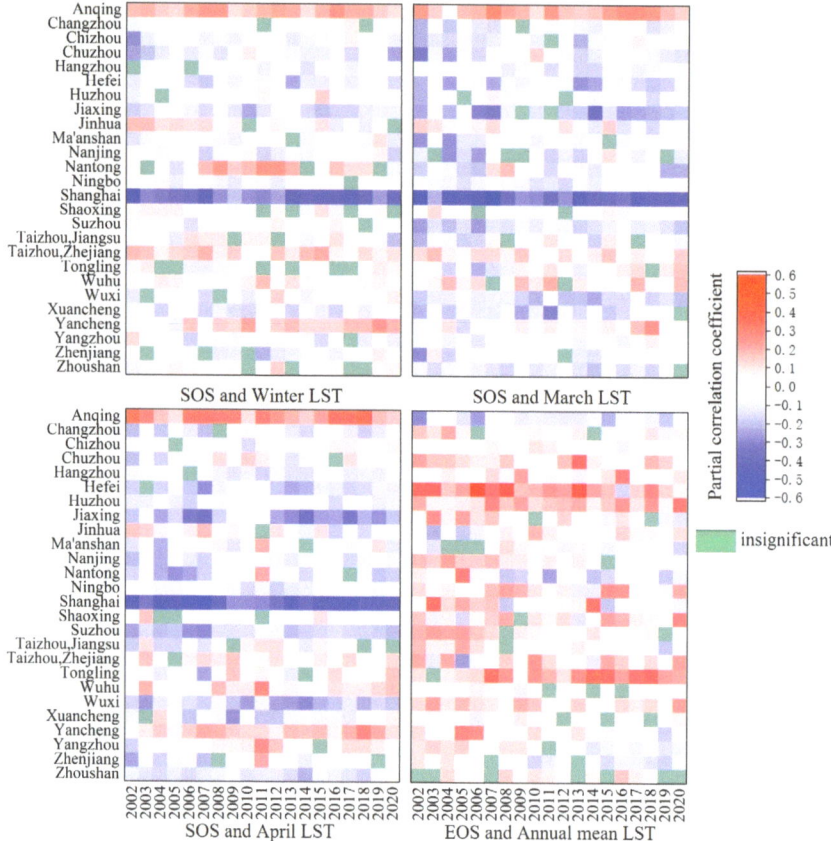

Figure 24. Partial correlation between vegetation phenology and LST in Cities of the Yangtze River Delta urban agglomeration.

4. Discussion

Vegetation phenology is highly sensitive to environmental changes, especially in areas of high urbanization, and environmental changes in these and surrounding areas due to intense human activities can have a significant impact on vegetation phenology [53]. Our results also found that there was a significant trend in increasing LST in the areas of concentrated impervious surfaces and that the vegetation phenology in these areas and nearby cropland changed significantly as a result of environmental changes, but the forest vegetation phenology near the areas of concentrated impervious surfaces changed relatively insignificantly. Scholars also found that the vegetation phenology of forests and shrubs was less sensitive to the surrounding environment than that of urban and cropland areas [28]. By analyzing the distribution of vegetation phenology and LST under different land covers, it was found that the vegetation phenology and LST of forests are less volatile than those of cropland and impervious surfaces, which may be related to the regulation of heat dissipation by forests themselves [54]. There is also a strong relationship between vegetation phenology and urban land cover and LST conditions in different cities; for example, cities with more forest in the southern part of the study area and more cropland in the northern part of the study area have longer GSL, while cities with more cropland in the western part of the riverine region have significantly delayed GSL due to high LST.

The results of the correlation analysis between vegetation phenology and LST showed that there were significant differences in the response of vegetation phenology to LST under different land covers, with forest and impervious surface SOS advancing with increasing LST but cropland SOS delaying with increasing LST; especially cropland near the area of concentration of impervious surfaces, SOS showed significant delay in LST increase due to urban heat island. This may be related to the following factors: (1) the urban heat island makes autumn and winter temperatures warmer and chilling accumulate insufficient, which may lead to a delay in SOS [55–57]; (2) differences in vegetation types, with annual herbaceous plants predominating in cropland areas and woody perennials predominating in forests and impervious surfaces, may be responsible for the different responses of vegetation phenology to temperature increases under different land covers [58]. We analyzed the partial correlation coefficients of vegetation phenology with LST and LST from 2002 to 2020 and retained the data in which the results of the linear fit of both were significant at the same time and found that neither the partial correlation coefficients of SOS with LST nor the fitted curves for LST met the conditions and that the results of the partial correlation coefficients of EOS with annual mean LST and the linear fit of annual mean LST in the three cities (Hefei, Shanghai, Yancheng) and impervious surfaces in the study area were simultaneously significant. As shown in Figure 25, the temporal variation of the correlation between vegetation phenology and LST showed that EOS exhibited a significant positive correlation with LST, but the intensity was gradually decreasing, while at the same time, the annual average LST in the study area showed a gradually increasing trend. Meanwhile, previous studies have found that the temperature sensitivity of vegetation phenology varies with temperature [59,60]. Therefore, it was judged that the elevated LST may have led to a decrease in the sensitivity of the EOS to the LST.

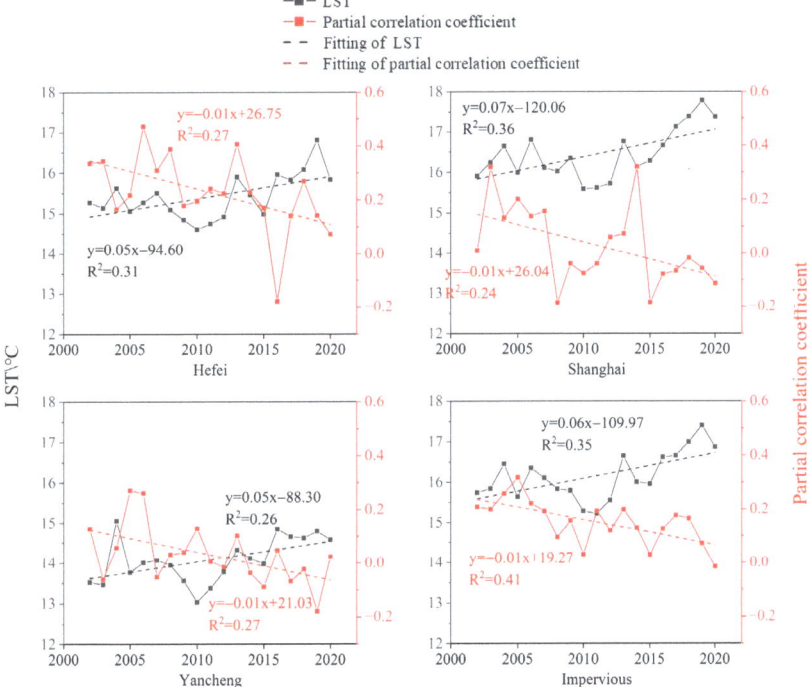

Figure 25. Trends in partial correlation coefficients of LST and EOS with LST under Hefei, Shanghai, Yancheng, and impervious surfaces in the Yangtze River Delta urban agglomeration, 2002–2020.

At the same time, there are some limitations to the content of our study. Firstly, although we have analyzed the response of vegetation phenology to LST in the Yangtze River Delta urban agglomeration, the urban agglomerations, as a region with strong and concentrated human activities, should not be neglected except for the factors of temperature and precipitation, as well as the humanistic factors such as the population density and the GDP, which are also likely to be responded to by the vegetation phenology. Secondly, previous studies have found that there can be a lag in the effect of meteorological factors on vegetation phenology [61], but we did not develop this in this study. In future studies, we hope to take into account the influence of human factors to improve the accuracy of our findings.

5. Conclusions

The main goal of the current study was to determine the spatial and temporal patterns of vegetation phenology and LST as well as the partial correlation between vegetation phenology and LST in the Yangtze River Delta urban agglomeration from 2002 to 2020 to investigate the characteristics of the spatial and temporal distributions and changes in vegetation phenology and LST under different land covers and cities over a long time series, as well as the response of vegetation phenology to LST. Accordingly, we analyzed the spatial and temporal variations of vegetation phenology and LST in the Yangtze River Delta urban agglomerations from 2002 to 2020 using vegetation phenology indicators extracted from the EVI data and the LST data and calculated the partial correlation coefficients of vegetation phenology and LST. The main conclusions were as follows:

1. Characteristics of spatial and temporal variation in vegetation phenology: (1) Cities located in the area of forests in the south and concentrated impervious surfaces had an earlier SOS, later EOS, and longer GSL, while those in the east and west along the river basins had a later SOS, earlier EOS, and shorter GSL. Analyzing the different land covers showed that forests had the earliest SOS, the latest EOS, and the longest GSL, along with the least volatility in phenological indicators. The cropland had the latest SOS, the earliest EOS, the shortest GSL, and the greatest volatility in vegetation phenology indicators; phenological indicators and the volatility of impervious surfaces are intermediate between forests and cropland. The reasons for these differences may be related to differences in thermal conditions, temperatures, and dominant vegetation types under different land covers. (2) The SOS of the study area from 2002 to 2020 showed a trend of advancement; the EOS showed a trend of postponement; and the GSL showed a trend of lengthening. This trend may have been caused by the gradual increase in LST in the study area.
2. Temporal and spatial variability characteristics of LST: (1) The LST in the study area at different times of the year generally showed a trend of gradual decrease from the southeast to the northwest. Cropland has the lowest LST in winter, March, and April; forest LST is higher in winter and March, but the annual mean LST is lower while it is the least volatile; and impervious surfaces have the highest April and annual mean LST, while they are the most volatile. This can be caused by impervious surfaces having more buildings, asphalt, and other surfaces. (2) From 2002 to 2020, winter, March, April, and annual mean LST in the study area showed an increased trend, the trend in elevated LST was most pronounced in March. This increased trend in LST may have been due to global warming and urbanization in the study area.
3. Vegetation phenology response to LST: (1) Cropland SOS was delayed with increased LST while forests and impervious surfaces were advanced; this may have been caused by differences in vegetation types under different land covers, as different vegetation responds differently to changes in LST; EOS was mainly delayed as LST increased. (2) Impervious surface vegetation phenology responded most strongly to LST while cropland and forests were less responsive; this may imply that vegetation in impervious surface areas responds more significantly to changes in LST. (3) Cropland responded more strongly to April LST while forests and impervious surfaces responded more strongly to March and April; this may indicate that the SOS response to LST in March and April will be more significant than the response to LST in winter. (4) The response of vegetation phenology to LST was variable, but the years of strong response were relatively concentrated.

Author Contributions: Conceptualization, L.Y.; methodology, L.Y.; software, Y.Y., X.F., R.S., X.W. and Y.L.; validation, Y.Y.; formal analysis, Y.Y.; investigation, Y.Y.; resources, Y.Y., X.F. and X.W.; data curation, Y.Y.; writing—original draft preparation, Y.Y.; writing—review and editing, L.Y. and R.S.; visualization, Y.Y.; supervision, L.Y., X.F., R.S., X.W. and Y.L.; project administration, L.Y.; funding acquisition, Y.L. All authors have read and agreed to the published version of the manuscript.

Funding: This research was funded by the National Natural Science Foundation of China (42171094) and the Natural Science Foundation of Shandong Province (ZR2021MD095; ZR2021QD093).

Data Availability Statement: Publicly available datasets were analyzed in this study.

Acknowledgments: The authors are very grateful to the editors and anonymous reviewers for their valuable time and advice on this manuscript.

Conflicts of Interest: The authors declare no conflicts of interest.

References

1. Xu, H.; Yang, S. Urban heat island effect and urban ecosystems. *J. Beijing Norm. Univ. (Nat. Sci.)* **2018**, *54*, 790–798. [CrossRef]
2. Liu, X.; Huang, Y.; Xu, X.; Li, X.; Li, X.; Ciais, P.; Lin, P.; Gong, K.; Ziegler, A.D.; Chen, A.; et al. High-spatiotemporal-resolution mapping of global urban change from 1985 to 2015. *Nat. Sustain.* **2020**, *3*, 564–570. [CrossRef]
3. Peng, J.; Jia, J.; Liu, Y.; Li, H.; Wu, J. Seasonal contrast of the dominant factors for spatial distribution of land surface temperature in urban areas. *Remote Sens. Environ.* **2018**, *215*, 255–267. [CrossRef]
4. Nayak, H.P.; Nandini, G.; Vinoj, V.; Landu, K.; Swain, D.; Mohanty, U.C.; Niyogi, D. Influence of urbanization on winter surface temperatures in a topographically asymmetric Tropical City, Bhubaneswar, India. *Comput. Urban Sci.* **2023**, *3*, 36. [CrossRef]
5. Kabano, P.; Lindley, S.; Harris, A. Evidence of urban heat island impacts on the vegetation growing season length in a tropical city. *Landsc. Urban Plan.* **2021**, *206*, 103989. [CrossRef]
6. Richardson, A.D.; Hollinger, D.Y.; Dail, D.B.; Lee, J.T.; Munger, J.W.; O'keefe, J. Influence of spring phenology on seasonal and annual carbon balance in two contrasting New England forests. *Tree Physiol.* **2009**, *29*, 321–331. [CrossRef]
7. Qiu, T.; Song, C.; Zhang, Y.; Liu, H.; Vose, J.M. Urbanization and climate change jointly shift land surface phenology in the northern mid-latitude large cities. *Remote Sens. Environ.* **2020**, *236*, 111477. [CrossRef]
8. Zhao, S.; Liu, S.; Zhou, D. Prevalent vegetation growth enhancement in urban environment. *Proc. Natl. Acad. Sci. USA* **2016**, *113*, 6313–6318. [CrossRef]
9. Li, X.; Zhou, Y.; Meng, L.; Asrar, G.; Sapkota, A.; Coates, F. Characterizing the relationship between satellite phenology and pollen season: A case study of birch. *Remote Sens. Environ.* **2019**, *222*, 267–274. [CrossRef]
10. Xia, C.; Li, J.; Liu, Q. Review of advances in vegetation phenology monitoring by remote sensing. *Natl. Remote Sens. Bull.* **2013**, *17*, 1–16.
11. Piao, S.; Mohammat, A.; Fang, J.; Cai, Q.; Feng, J. NDVI-based increase in growth of temperate grasslands and its responses to climate changes in China. *Glob. Environ. Chang.* **2006**, *16*, 340–348. [CrossRef]
12. Cleland, E.E.; Chuine, I.; Menzel, A.; Mooney, H.A.; Schwartz, M.D. Shifting plant phenology in response to global change. *Trends Ecol. Evol.* **2007**, *22*, 357–365. [CrossRef]
13. Jeong, S.-J.; Ho, C.-H.; Gim, H.-J.; Brown, M.E. Phenology shifts at start vs. end of growing season in temperate vegetation over the Northern Hemisphere for the period 1982–2008. *Glob. Chang. Biol.* **2011**, *17*, 2385–2399. [CrossRef]
14. Keenan, T.F.; Gray, J.; Friedl, M.A.; Toomey, M.; Bohrer, G.; Hollinger, D.Y.; Munger, J.W.; O'Keefe, J.; Schmid, H.P.; Wing, I.S. Net carbon uptake has increased through warming-induced changes in temperate forest phenology. *Nat. Clim. Chang.* **2014**, *4*, 598–604. [CrossRef]
15. Richardson, A.D.; Andy Black, T.; Ciais, P.; Delbart, N.; Friedl, M.A.; Gobron, N.; Hollinger, D.Y.; Kutsch, W.L.; Longdoz, B.; Luyssaert, S.; et al. Influence of spring and autumn phenological transitions on forest ecosystem productivity. *Philos. Trans. R. Soc. Lond. Ser. B Biol. Sci.* **2010**, *365*, 3227–3246. [CrossRef]
16. Piao, S.; Ciais, P.; Friedlingstein, P.; Peylin, P.; Reichstein, M.; Luyssaert, S.; Margolis, H.; Fang, J.; Barr, A.; Chen, A.; et al. Net carbon dioxide losses of northern ecosystems in response to autumn warming. *Nature* **2008**, *451*, 49–52. [CrossRef]
17. Zeng, H.; Jia, G.; Epstein, H. Recent changes in phenology over the northern high latitudes detected from multi-satellite data. *Environ. Res. Lett.* **2011**, *6*, 045508. [CrossRef]
18. Dragoni, D.; Rahman, A.F. Trends in fall phenology across the deciduous forests of the Eastern USA. *Agric. For. Meteorol.* **2012**, *157*, 96–105. [CrossRef]
19. Aono, Y.; Kazui, K. Phenological data series of cherry tree flowering in Kyoto, Japan, and its application to reconstruction of springtime temperatures since the 9th century. *Int. J. Climatol.* **2008**, *28*, 905–914. [CrossRef]
20. Annette, M.; Ye, Y.; Michael, M.; Tim, S.; Helfried, S.; Regula, G.; Nicole, E. Climate change fingerprints in recent European plant phenology. *Glob. Chang. Biol.* **2020**, *26*, 2599–2612. [CrossRef]
21. Jochner, S.; Menzel, A. Urban phenological studies—Past, present, future. *Environ. Pollut.* **2015**, *203*, 250–261. [CrossRef]
22. Zhang, X.; Friedl, M.A.; Schaaf, C.B.; Strahler, A.H.; Schneider, A. The footprint of urban climates on vegetation phenology. *Geophys. Res. Lett.* **2004**, *31*, L122091–L122094. [CrossRef]

23. Zhou, D.; Zhao, S.; Zhang, L.; Liu, S. Remotely sensed assessment of urbanization effects on vegetation phenology in China's 32 major cities. *Remote Sens. Environ.* **2016**, *176*, 272–281. [CrossRef]
24. Ji, Y.; Zhan, W.; Du, H.; Wang, S.; Li, L.; Xiao, J.; Liu, Z.; Huang, F.; Jin, J. Urban-rural gradient in vegetation phenology changes of over 1500 cities across China jointly regulated by urbanization and climate change. *ISPRS J. Photogramm. Remote Sens.* **2023**, *205*, 367–384. [CrossRef]
25. Zhu, E.; Fang, D.; Chen, L.; Qu, Y.; Liu, T. The Impact of Urbanization on Spatial–Temporal Variation in Vegetation Phenology: A Case Study of the Yangtze River Delta, China. *Remote Sens.* **2024**, *16*, 914. [CrossRef]
26. Jia, W.; Zhao, S.; Zhang, X.; Liu, S.; Henebry, G.M.; Liu, L. Urbanization imprint on land surface phenology: The urban–rural gradient analysis for Chinese cities. *Glob. Chang. Biol.* **2021**, *27*, 2895–2904. [CrossRef]
27. Yang, J.; Luo, X.; Jin, C.; Xiao, X.; Xia, J. Spatiotemporal patterns of vegetation phenology along the urban–rural gradient in Coastal Dalian, China. *Urban For. Urban Green.* **2020**, *54*, 126784. [CrossRef]
28. Zhang, Y.; Yin, P.; Li, X.; Niu, Q.; Wang, Y.; Cao, W.; Huang, J.; Chen, H.; Yao, X.; Yu, L.; et al. The divergent response of vegetation phenology to urbanization: A case study of Beijing city, China. *Sci. Total Env.* **2022**, *803*, 150079. [CrossRef]
29. Zhou, D.; Zhao, S.; Zhang, L.; Sun, G.; Liu, Y. The footprint of urban heat island effect in China. *Sci. Rep.* **2015**, *5*, 11160. [CrossRef]
30. Wang, M.; Luo, Y.; Zhang, Z.; Xie, Q.; Wu, X.; Ma, X. Recent advances in remote sensing of vegetation phenology: Retrieval algorithm and validation strategy. *Natl. Remote Sens. Bull.* **2022**, *26*, 431–455. [CrossRef]
31. Justice, C.O.; Vermote, E.; Townshend, J.R.; Defries, R.; Roy, D.P.; Hall, D.K.; Salomonson, V.V.; Privette, J.L.; Riggs, G.; Strahler, A.; et al. Moderate Resolution Imaging Spectroradiometer (MODIS): Land remote sensing for global change research. *IEEE Trans. Geosci. Remote Sens.* **1998**, *36*, 1228–1249. [CrossRef]
32. Rizvi, S.H.; Fatima, H.; Iqbal, M.J.; Alam, K. The effect of urbanization on the intensification of SUHIs: Analysis by LULC on Karachi. *J. Atmos. Sol. Terr. Phys.* **2020**, *207*, 105374. [CrossRef]
33. Siddiqui, A.; Kushwaha, G.; Nikam, B.; Srivastav, S.K.; Shelar, A.; Kumar, P. Analysing the day/night seasonal and annual changes and trends in land surface temperature and surface urban heat island intensity (SUHII) for Indian cities. *Sustain. Cities Soc.* **2021**, *75*, 103374. [CrossRef]
34. Yin, P.; Li, X.; Zhou, Y.; Mao, J.; Fu, Y.H.; Cao, W.; Gong, P.; He, W.; Li, B.; Huang, J.; et al. Urbanization effects on the spatial patterns of spring vegetation phenology depend on the climatic background. *Agric. For. Meteorol.* **2024**, *345*, 109718. [CrossRef]
35. Tang, W.; Zhou, J.; Ma, J.; Wang, Z.; Ding, L.; Zhang, X.; Zhang, X. TRIMS LST: A daily 1-km all-weather land surface temperature dataset for the Chinese landmass and surrounding areas (2000–2021). *Earth Syst. Sci. Data Discuss.* **2023**, *2023*, 1–34. [CrossRef]
36. Zhou, J.; Zhang, X.; Zhan, W.; Gottsche, F.M.; Liu, S.; Olesen, F.S.; Hu, W.; Dai, F. A Thermal Sampling Depth Correction Method for Land Surface Temperature Estimation From Satellite Passive Microwave Observation Over Barren Land. *IEEE Trans. Geosci. Remote Sens.* **2017**, *55*, 4743–4756. [CrossRef]
37. Zhang, X.; Zhou, J.; Göttsche, F.-M.; Zhan, W.; Liu, S.; Cao, R. A Method Based on Temporal Component Decomposition for Estimating 1-km All-Weather Land Surface Temperature by Merging Satellite Thermal Infrared and Passive Microwave Observations. *IEEE Trans. Geosci. Remote Sens. A Publ. IEEE Geosci. Remote Sens. Soc.* **2019**, *57*, 4670–4691. [CrossRef]
38. Zhang, X.; Zhou, J.; Liang, S.; Wang, D. A practical reanalysis data and thermal infrared remote sensing data merging (RTM) method for reconstruction of a 1-km all-weather land surface temperature. *Remote Sens. Environ.* **2021**, *260*, 112437. [CrossRef]
39. Ji, Y.; Jin, J.; Zhan, W.; Guo, F.; Yan, T. Quantification of Urban Heat Island-Induced Contribution to Advance in Spring Phenology: A Case Study in Hangzhou, China. *Remote Sens.* **2021**, *13*, 3684. [CrossRef]
40. Polgar, C.; Gallinat, A.; Primack, R.B. Drivers of leaf-out phenology and their implications for species invasions: Insights from Thoreau's Concord. *New Phytol.* **2014**, *202*, 106–115. [CrossRef]
41. Gou, J.; Miao, C.; Samaniego, L.; Xiao, M.; Wu, J.; Guo, X. CNRD v1.0: A High-Quality Natural Runoff Dataset for Hydrological and Climate Studies in China. *Bull. Am. Meteorol. Soc.* **2021**, *102*, E929–E947. [CrossRef]
42. Miao, C.; Gou, J.; Fu, B.; Tang, Q.; Duan, Q.; Chen, Z.; Lei, H.; Chen, J.; Guo, J.; Borthwick, A.G.L.; et al. High-quality reconstruction of China's natural streamflow. *Sci. Bull.* **2022**, *67*, 547–556. [CrossRef]
43. Han, J.Y.; Miao, C.Y.; Gou, J.J.; Zheng, H.Y.; Zhang, Q.; Guo, X.Y. A new daily gridded precipitation dataset for the Chinese mainland based on gauge observations. *Earth Syst. Sci. Data* **2023**, *15*, 3147–3161. [CrossRef]
44. Yang, J.; Huang, X. The 30 m annual land cover datasets and its dynamics in China from 1985 to 2022. *Earth Syst. Sci. Data* **2021**, *13*, 3907–3925. [CrossRef]
45. Chen, J.; Jönsson, P.; Tamura, M.; Gu, Z.; Matsushita, B.; Eklundh, L. A simple method for reconstructing a high-quality NDVI time-series data set based on the Savitzky–Golay filter. *Remote Sens. Environ.* **2004**, *91*, 332–344. [CrossRef]
46. Jönsson, P.; Eklundh, L. Seasonality extraction by function fitting to time-series of satellite sensor data. *IEEE Trans. Geosci. Remote Sens.* **2002**, *40*, 1824–1832. [CrossRef]
47. Beck, P.S.A.; Atzberger, C.; Høgda, K.A.; Johansen, B.; Skidmore, A.K. Improved monitoring of vegetation dynamics at very high latitudes: A new method using MODIS NDVI. *Remote Sens. Environ.* **2006**, *100*, 321–334. [CrossRef]
48. Song, C.; Linggong, K.E.; Liu, S.; Liu, G.; Zhong, X. Comparison of Three NDVI Timeseries Fitting Methods based on TIMESAT—Taking the Grassland in Northern Tibet as Case. *Remote Sens. Technol. Appl.* **2011**, *26*, 147–155. [CrossRef]
49. Liang, L.; Schwartz, M.D.; Fei, S. Validating satellite phenology through intensive ground observation and landscape scaling in a mixed seasonal forest. *Remote Sens. Environ.* **2011**, *115*, 143–157. [CrossRef]

50. Ganguly, S.; Friedl, M.A.; Tan, B.; Zhang, X.; Verma, M. Land surface phenology from MODIS: Characterization of the Collection 5 global land cover dynamics product. *Remote Sens. Environ.* **2010**, *114*, 1805–1816. [CrossRef]
51. He, Y.; Fan, G.; Zhang, X.; Li, Z.; Gao, D. Vegetation Phenological Variation and Its Response to Climate Changes in Zhejiang Province. *J. Nat. Resour.* **2013**, *28*, 220–233. [CrossRef]
52. Song, C.; You, S.; Ke, L.; Liu, G.; Zhong, X. Phenological variation of typical vegetation types in northern Tibet and its response to climate changes. *Acta Ecol. Sin.* **2012**, *32*, 1045–1055. [CrossRef]
53. Yuan, H.; Yan, J.; Zang, J.; Wang, Z.; Xu, W.; Zhang, H. Influences of climate change and human activities on vegetation phenology of Shanghai. *Acta Ecol. Sin.* **2023**, *43*, 8803–8815. [CrossRef]
54. Yosef, R.; Rakholia, S.; Mehta, A.; Bhatt, A.; Kumbhojkar, S. Land Surface Temperature Regulation Ecosystem Service: A Case Study of Jaipur, India, and the Urban Island of Jhalana Reserve Forest. *Forests* **2022**, *13*, 1101. [CrossRef]
55. Heide, O.M. High autumn temperature delays spring bud burst in boreal trees, counterbalancing the effect of climatic warming. *Tree Physiol.* **2003**, *23*, 931–936. [CrossRef]
56. Chuine, I.; Morin, X.; Bugmann, H. Warming, photoperiods, and tree phenology. *Science* **2010**, *329*, 277–278. [CrossRef]
57. Chung, U.; Jung, J.-e.; Seo, H.-c.; Yun, J.I. Using urban effect corrected temperature data and a tree phenology model to project geographical shift of cherry flowering date in South Korea. *Clim. Chang.* **2009**, *93*, 447–463. [CrossRef]
58. Lai, X.; Li, M.; Liu, C.; Zhong, Y.; Lin, L.; Wang, H. The phenological responses of plants to the heat island effect in the main urban area of Chongqing. *Acta Ecol. Sin.* **2019**, *39*, 7025–7034. [CrossRef]
59. Gusewell, S.; Furrer, R.; Gehrig, R.; Pietragalla, B. Changes in temperature sensitivity of spring phenology with recent climate warming in Switzerland are related to shifts of the preseason. *Glob. Chang. Biol.* **2017**, *23*, 5189–5202. [CrossRef]
60. Wang, L.; De Boeck, H.J.; Chen, L.; Song, C.; Chen, Z.; McNulty, S.; Zhang, Z. Urban warming increases the temperature sensitivity of spring vegetation phenology at 292 cities across China. *Sci. Total Environ.* **2022**, *834*, 155154. [CrossRef]
61. Deng, C.; Bai, H.; Gao, S.; Huang, X.; Meng, Q.; Zhao, T.; Zhang, Y.; Su, K.; Guo, S. Comprehensive effect of climatic factors on plant phenology in Qinling Mountains region during 1964–2015. *Acta Ecol. Sin.* **2018**, *73*, 917–931. [CrossRef]

Disclaimer/Publisher's Note: The statements, opinions and data contained in all publications are solely those of the individual author(s) and contributor(s) and not of MDPI and/or the editor(s). MDPI and/or the editor(s) disclaim responsibility for any injury to people or property resulting from any ideas, methods, instructions or products referred to in the content.

Article

Vegetation Dynamics of Sub-Mediterranean Low-Mountain Landscapes under Climate Change (on the Example of Southeastern Crimea)

Vladimir Tabunshchik, Roman Gorbunov, Tatiana Gorbunova * and Mariia Safonova

A.O. Kovalevsky Institute of Biology of the Southern Seas of RAS, 299011 Sevastopol, Russia; tabunshchyk@ya.ru (V.T.); karadag_station@mail.ru (R.G.); malashina@ibss-ras.ru (M.S.)
* Correspondence: gorbunovatyu@gmail.com

Abstract: In the context of a changing environment, understanding the interaction between vegetation and climate is crucial for assessing, predicting, and adapting to future changes in different vegetation types. Vegetation exhibits high sensitivity to external environmental factors, making this understanding particularly significant. This study utilizes geospatial analysis techniques, such as geographic information systems, to investigate vegetation dynamics based on remote sensing data and climatic variables, including annual air temperature, annual precipitation, and annual solar radiation. The research methodology encompasses data collection, processing, and analysis, incorporating multispectral imagery and multilayered maps of various parameters. The calculation of the normalized difference vegetation index serves to evaluate changes in vegetation cover, identify areas experiencing variations in green biomass, and establish strategies for the future development of different vegetation types. During the period from 2001 to 2022, the average normalized difference vegetation index value in the Southeastern Crimea region amounted to 0.443. The highest average values were recorded in the year 2006, reaching a magnitude of 0.469. Conversely, the lowest values were observed in the years 2001–2022, constituting 0.397. It has been ascertained that an overarching positive trend in the evolution of NDVI values from 2001 to 2022 is apparent, thus implying a notable augmentation in vegetative biomass. However, adversarial trends manifest in discrete locales adjacent to the cities of Sudak and Feodosia, along with the coastal stretches of the Black Sea. Correlation analysis is employed to establish relationships between vegetation changes and climatic indicators. The findings contribute to our understanding of the vulnerability of various vegetation types and ecosystems in the Southeastern Crimea region. The obtained data provide valuable insights for the development of sustainable vegetation resource management strategies and climate change adaptation in the region.

Keywords: forest; change; ecosystems; GIS; remote sensing; NDVI; Crimean Peninsula; air temperature; precipitation; solar radiation; multispectral imagery

Citation: Tabunshchik, V.; Gorbunov, R.; Gorbunova, T.; Safonova, M. Vegetation Dynamics of Sub-Mediterranean Low-Mountain Landscapes under Climate Change (on the Example of Southeastern Crimea). *Forests* **2023**, *14*, 1969. https://doi.org/10.3390/f14101969

Academic Editors: Cate Macinnis-Ng, Jianping Wu, Zhongbing Chang and Xin Xiong

Received: 21 July 2023
Revised: 15 September 2023
Accepted: 27 September 2023
Published: 28 September 2023

Copyright: © 2023 by the authors. Licensee MDPI, Basel, Switzerland. This article is an open access article distributed under the terms and conditions of the Creative Commons Attribution (CC BY) license (https://creativecommons.org/licenses/by/4.0/).

1. Introduction

Vegetation change is considered a key indicator of ecosystem response to environmental factors and conditions [1,2]. Global vegetation change has become a pressing issue in recent years, posing a significant threat [3,4]. Consequently, all countries are actively involved in addressing the drastic reduction of global vegetation cover, as evidenced by the inclusion of this topic in various regulatory documents worldwide [5,6]. Climate factors exert a profound influence on vegetation change [7–10], while endogenous catastrophic processes, such as earthquakes [11], volcanic eruptions [12], fires [13], erosion-induced loss of topsoil fertility [14,15], floods [16], adverse atmospheric phenomena (e.g., hurricanes, typhoons) [17], and plant diseases caused by fungi, lichens, insects, and other agents [18,19], further contribute to the transformation of vegetation. Moreover, anthropogenic activities, encompassing both complex processes and intentional clearing of vegetation for various purposes [20–22], significantly impact vegetation change. Deforestation is particularly acute in large timber-rich countries such as Brazil, Canada, China, Russia, and others. Various approaches are employed to assess vegetation change, including computer modeling, remote sensing methods [23–25], changes in land cover composition [26–28], alterations in vegetation index characteristics [29–31], and the utilization of multispectral satellite imagery for classification purposes [32,33].

Li et al. [34] point out that traditional vegetation dynamics monitoring based on field-sampled data has limitations due to the intricate data collection process, which presents challenges in analyzing long-term changes in vegetation. Consequently, the application of remote sensing methods addresses many challenges when studying vegetation dynamics. In recent years, the use of remote sensing data has enabled the near real-time monitoring of vegetation change, including its qualitative characteristics. The analysis of NDVI dynamics has gained prominence as a widely employed method for assessing vegetation change [30,34,35]. This can be attributed to the simplicity of NDVI calculation and the availability of extensive archives of high-resolution multispectral satellite imagery, such as MODIS, Landsat, Sentinel, and others. Traditionally, NDVI has found practical applications in agriculture [36] for crop condition analysis and the calculation of norms for various land improvement operations. NDVI has been actively applied in recent years to evaluate vegetation change in diverse regions worldwide, including China [37–40], India [41,42], the United States [43], Russia [41], Bangladesh [28], Argentina [35], Iran [44], Pakistan [31], among others. Gandhi et al. [30] demonstrated the potential of employing NDVI in analyzing vegetation change in the Vellore District, India. They found that forest or shrub land and barren land cover types decreased by approximately 6% and 23%, respectively, from 2001 to 2006. In contrast, agricultural land, built-up areas, and water areas increased by approximately 19%, 4%, and 7%, respectively. Jiang et al. [38], utilizing NDVI calculations, revealed a significant increase in vegetation NDVI in Tibet from 2001 to 2020, with the annual mean NDVI fluctuating between 0.31 and 0.34. Han et al. [37] demonstrated that the NDVI values in Anhui Province ranged between 0.5 and 0.58, with a multi-year annual mean of 0.55. Vegetation cover in Anhui Province gradually improved from 2001 to 2019. Johnson et al. [43] assessed crop productivity in the United States using MODIS NDVI.

It is essential to recognize that forest landscapes hold significant value within the spectrum of vegetation cover due to their crucial contributions, as highlighted by numerous researchers [45–47]. Forests play a vital role in carbon sequestration, exhibiting the highest potential in this regard [48]. They also function as complex ecosystems that influence global substance and energy transformation cycles [49,50], while providing recreational benefits [51]. Among the most vulnerable and susceptible ecosystems are the forests of the Amazon [52,53], Equatorial Africa [54,55], Vietnam [56], and Siberia [57]. Additionally, forests situated at the boundaries of their natural distribution range, subject to negative impacts from both natural and anthropogenic factors, warrant special attention. However, the study of forest landscape functioning often receives insufficient attention.

The Crimean Peninsula, particularly Southeastern Crimea, represents a typical region characterized by vulnerable forest landscapes. Moreover, Southeastern Crimea marks the northern limit of downy oak forests [58]. In addition to forests, the region encompasses a limited number of steppe landscapes, which also respond to climate change.

This study aims to: (1) assess changes in NDVI values within different vegetation communities in Southeastern Crimea from 2001 to 2022; (2) analyze climatic changes in Southeastern Crimea during the same period; (3) establish relationships between climatic changes and vegetation dynamics in Southeastern Crimea from 2001 to 2022; and (4) evaluate trends in vegetation change in Southeastern Crimea from 2001 to 2022.

Section 1 addresses foundational theoretical inquiries concerning the feasibility of investigating vegetation dynamics through the application of the normalized difference vegetation index (NDVI) in conjunction with geoinformatics methodologies. Elaborate scrutiny is directed towards discerning the multifarious factors that exert potential influence upon the alterations within the vegetative canopy. Of particular emphasis is the intricate topic of global forest dynamics and transformation. The inaugural section endeavors to posit the proposition that the assimilation of the NDVI and the employment of remote sensing techniques, including the dissection of satellite-derived imagery and the computation of pertinent vegetation indices, confer the analytical capacity to evaluate the intricate trajectory of vegetation dynamics, encompassing sylvan ecosystems, within the expanse of the delimited study area. Section 2, elucidates the geographical parameters of the research locale, encapsulating its physiogeographic attributes and the botanical composition indigenous to the study domain. Concomitantly, it expounds upon the research methodology grounded in the exploitation of spatiotemporal variations intrinsic to the NDVI. This methodological framework encompasses the computation of trend analyses, coefficient of variation assessments, the application of the Hurst index, and the discernment of climatic influencers shaping vegetational metamorphosis. Within Section 3, a compendium of cartographic representations and graphical depictions, conceived through the implementation of the methodology outlined in Section 2, is presented. Section 4 provides a comprehensive evaluation of the obtained results, including a detailed exploration of their implications and significance. Additionally, this section outlines the inherent limitations that have constrained the scope and applicability of the conducted research. Section 5 encapsulates the definitive postulates derived from the research endeavor. Furthermore, it undertakes a discourse on the potential trajectories for future investigations in this domain.

2. Materials and Methods

2.1. Study Area

Southeastern Crimea is situated in Eastern Europe, in the southeastern part of the Crimean Peninsula (Figure 1). Its geographical coordinates range from 34°45–35°25′ E and 44°45–44°55′ N. The area of Southeastern Crimea is approximately 568 square kilometers. The boundary of Southeastern Crimea is defined according to [59]. The region is characterized by a complex and rugged terrain, with limited surface water resources. A detailed geographic description of the study area is provided in [58]. The territory of Southeastern Crimea is characterized by Mediterranean climatic features. The average annual air temperature within the Southeastern Crimea area, which varies from the northwest to the southeast, ranges from +9° to +13°. Southeastern Crimea experiences between 100 and 300 mm of precipitation during the winter period. The precipitation field decreases from west to east, with the greatest amount falling in the northwestern part of the study area. In summer, 80 to 160 mm of precipitation falls over the territory of Southeastern Crimea, with the precipitation field decreasing from the northwest to the southeast. Both in winter and in summer, the coastal areas of Southeastern Crimea are the most arid. The annual distribution of precipitation varies from 700 to 350 mm in the west-to-east and northwest-to-northeast directions [58].

2.2. Data

2.2.1. Vegetation of Southeastern Crimea

The vegetation data for Southeastern Crimea were obtained from the vegetation map presented in [60]. This map served as the primary source of information on the vegetation cover within the study area (Figure 2).

According to [60], the spatial distribution of vegetation in the study region is influenced by elevation gradients. However, the elevation zones are not continuous belts due to various local orographic factors. From north to south, as the absolute elevation decreases, a succession of vegetation types can be observed. These include beech forests with a mixture of Stephen maple, durmast oak forests with a blend of hornbeam and ash, pubescent oak forests, pubescent oak light forest in the complex with tomillares- and savannoids-like elements, and forb-feather grass true submontane steppes of the Crimean Mountains. The flatter regions and lower coastal areas of Southeastern Crimea are predominantly occupied by agricultural lands featuring orchards, vineyards, and cultivated fields. Along the coastline, juniper forests can be found. Urban communities are widespread across the region. Table 1 provides an overview of the distribution of the major vegetation types within Southeastern Crimea.

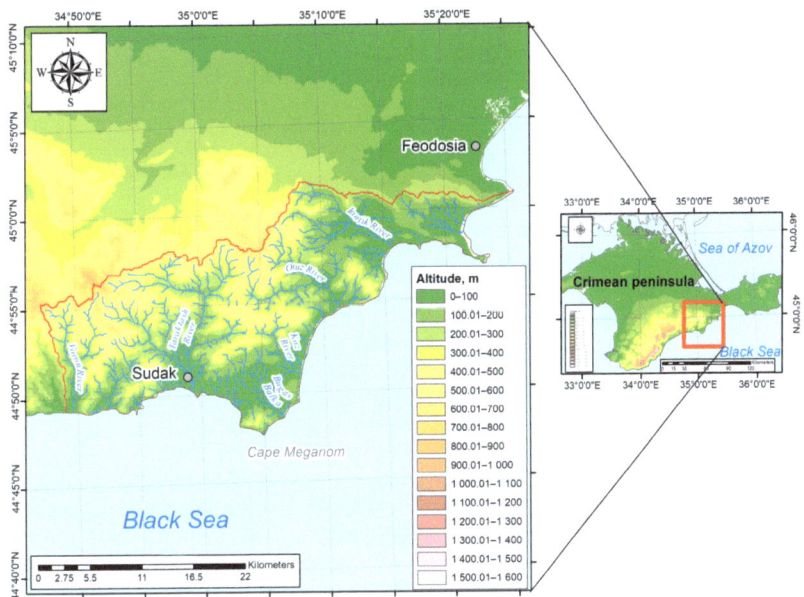

Figure 1. Geographical location of the study area [58].

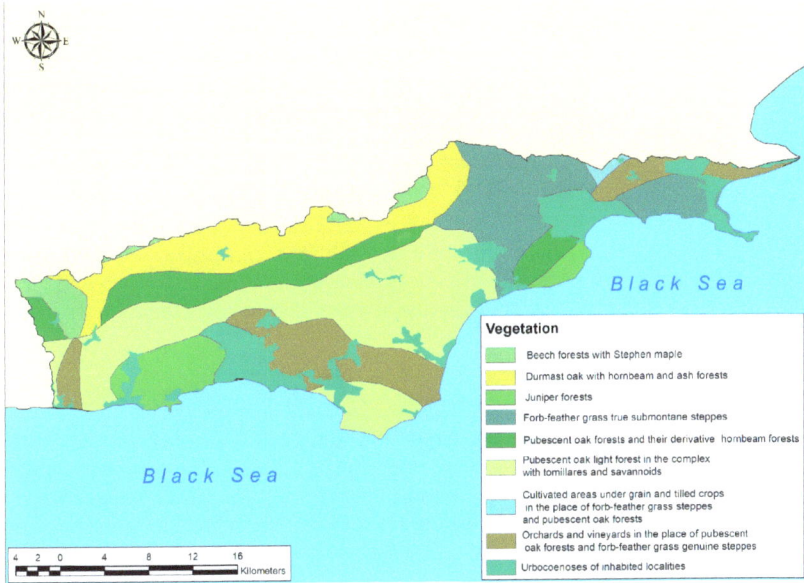

Figure 2. Vegetation of Southeastern Crimea (adapted from [60]).

Table 1. Major vegetation types in Southeastern Crimea.

Plant Community	Area, km^2
Juniper forests	38.42
Beech forests with Stephen maple	18.11
Durmast oak with hornbeam and ash forests	80.54
Pubescent oak forests and their derivative hornbeam forests	61.28
Pubescent oak light forest in the complex with tomillares and savannoids	169.03
Forb-feather grass true submontane steppes	85.39
Orchards and vineyards in the place of pubescent oak forests and forb-feather grass genuine steppes	84.31
Cultivated areas under grain and tilled crops in the place of forb-feather grass steppes and pubescent oak forests	2.58

2.2.2. Satellite Data

The study utilized MODIS satellite imagery covering the period from 2001 to 2022. The use of MODIS imagery (500 m/pixel, 8-day composite) was primarily motivated by its broader temporal coverage, including the acquisition of cloud-free and sparsely clouded satellite images within the study area, in comparison with competitors such as Landsat (30 m/pixel) and Sentinel-2 (10 m/pixel), which may offer higher spatial resolution. MODIS has a higher frequency of Earth observation within the study area. Processing and computation of NDVI values from MODIS satellite imagery were conducted using the Google Earth Engine (GEE) cloud computing platform.

2.2.3. Temperature, Precipitation, and Solar Radiation Data

In recent years, extensive analysis has been conducted on various meteorological databases [61,62]. Spatial and temporal distribution data of air temperature, precipitation, and solar radiation fields were obtained from publicly available databases and published works by other researchers. Air temperature data within the Southeastern Crimea region were sourced from the ClimateEU database [63]. Precipitation and solar radiation data were retrieved using the Google Earth Engine platform from the CHIRPS [62] and ERA5-Land [64] datasets, respectively.

2.3. Methods

2.3.1. NDVI Trend Analysis

The analysis of NDVI trend changes within Southeastern Crimea was conducted for the entire study area and for each pixel of the NDVI raster. The linear regression model was extensively applied to assess the trend changes [38,65,66]. The assessment of NDVI changes over time was performed using the formula:

$$Slope = \frac{n * NDVI_i * \sum_{i=1}^{n} i - \sum_{i=1}^{n} i * \sum_{i=1}^{n} i}{n * \sum_{i=1}^{n} i^2 - (\sum_{i=1}^{n} i)^2} \quad (1)$$

where i—year; n—the number of years of observation; $NDVI_i$—NDVI value for year i.

Negative slope values indicate a decrease in NDVI values over time, while positive slope values indicate an increase in NDVI values over time. Additionally, the *p*-values were evaluated using the R Studio software (Version 2023.09.0+463; Posit PBC, Boston, MA, USA).

2.3.2. Coefficient of Variation

In order to evaluate the stability of NDVI values changes in the Southeastern Crimea region from 2001 to 2022, the coefficient of variation was calculated. The coefficient of variation was computed using the following formula:

$$CV = \frac{\sqrt{\frac{1}{n}\sum_{i=1}^{n}(NDVI_i - NDVI)^2}}{NDVI} \qquad (2)$$

where CV—coefficient of variation; n—the number of years in the study period; $NDVI_i$—NDVI value in year i; $NDVI$—the mean NDVI value for the entire study period [38].

The assessment of the coefficient of variation was performed using the R Studio software package for each pixel within the raster in the study region. Data values from the raster were extracted using the Quantum GIS software package (Version 3.16.16; Open Source Geospatial Foundation (OSGeo), Beaverton, OR, USA). To gauge the stability of changes, the coefficient of variation values was categorized into four classes based on the guidelines provided in [37]: very stable ($CV \leq 0.04$), stable ($0.04 < CV \leq 0.08$), slightly changed ($0.08 < CV \leq 0.12$), and significantly changed ($CV > 0.12$) for each pixel in the analyzed image.

2.3.3. Hurst Index

In order to assess strategies for the development of forest ecosystems, the Hurst index was calculated as a means to forecast the evolution of temporal trends. The Hurst index is an effective method for identifying long-term dependencies in time series [29]. Detailed descriptions and calculations of the Hurst index can be found in several works [29,37,38]. The R Studio software package was utilized to simplify the calculations and determine the Hurst index for each pixel within the study region. Data values were obtained from the raster using the Quantum GIS software package. To calculate the Hurst index, data for 8-day periods of measurements of the MODIS space satellite were used. The calculation of the Hurst index was conducted using the R Studio programming environment, facilitated by the «pracma» library. This method involves the computation of the Hurst index through R/S analysis. To execute the R/S analysis, temporal series spanning the years 2001 to 2022 were individually subjected to a comprehensive examination for stationarity or non-stationarity. This evaluation was undertaken for each spatial cell. The Augmented Dickey–Fuller test, available within the R Studio environment and facilitated by the «tseries» library, was employed for this purpose. In instances where non-stationary time series were encountered, a sequence of transformations was applied to render them stationary. The R Studio program was employed for this transformation process, utilizing techniques such as logarithmization and differencing. These methods were strategically utilized to normalize variance and mitigate the presence of trends within the data. Upon the attainment of stationary time series, the computation of the Hurst index was undertaken through the utilization of the «pracma» library within R Studio. This index is derived from the R/S analysis and serves as an indicator of long-range dependence or persistence within the data. The calculated Hurst index values offer insights into the underlying temporal dynamics of the analyzed variables across different spatial cells. It is worth noting that this methodological approach aligns with a comprehensive workflow involving data preprocessing, statistical analysis, and computational procedures, all orchestrated within the R Studio environment.

2.3.4. Correlation Analysis

A correlation analysis was conducted to assess the influence of climatic factors, including annual precipitation, annual air temperatures, and annual solar radiation, on the average annual NDVI values within the Southeastern Crimea region. Correlation analysis is a widely used technique for examining the relationships between climate factors and NDVI values in various research regions worldwide [38,67,68]. Data on average annual

NDVI values, air temperatures, precipitation, and solar radiation were obtained for a grid of points using the Quantum GIS software package (specifically, the Vector–Raster SAGA toolset). Each point within the presented grid of points functions as the central reference for a square, measuring 500 by 500 m. Consequently, these points are uniformly separated by a distance of 500 m from one another. The correlation calculation was performed according to the following formula:

$$R_{xy} = \frac{\sum_{i=1}^{n}[(x_i - x)(y_i - y)]}{\sqrt{\sum_{i=1}^{n}(x_i - x)^2 \sum_{i=1}^{n}(y_i - y)}} \tag{3}$$

where R_{xy}—the correlation coefficient; n—the number of years in the study period; x and y—the factors used for the correlation analysis, representing the sample means of the variables [38].

The correlation assessment was performed using the R Studio software package. For the purpose of conducting correlation analysis, NDVI and climatic datasets spanning the years 2001 to 2022 were subjected to rigorous examination in terms of their stationarity or non-stationarity. This examination was undertaken utilizing the Augmented Dickey–Fuller test, a statistical method, within the computational environment of R Studio, specifically leveraging the «tseries» library. Time series exhibiting non-stationarity underwent a process of transformation into stationary series, accomplished through the application of logarithmic and differencing techniques. These techniques were instrumental in homogenizing variance across the temporal dimension and effecting the elimination of underlying trends. Subsequent to the attainment of stationary time series, correlation coefficients were computed individually for each temporal data point, a crucial step in elucidating the relationships between the NDVI and climatic variables. Upon generation of these correlation datasets, a seamless transition was effected into the ArcGIS software package (Version 10.8; ESRI, Redlands, CA, USA). Within this geospatial environment, an interpolation procedure, employing the well-regarded «Spline» method, was executed. This interpolation operation facilitated the derivation of intermediate values between discrete data points. Subsequently, cartographic representations were generated to visually articulate the spatial patterns emerging from the interpolated data, providing valuable insights into the dynamics of the studied variables across the geographical extent.

3. Results

3.1. NDVI Dynamics in Southeastern Crimea

A consistent increase in the NDVI values has been observed in Southeastern Crimea from 2001 to 2022 (Figures 3 and 4).

During the period from 2001 to 2022, the average NDVI value in Southeastern Crimea was 0.443. The highest average values were recorded in 2006, reaching 0.469, while the lowest values were observed in 2001–2002, at 0.397. In this context, the spatial distribution of the NDVI values within the investigated area from 2001 to 2022 exhibits a range of fluctuation spanning from 0.1 to 0.62. When comparing the annual average values for each year with the long-term mean, a clear trend of dividing the study period into two periods becomes apparent: before 2014, when the annual average NDVI values were consistently lower than the long-term mean, and after 2014, when the annual average NDVI values exceeded the long-term mean.

Figure 4 presents the dynamics of annual average NDVI values within the major vegetation communities in Southeastern Crimea.

Within the juniper forests, the average NDVI value was 0.43; within pubescent oak forests and their derivative hornbeam forests, it was 0.57; within the pubescent oak light forest in the complex with tomillares and savannoids, it was 0.54; and within the pubescent oak light forest in the complex with tomillares and savannoids, it was 0.45. Positive trends of increasing NDVI values are evident, particularly within the oak forests. From Figures 3 and 4, it is evident that there is a trend of increasing NDVI values, indicating overall vegetation growth in Southeastern Crimea from 2001 to 2022. The distribution of annual average NDVI values within the major vegetation communities is presented in more detail in Figure 4, where the durmast oak with hornbeam and ash forests exhibit the highest NDVI values, while pubescent oak light forest in the complex with tomillares and savannoids show the lowest values.

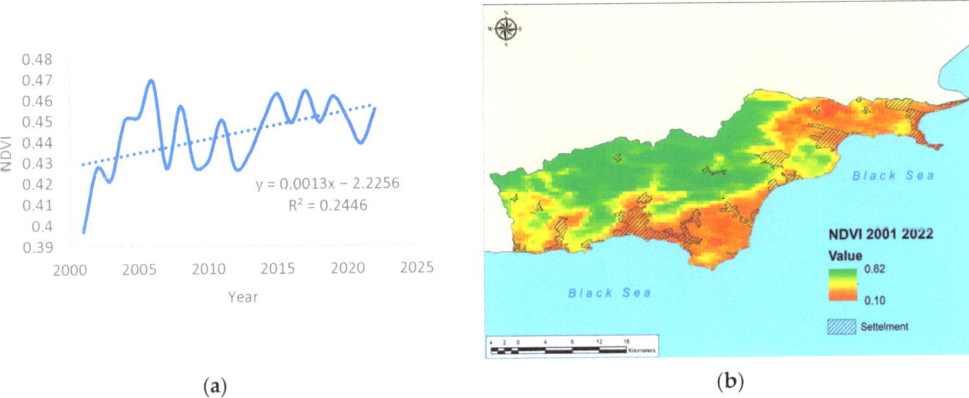

Figure 3. Dynamics of annual average NDVI values from 2001 to 2022 in Southeastern Crimea: (**a**) annual average values; (**b**) map of average NDVI values.

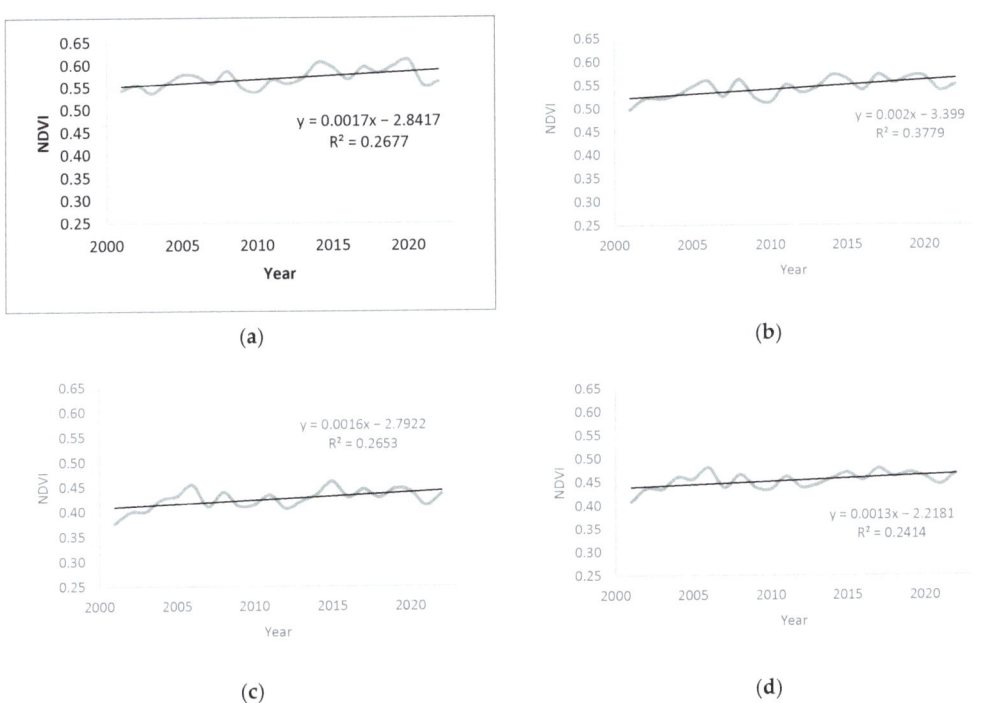

Figure 4. Dynamics of annual average NDVI values from 2001 to 2022 within the major vegetation communities in Southeastern Crimea: (**a**) durmast oak with hornbeam and ash forests; (**b**) pubescent oak forests and their derivative hornbeam forests; (**c**) juniper forests; (**d**) pubescent oak light forest in the complex with tomillares and savannoids.

In the southeastern region of Crimea, it is noteworthy that the peak of vegetation activity predominantly occurs in August. Figure 5 illustrates the monthly averages as well as the maximum and minimum values of the NDVI (Normalized Difference Vegetation Index) for the month of August, covering the period from 2001 to 2022.

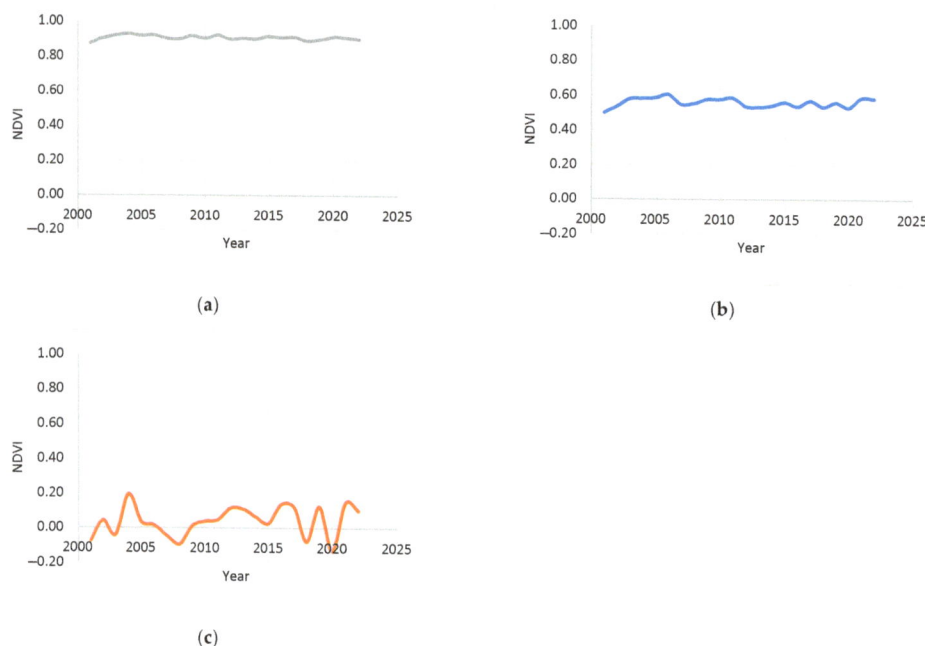

Figure 5. NDVI values in southeastern Crimea in August: (**a**) maximum values; (**b**) average values; (**c**) minimum values.

Figure 6 presents the minimum, maximum, and average values of the NDVI vegetation index for various vegetation communities in Southeastern Crimea during August.

The graph illustrating maximum NDVI values across the extensive study area demonstrates a pronounced smoothing effect, thus limiting its ability to discern spatial and temporal differentiations within the analyzed region. Conversely, as we narrow down the spatial units under scrutiny and reduce their corresponding areas, the distinct differences become more apparent, rendering maximum data more meaningful. It is worth noting, however, that the practical application of maximum values should be exercised in light of the total count of maximum pixels within the designated study area. Meanwhile, the graph illustrating average NDVI values effectively captures discrepancies and offers the analytical and interpretive potential that underpins its utility. The smoothing effect evident in the graph of maximum values is an outcome of the representation of peak values of individual pixels, which can be spatially dispersed across various segments of the studied region, thereby rendering an incomplete reflection of the overarching patterns of change. A comparable circumstance applies to the distribution of minimum NDVI values in August, albeit, here, the scenario is characterized by a substantial dispersion of values, owing to certain pixels yielding negative NDVI values. Consequently, the application of analysis to both maximum and minimum values is, to a substantial extent, constrained. However, the validity of such analyses should be predicated on the proportional prevalence of minimum and maximum pixels within the examined region. We initially abstained from immediate utilization of minimum values due to the region's characteristics, wherein minimum values could potentially encompass negative NDVI values attributed to the presence of water bodies and the dynamic sea coastline. Through comprehensive analysis of all NDVI values within the study area, we determined that for certain years, negative minimum values accounted for less than 0.1% of all data. A comparable pattern emerged in the distribution of maximum values. Specifically, we scrutinized NDVI values > 0.9 and ascertained that, for the majority of years, these values were either absent or constituted less than 0.5% of the complete dataset. However, while information about minimum and maximum values can serve to pinpoint localized growth or decline patterns, it is crucial to recognize that these values do not offer a holistic depiction of vegetation dynamics or the comprehensive operational landscape. Consequently, our study primarily relied upon the utilization of average NDVI values. This preference is rooted in the fact that minimum and maximum values, being isolated occurrences and accounting for a minute portion of the analyzed dataset, cannot adequately encompass the multifaceted character of the studied territory in Southeastern Crimea.

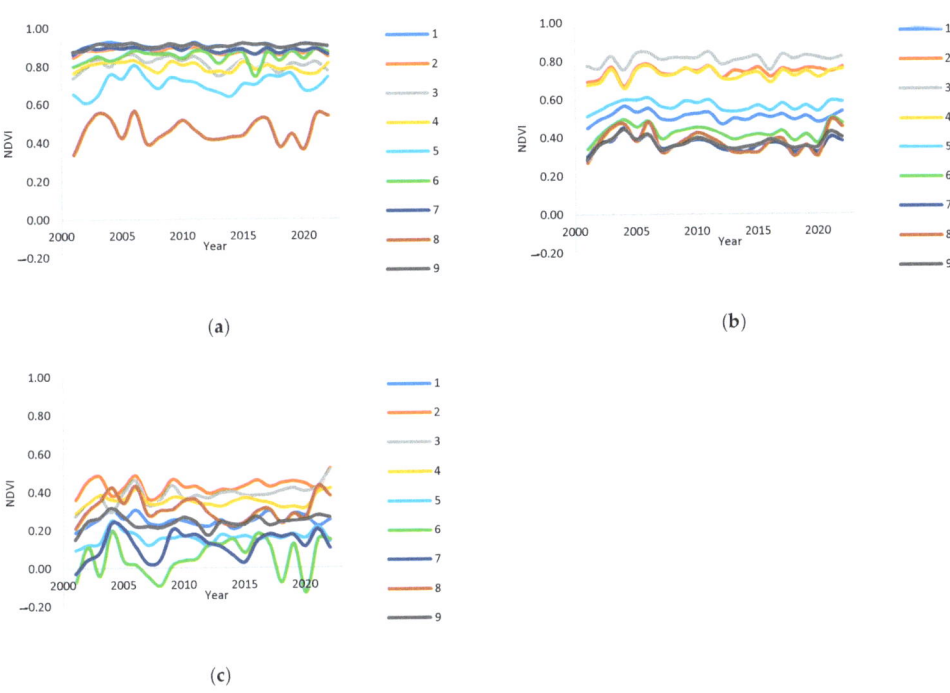

Figure 6. NDVI values within vegetation communities of Southeastern Crimea in August: (**a**) maximum values; (**b**) average values; (**c**) minimum values (numerical indicators on the graphs denoted as: 1—juniper forests; 2—beech forests with Stephen maple; 3—durmast oak with hornbeam and ash forests; 4—pubescent oak forests and their derivative hornbeam forests; 5—pubescent oak light forest in the complex with tomillares and savannoids; 6—forb-feather grass true submontane steppes; 7—orchards and vineyards in the place of pubescent oak forests and forb-feather grass genuine steppes; 8—cultivated areas under grain and tilled crops in the place of forb-feather grass steppes and pubescent oak forests; 9—urbocoenoses of inhabited localities.

3.2. NDVI Trends

Significant spatial and temporal differentiation is observed within the territory of Southeastern Crimea, not only in the NDVI values themselves but also in the direction of their trends. Figure 7 illustrates the analysis of NDVI trend changes for the entire period from 2001 to 2022, as well as for five-year periods.

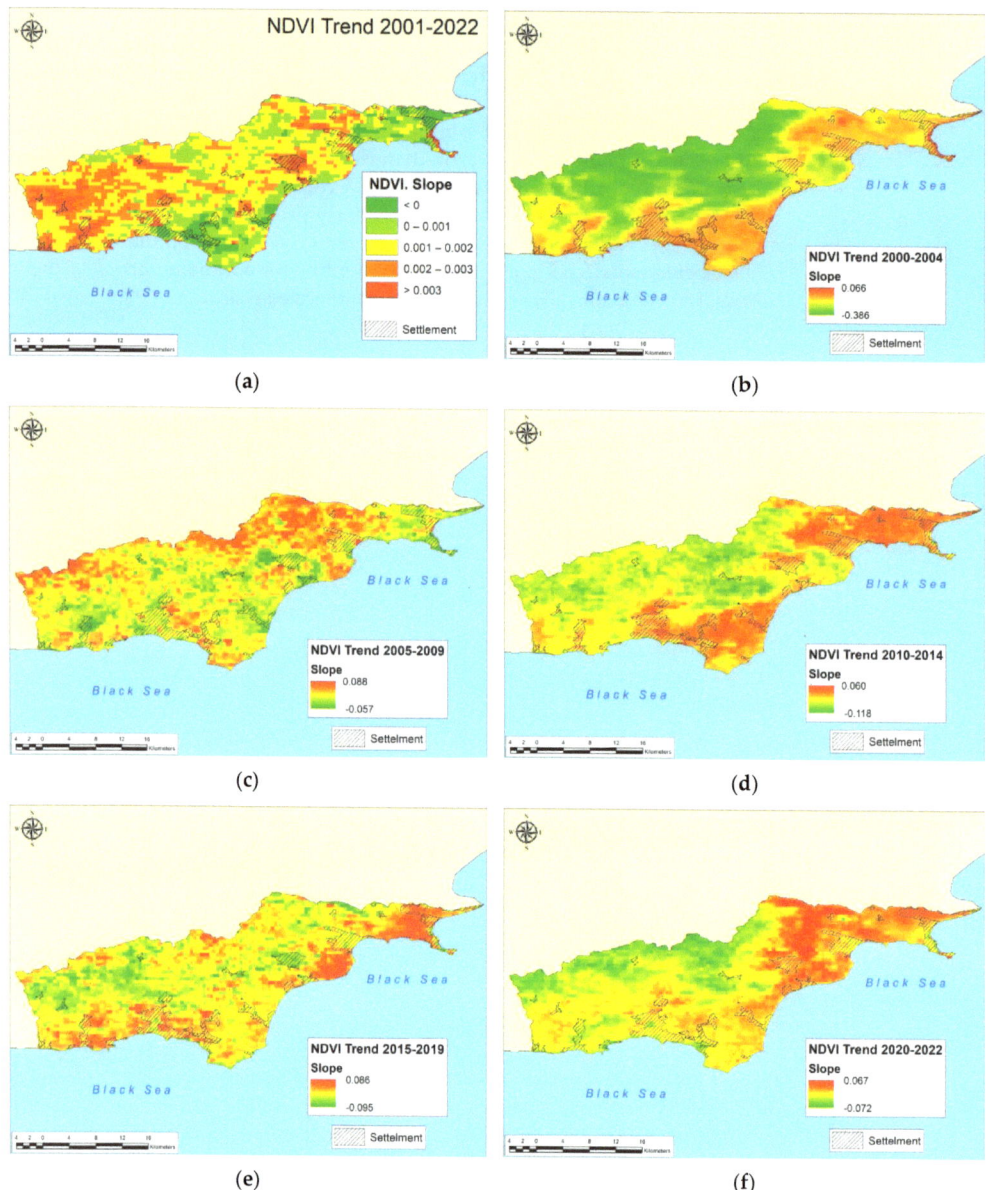

Figure 7. NDVI trend analysis: (**a**) 2001–2022; (**b**) 2001–2004; (**c**) 2005–2009; (**d**) 2010–2014; (**e**) 2015–2019; (**f**) 2020–2022.

As shown in Figure 7a, there is an overall positive trend of NDVI values from 2001 to 2022, indicating an increase in vegetation biomass. However, negative trends are observed in specific areas near the cities of Sudak and Feodosia. When examining the spatial–temporal dynamics for five-year periods (Figure 7b–f), a complex pattern emerges within each spatial cell (pixel). During the period from 2001 to 2004, a noticeable decline in vegetation biomass and a decrease in NDVI values are observed in the central, northern, and northwestern parts of Southeastern Crimea. From 2005 to 2009, favorable conditions for vegetation growth are established in these areas, as indicated by positive values of the NDVI trend (Slope NDVI). Conversely, in areas where environmental conditions for vegetation growth were favorable from 2001 to 2004, conditions leading to a decrease in vegetation biomass are observed from 2005 to 2009. The situation in 2010–2014 is similar to that in 2001–2004. From 2015 to 2019, a complex pattern of growth and decline in vegetation is observed, yet the overall situation resembles the period of 2005–2009. Notably, starting from 2015, negative trends in NDVI values are observed in the western and northwestern parts of Southeastern Crimea, indicating a decrease in vegetation biomass. Therefore, quasi-five-year cycles can be identified, reflecting changes in vegetation cover in Southeastern Crimea.

If we consider the average values of the NDVI trend within the major vegetation communities, it can be observed that there is generally either an increase or a relatively stable trend in vegetation biomass (Table 2).

Table 2. Average multi-year trend values within the vegetation communities of Southeastern Crimea from 2001 to 2022.

Variation Trend	Trend Prediction		
	Minimum	Maximum	Average
Juniper forests	−0.0012	0.0060	0.0015
Beech forests with Stephen maple	0.0000	0.0040	0.0014
Durmast oak with hornbeam and ash forests	−0.0007	0.0050	0.0010
Pubescent oak forests and their derivative hornbeam forests	0.0000	0.0040	0.0014
Pubescent oak light forest in the complex with tomillares and savannoids	−0.0020	0.0050	0.0010
Forb-feather grass true submontane steppes	−0.0020	0.0103	0.0010
Orchards and vineyards in the place of pubescent oak forests and forb-feather grass genuine steppes	−0.0030	0.0050	0.0003
Cultivated areas under grain and tilled crops in the place of forb-feather grass steppes and pubescent oak forests	−0.0010	0.0030	0.0011
Urbocoenoses of inhabited localities	−0.0020	0.0050	0.0010

Anthropogenically created vegetation communities, such as gardens and vineyards, exhibit the least variability as they are artificially maintained throughout their existence due to human activities.

3.3. Coefficient of Variation

Let us now delve into a more detailed analysis of the coefficient of variation (CV) of NDVI values in Southeastern Crimea (Figure 8).

By assessing the CV of NDVI values, we were able to identify the most stable and highly variable areas within Southeastern Crimea. As shown in Figure 8, a significant portion of the study area exhibits a stable distribution of NDVI values. However, minor and significant fluctuations are predominantly observed in the southern and southeastern regions.

Table 3 presents the changes in the CV of NDVI within the vegetation communities of Southeastern Crimea from 2001 to 2022.

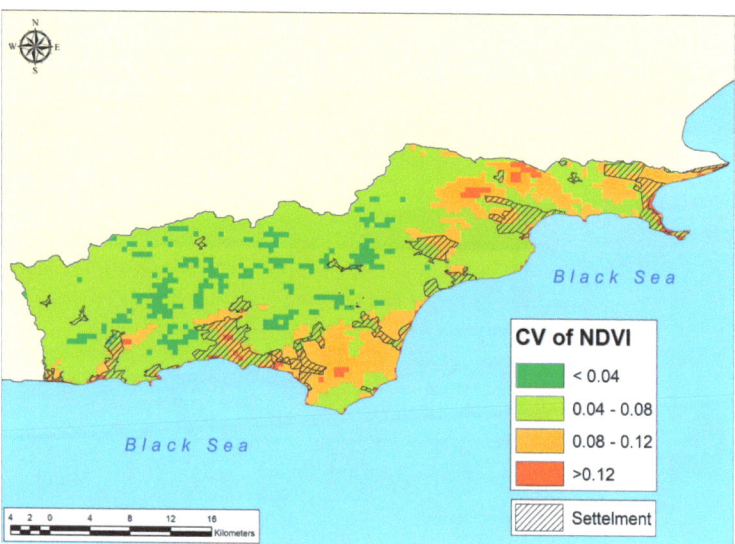

Figure 8. Coefficient of variation of NDVI in Southeastern Crimea.

Table 3. Average multi-year changes in trend values within vegetation communities of Southeastern Crimea (2001–2022).

Variation Trend	CV Minimum	CV Maximum	Average
Juniper forests	0.04	0.21	0.06
Beech forests with Stephen maple	0.04	0.07	0.05
Durmast oak with hornbeam and ash forests	0.03	0.09	0.05
Pubescent oak forests and their derivative hornbeam forests	0.03	0.09	0.05
Pubescent oak light forest in the complex with tomillares and savannoids	0.03	4.43	0.06
Forb-feather grass true submontane steppes	0.04	2.90	0.08
Orchards and vineyards in the place of pubescent oak forests and forb-feather grass genuine steppes	0.03	0.44	0.08
Cultivated areas under grain and tilled crops in the place of forb-feather grass steppes and pubescent oak forests	0.06	0.13	0.09
Urbocoenoses of inhabited localities	0.04	0.19	0.08

As can be observed from Table 3 and the Figure 8, the forest ecosystems exhibit greater stability compared with the steppe ecosystems and anthropogenically created agricultural lands and populated areas.

3.4. Hurst Index

The Hurst index provides a comprehensive assessment of vegetation variability and offers insights into forecasted changes. Figure 9 illuminates the spatial differentiation of Hurst index values in Southeastern Crimea, while Table 4 furnishes a comprehensive account of the minimum, maximum, and mean Hurst index values pertaining to the principal vegetation communities within the same region.

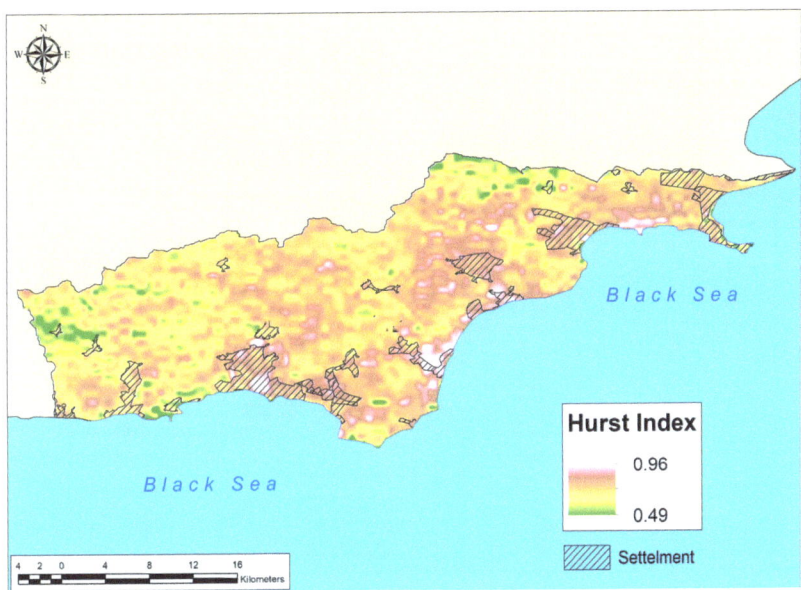

Figure 9. Hurst index of NDVI in Southeastern Crimea.

Table 4. Hurst index values within vegetation communities of Southeastern Crimea (2001–2022).

	Hurst Index		
	Minimum	Maximum	Average
Juniper forests	0.49	0.85	0.73
Beech forests with Stephen maple	0.58	0.84	0.71
Durmast oak with hornbeam and ash forests	0.58	0.87	0.74
Pubescent oak forests and their derivative hornbeam forests	0.59	0.88	0.74
Pubescent oak light forest in the complex with tomillares and savannoids	0.56	0.93	0.76
Forb-feather grass true submontane steppes	0.54	0.94	0.76
Orchards and vineyards in the place of pubescent oak forests and forb-feather grass genuine steppes	0.59	0.93	0.77
Cultivated areas under grain and tilled crops in the place of forb-feather grass steppes and pubescent oak forests	0.65	0.87	0.74
Urbocoenoses of inhabited localities	0.61	0.94	0.78

It has been determined that the range of Hurst index values within Southeastern Crimea varies from 0.49 to 0.96, with a calculated mean value of 0.75. In the broader context of the Southeastern Crimea region, it is evident that the range of Hurst index values, calculated for NDVI data, exhibits pronounced spatial heterogeneity. The lowest Hurst index values are predominantly observed in the northernmost and northwestern sectors of the research area, while the highest values are consistently recorded in the coastal, southern, and southeastern regions. Notably, elevated Hurst index values are particularly prominent in proximity to urban settlements such as Shchebetovka, Kurortnoe, Sudak, Solnechnaya Dolina, and others.

3.5. Influence of Climatic Factors on NDVI Changes in Southeastern Crimea

To assess the impact of climatic factors on NDVI changes, an examination of the temporal dynamics of annual mean air temperature, precipitation, and solar radiation within Southeastern Crimea was conducted. Additionally, correlation coefficients were computed to determine the relationship between these factors and NDVI values (Figure 10).

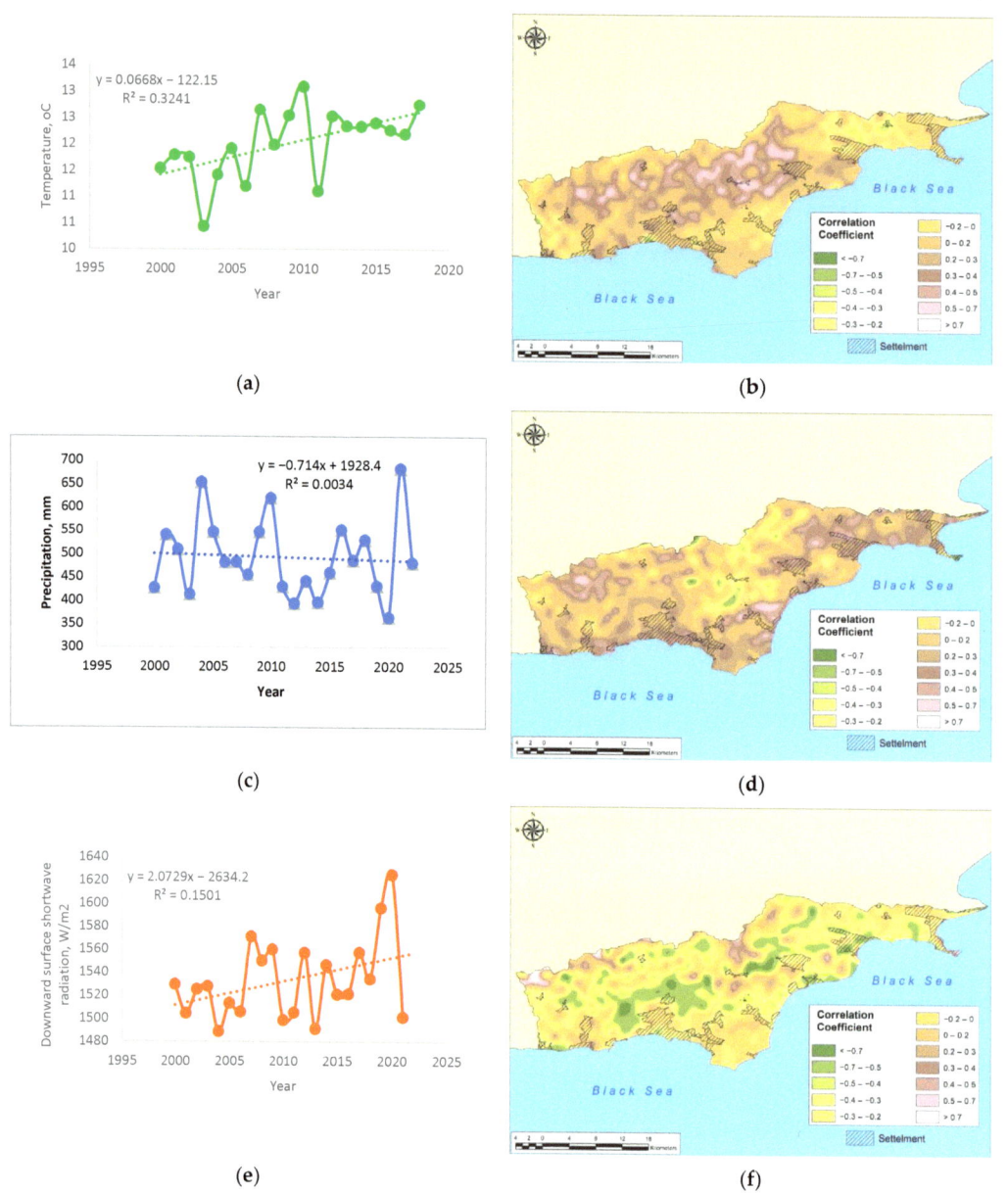

Figure 10. Relationship between air temperature, precipitation, solar radiation, and NDVI: (**a**) dynamics of annual mean air temperature in Southeastern Crimea; (**b**) correlation coefficient between annual mean NDVI values and air temperature; (**c**) dynamics of annual mean precipitation in Southeastern Crimea; (**d**) correlation coefficient between annual mean NDVI values and precipitation; (**e**) dynamics of annual mean solar radiation in Southeastern Crimea; (**f**) correlation coefficient between annual mean NDVI values and solar radiation.

The findings from Figure 10 reveal an upward trend in air temperature and solar radiation, accompanied by a slight reduction in precipitation levels in Southeastern Crimea. Notably, despite the modest decrease in precipitation, there exists a significant correlation between precipitation and NDVI values. Moreover, the correlation coefficients between NDVI and air temperature, as well as NDVI and solar radiation, indicate a moderate level of significance.

4. Discussion

Studying vegetation changes is a critical task as vegetation responds rapidly to various environmental factors. This is particularly important in regions with forests, which are valuable resources for multiple sectors. In this study, we analyzed the dynamics of NDVI and Hurst index values within the major vegetation communities of Southeastern Crimea.

Contrary to several published works [29,37,69], Southeastern Crimea does not exhibit significant variability in NDVI values. This can be attributed to the smaller study area compared with previous studies [69,70] and the prevalence of natural vegetation cover rather than contrasting or absent vegetation cover.

Although there are studies on NDVI dynamics for the Crimean Peninsula and its parts [42,43,71–73], Southeastern Crimea remains understudied. Notably, there are works assessing vegetation dynamics [74]. Fan et al. [70] calculated NDVI changes in the Crimean Peninsula within the Belt and Road Initiative region from 1982 to 2015. However, comparing their data with ours is challenging due to differences in spatial scales.

Comparing the Hurst index values of Southeastern Crimea with other regions worldwide, values below 0.5 prevail, similar to the Tibet Autonomous Region (China) [69] and Inner Mongolia (China) [29]. However, the Hurst index values (<0.4) indicate isolated centers of anti-sustainability within the downy oak forests of the Karadag Nature Reserve, supporting our previous findings [58]. Conversely, centers of instability predominantly occur at the boundaries of urban areas due to negative anthropogenic impact. Overall, studying NDVI dynamics and trends helps identify the most and least susceptible land and forest ecosystems. However, defining classes and boundaries presents challenges compared with previous studies [37,38], and alternative classification variants from [37,38] are not applicable in our research.

Forest ecosystems exhibit the highest correlation between vegetation cover and air temperature, precipitation, solar radiation, indicating favorable conditions for forest growth. However, the downy oak forests, located at the edge of their range, face unfavorable environmental conditions and cannot achieve their full potential in terms of green biomass. This underscores the presence of valuable and less vulnerable ecosystems within the Crimean Mountains.

Our findings are closely related to the data obtained by Han et al. [37] in Anhui Province (China), which indicates that the period around 2014–2015 marks a turning point in NDVI trend changes. However, unlike Han et al. [37], who attribute this to increased catastrophic natural phenomena, such as landslides and avalanches, this does not apply to Southeastern Crimea, where anthropogenic factors and global circulation processes play a significant role.

Considering vegetation-covered regions, NDVI values cannot be negative (unlike water surfaces), allowing us to analyze average values. However, due to pixel size limitations in satellite imagery, water bodies may be included, which are subject to boundary changes due to natural and anthropogenic factors, particularly relevant in the Crimean Peninsula [75]. Nevertheless, the influence on calculated NDVI values in our study region is minor since it belongs to water-deficient areas [58] and comprises numerous natural landscapes devoid of natural or human-created water bodies. Analyzing average annual NDVI values, as presented in Section 3, does not provide a clear understanding of change trends, necessitating more complex indicators to assess vegetation changes. The calculation of Hurst index and trend prediction values effectively addresses this research objective. Thus, the Hurst index proves to be a useful tool for analyzing future changes in regional

NDVI, determining the stability or instability of trends, and predicting long-term vegetation cover changes.

The question of selecting the primary factor or combination of factors influencing vegetation growth within a specific research region remains unresolved. While temperature, precipitation, and solar radiation significantly contribute to the functioning of distinct vegetation communities in Southeastern Crimea, they are not the sole factors. Identifying and analyzing various environmental factors that influence vegetation functioning represent a promising research direction. The utilization of G.E. Hutchinson's concept of multidimensional ecological niche [76] is highly relevant for studying vegetation dynamics. This framework provides valuable insights into the ecological requirements and responses of species within their respective environments. Simultaneously, our analysis has exclusively examined the impact of three climatic factors on vegetation change. Recognizing the intricate influence of climatic variables on the development and functioning of ecosystems and landscapes, it is imperative for future investigations to encompass a more comprehensive array of external factors. This expansion should not only encompass climatic variables but also extend to various other elements within the external environment.

When addressing the matter of selecting vegetation indices appropriate for the evaluation of vegetative landscape dynamics, the adoption of the NDVI demonstrates markedly heightened utility and applicability when compared with alternative vegetation indices. The preference for the NDVI vegetation index can primarily be attributed to its extensive prevalence and its capacity for facilitating inter-comparability among datasets generated by researchers. Esteemed scholars, including those referenced in sources [37,38,77], affirm that the NDVI possesses an enhanced capacity for delineating the growth status of vegetation, engendered by its heightened sensitivity to vegetative constituents. Accordingly, it is conventionally embraced as an efficacious evaluative metric for the surveillance of regional ecological systems. In the words of Li et al. [78], amongst the entire gamut of vegetation indices, the NDVI evinces a robust correlation with net primary productivity (NPP), canopy extent, and biomass. This confluence of attributes enables it to aptly encapsulate and gauge the trajectory of vegetative growth. Consequently, it assumes prominence within investigations into the temporal vicissitudes of regional vegetative ecosystems.

Concurrently, an extensive array of vegetation indices, amenable to computation based on satellite-derived imagery, exists. For instance, the ratio vegetation index (RVI) [79], infrared percentage vegetation index (IPVI) [80], and transformed vegetation index (TVI) [81], while relying on identical satellite image channels as NDVI for computation, are integrally linked to NDVI. Their autonomous consideration, dissociated from the NDVI framework, is bereft of practical import. Notably, NDVI's ascendancy is underscored by its scalar range, spanning from -1 to 1, unlike the ratio vegetation index (RVI) [79] or the difference vegetation index (DVI) [82], both characterized by dimensionality sans constraints, thereby exacerbating the intricacies attendant upon inter-data comparison and interpretation.

Pioneers in the field, including Huete et al. [83] and Elvidge and Lyon [84], have underscored the substantive influence of soil cover upon vegetation indices. However, it is germane to apprehend that the sensitivity of all vegetation indices to the overprint of soil cover and unadorned tracts bereft of vegetation is ubiquitous. In the context of an examined image pixel, NDVI is amenable to calculation, encapsulating the entwined characteristics of soil and aquatic features. Yet, with the burgeoning of vegetative constituents within the confines of this pixel, the NDVI amplitude burgeons in tandem, reflective of the enshrouding of soil domains or the supplanting of erstwhile aquatic expanses. Moreover, certain vegetation indices encompass considerations of both soil and ground surface influence. Instances thereof comprise the transformed soil adjusted vegetation index (TSAVI) [85], modified soil adjusted vegetation index (MSAVI) [86], and enhanced vegetation index (EVI) [87]. Their deployment, however, is confined to locales characterized by a luxuriant mantle of vegetation, encompassing the studied region. This confinement emanates from the intricate constraints impeding precise delimitation of soil constants, as delineated extensively in the monograph [88]. Under these circumstances, the predilection for NDVI prevails, owing to its optimized resonance with vegetative profusion, emblematic of Southeastern Crimea's ecological tapestry. Although the integration of the enhanced vegetation index (EVI) was contemplated during the preparatory stages of inquiry, its notable proclivity for inaccuracy precluded its integration into the investigative paradigm. Furthermore, as underscored in the discourse [42], EVI values evince marginal alignment with alternative vegetation index metrics (most notably NDVI). The calculus of EVI betrayed an elevated susceptibility to terrain topography, a consideration that becomes particularly salient in the backdrop of the study area's intricate and variegated topographical terrain, thereby giving rise to pronounced disparities [89]. Martín-Ortega et al. [90] elucidate the fact that the enhanced vegetation index (EVI) displays enhanced sensitivity to biophysical parameters such as leaf area index (LAI), and is notably impacted by atmospheric conditions to a greater extent than the conventional normalized difference vegetation index (NDVI), owing to the inherent non-ratio nature of EVI. Moreover, empirical evidence underscores that EVI exhibits a heightened degree of responsiveness, approximately five times greater than NDVI, towards fluctuations in near-infrared reflectance (NIR). In contrast, Martín-Ortega et al. [90] observe that the ratio-based formulation of NDVI confers the ability to effectively ameliorate a substantial quantum of perturbations stemming from dynamic solar angles, topographical variations, cloud-induced interference, and shading effects. This inherent property endows NDVI with enhanced robustness against alterations in luminous conditions.

The incorporation of vegetation indices calibrated to ameliorate the influence of soil cover necessitates a nuanced incorporation of regional idiosyncrasies, concomitant with the integration of sundry correction coefficients. These coefficients, inevitably variably distributed across raster cells due to the heterogeneity characterizing soil and terrestrial substrate, preclude universal homogenization. Ergo, the electivity for NDVI culminates in its conspicuously salient relevance for the entire Crimea Peninsula, and Southeastern Crimea, specifically, as manifestly chronicled in scholarly contributions [73,74,91]. The ramifications of soil characteristics at localized and regional research junctures command prodigious endeavor and temporal investment in the context of field and laboratory spectroscopic evaluations of soil. Analogous undertakings, frequently conducted to fine-tune global models and extricate regional particularities across assorted scientific disciplines leveraging remote terrestrial sensing datasets, are amply discernible in the extant corpus [92,93].

Another crucial aspect to consider in the analysis of NDVI value changes and the calculation of the Hurst index was the transformation of time series into a stationary form. This was necessitated by the fact that one of the most widely used methods for calculating the Hurst exponent is the R/S analysis. Holl et al. [94] point out that in the natural world, real-life data often contains inherent trends that render the series nonstationary, thereby rendering the R/S analysis inappropriate. This phenomenon arises from the fact that the R/S analysis can be applied to series that exhibit a degree of stationarity on mean [95]. Furthermore, as indicated by reference [96], it is imperative to consider that in order to compute the Hurst index, the length of the observational series should encompass a minimum of 256 measurements.

Another limitation of this study is the paucity of data and the challenges associated with the geospatial processing of climatic characteristics within Southeastern Crimea. We concur with the findings of Han et al. [37], who highlighted that different interpolation methods for climatic data can lead to divergent raster fields of climate factors.

Furthermore, it is important to recognize the potential use of more detailed satellite imagery with higher spatial resolution (e.g., Sentinel-2 with a resolution of 10 m/pixel), in contrast to the MODIS data employed in this study. Nevertheless, the use of MODIS satellite imagery was primarily driven by its extensive spatial and temporal coverage, despite its lower resolution. As satellite imaging frequency increases in the future, endeavors should be made to enhance data quality and obtain higher-resolution datasets. Additionally, it is imperative to improve the quality of available open climatic data, which currently can only be spatially correlated with MODIS satellite imagery. Another pivotal constraint inherent to this research pertains to the observation duration, a parameter dictated by the accessibility of data procured from orbiting satellites. In the case of the MODIS satellite, as delineated within our investigation, the dataset spans from 2001 onwards. In the event alternative spaceborne satellites are employed for the acquisition of requisite multispectral satellite imagery, instrumental for NDVI computations, the temporal extent of observational records could conceivably expand (as in the case of Landsat) or conversely contract (as exemplified by Sentinel-2). Another salient consideration in research lies in the recognition that the computation of NDVI values, executed across diverse satellite platforms and software suites (such as MODIS, Landsat, Sentinel-2, among others), inherently yields variations due to discrepant imaging epochs. The challenge of coherently aligning images for a specific temporal point compounds this variation. Moreover, even in instances where alignment is achieved, the fact remains that MODIS, Landsat, Sentinel-2, and analogous satellite sensors capture imagery within distinct spectral ranges. Consequently, the comparative analysis of resultant data emerges as an intricate endeavor. It is imperative to duly acknowledge this intricacy as a prospective stipulation shaping the contours of the study's limitations.

Addressing these limitations and incorporating advanced data sources and analytical methods will contribute to a more comprehensive understanding of vegetation dynamics and the underlying environmental factors in Southeastern Crimea.

5. Conclusions

Vegetation cover serves as a crucial indicator of the environmental condition and offers valuable insights into ecosystem health. It plays a significant role in assessing environmental parameters, monitoring anthropogenic activities, evaluating ecosystem services, and understanding forest landscape functions. Understanding the dynamics of vegetation change is essential for effective conservation planning, species preservation, sustainable forest management, and protection. However, it is crucial to recognize the conflicting interests between economic exploitation of forests and environmental conservation efforts.

To achieve a balance between economic development and environmental preservation, it is necessary to comprehend the impact of human activities and climate change on vegetation cover. Future research should focus on comprehensive analysis and comparison of various vegetation indices beyond NDVI, exploring functional characteristics of vegetation such as primary productivity and carbon sequestration, integrating NDVI calculations with other remote sensing techniques such as unmanned aerial vehicle (UAV) imagery for precise assessments of canopy structure and three-dimensional characteristics, and investigating localized redistribution of key meteorological parameters. An auspicious avenue of research lies in the utilization of detrended fluctuation analysis techniques for the assessment of NDVI dynamics.

The utilization of geospatial analysis and remote sensing techniques enables the acquisition of extensive spatial information regarding vegetation dynamics and its correlation with climate change. This knowledge facilitates improved environmental planning, decision making, and the implementation of sustainable practices for conservation and development.

Studying vegetation dynamics provides valuable insights into environmental changes and plays a pivotal role in preserving natural ecosystems, managing resources, and striving towards sustainable development objectives.

Author Contributions: Conceptualization, V.T. and R.G.; methodology, V.T. and R.G.; validation, V.T. and R.G.; formal analysis, V.T., R.G., T.G. and M.S.; investigation, V.T. and T.G.; writing—original draft preparation, V.T., T.G., M.S. and R.G.; writing—review and editing, V.T., T.G., M.S. and R.G.; visualization, T.G. and V.T.; supervision, R.G.; project administration, R.G. All authors have read and agreed to the published version of the manuscript.

Funding: This research was funded by Russian Science Foundation, grant number 22-27-00579.

Data Availability Statement: Not applicable.

Acknowledgments: The authors are grateful to Kelip A. and Kalinowski P. for technical support of the study.

Conflicts of Interest: The authors declare no conflict of interest. The funders had no role in the design of the study; in the collection, analyses, or interpretation of data; in the writing of the manuscript; or in the decision to publish the results.

References

1. Miles, J. *Vegetation Dynamics*; Springer Science & Business Media: Dordrecht, The Netherlands, 1979; 80p. [CrossRef]
2. Zhou, Z.; Ding, Y.; Shi, H.; Cai, H.; Fu, Q.; Liu, S.; Li, T. Analysis and prediction of vegetation dynamic changes in China: Past, present and future. *Ecol. Indic.* **2020**, *117*, 106642. [CrossRef]
3. Piao, S.; Fang, J.; Zhou, L.; Zhu, B.; Tan, K.; Tao, S. Changes in vegetation net primary productivity from 1982 to 1999 in China. *Glob. Biogeochem. Cycles* **2005**, *19*, 2. [CrossRef]
4. Mayeux, H.S.; Johnson, H.B.; Polley, H.W. Global change and vegetation dynamics. In *Noxious Range Weeds*, 1st ed.; CRC Press: Boca Raton, FL, USA, 1992; pp. 62–74. [CrossRef]
5. Brancalion, P.H.; Garcia, L.C.; Loyola, R.; Rodrigues, R.R.; Pillar, V.D.; Lewinsohn, T.M. A critical analysis of the Native Vegetation Protection Law of Brazil (2012): Updates and ongoing initiatives. *Nat. Conserv.* **2016**, *14*, 1–15. [CrossRef]
6. Opoku, A. Biodiversity and the built environment: Implications for the Sustainable Development Goals (SDGs). *Resour. Conserv. Recycl.* **2019**, *141*, 1–7. [CrossRef]
7. Li, Z.; Chen, Y.; Li, W.; Deng, H.; Fang, G. Potential impacts of climate change on vegetation dynamics in Central Asia. *J. Geophys. Res. Atmos.* **2015**, *120*, 12345–12356. [CrossRef]
8. Jiang, L.; Bao, A.; Guo, H.; Ndayisaba, F. Vegetation dynamics and responses to climate change and human activities in Central Asia. *Sci. Total Environ.* **2017**, *599*, 967–980. [CrossRef]
9. Verrall, B.; Pickering, C.M. Alpine vegetation in the context of climate change: A global review of past research and future directions. *Sci. Total Environ.* **2020**, *748*, 141344. [CrossRef]
10. Kalisa, W.; Igbawua, T.; Henchiri, M.; Ali, S.; Zhang, S.; Bai, Y.; Zhang, J. Assessment of climate impact on vegetation dynamics over East Africa from 1982 to 2015. *Sci. Rep.* **2019**, *9*, 16865. [CrossRef]
11. Shen, P.; Zhang, L.M.; Fan, R.L.; Zhu, H.; Zhang, S. Declining geohazard activity with vegetation recovery during first ten years after the 2008 Wenchuan earthquake. *Geomorphology* **2020**, *352*, 106989. [CrossRef]
12. Malawani, M.N.; Lavigne, F.; Gomez, C.; Mutaqin, B.W.; Hadmoko, D.S. Review of Local and Global Impacts of Volcanic Eruptions and Disaster Management Practices: The Indonesian Example. *Geosciences* **2021**, *11*, 109. [CrossRef]
13. Thonicke, K.; Venevsky, S.; Sitch, S.; Cramer, W. The role of fire disturbance for global vegetation dynamics: Coupling fire into a Dynamic Global Vegetation Model. *Glob. Ecol. Biogeogr.* **2001**, *10*, 661–677. [CrossRef]
14. Touré, A.A.; Tidjani, A.D.; Rajot, J.L.; Marticorena, B.; Bergametti, G.; Bouet, C.; Garba, Z. Dynamics of wind erosion and impact of vegetation cover and land use in the Sahel: A case study on sandy dunes in southeastern Niger. *Catena* **2019**, *177*, 272–285. [CrossRef]
15. Tang, C.; Liu, Y.; Li, Z.; Guo, L.; Xu, A.; Zhao, J. Effectiveness of vegetation cover pattern on regulating soil erosion and runoff generation in red soil environment, southern China. *Ecol. Indic.* **2021**, *129*, 107956. [CrossRef]
16. Chen, J.; Shao, Z.; Huang, X.; Cai, B.; Zheng, X. Assessing the impact of floods on vegetation worldwide from a spatiotemporal perspective. *J. Hydrol. Hydrol.* **2023**, *622*, 129715. [CrossRef]
17. Hu, T.; Smith, R.B. The Impact of Hurricane Maria on the Vegetation of Dominica and Puerto Rico Using Multispectral Remote Sensing. *Remote Sens.* **2018**, *10*, 827. [CrossRef]
18. Olivero, J.; Fa, J.E.; Real, R.; Márquez, A.L.; Farfán, M.A.; Vargas, J.M.; Gaveau, D.; Salim, M.A.; Park, D.; Suter, J.; et al. Recent loss of closed forests is associated with Ebola virus disease outbreaks. *Sci. Rep.* **2017**, *7*, 14291. [CrossRef]
19. Romero-Alvarez, D.; Escobar, L.E. Vegetation loss and the 2016 Oropouche fever outbreak in Peru. *Mem. Inst. Oswaldo Cruz* **2017**, *112*, 292–298. [CrossRef]
20. McAlpine, C.A.; Fensham, R.J.; Temple-Smith, D.E. Biodiversity conservation and vegetation clearing in Queensland: Principles and thresholds. *Rangel. J.* **2002**, *24*, 36–55. [CrossRef]
21. Nogueira, E.M.; Yanai, A.M.; Fonseca, F.O.R.; Fearnside, P.M. Carbon stock loss from deforestation through 2013 in Brazilian Amazonia. *Glob. Chang. Biol.* **2015**, *21*, 1271–1292. [CrossRef]
22. Mederski, P.; Jakubowski, M.; Karaszewski, Z. The Polish landscape changing due to forest policy and forest management. *iForest Biogeosciences For.* **2009**, *2*, 140–142. [CrossRef]
23. Yurova, A.Y.; Volodin, E.M. Coupled simulation of climate and vegetation dynamics. *Izv. Atmos. Ocean. Phys.* **2011**, *47*, 531–539. [CrossRef]
24. Hobbs, R.J. Remote Sensing of Spatial and Temporal Dynamics of Vegetation. In *Remote Sensing of Biosphere Functioning*; Hobbs, R.J., Mooney, H.A., Eds.; Springer: New York, NY, USA, 1990; pp. 203–219. [CrossRef]
25. Sun, G.Q.; Li, L.; Li, J.; Liu, C.; Wu, Y.P.; Gao, S.; Wang, Z.; Feng, G.L. Impacts of climate change on vegetation pattern: Mathematical modelling and data analysis. *Phys. Life Rev.* **2022**, *43*, 239–270. [CrossRef] [PubMed]
26. Potapov, P.; Hansen, M.C.; Kommareddy, I.; Kommareddy, A.; Turubanova, S.; Pickens, A.; Adusei, B.; Tyukavina, A.; Ying, Q. Landsat Analysis Ready Data for Global Land Cover and Land Cover Change Mapping. *Remote Sens.* **2020**, *12*, 426. [CrossRef]
27. Liu, B.; Pan, L.; Qi, Y.; Guan, X.; Li, J. Land Use and Land Cover Change in the Yellow River Basin from 1980 to 2015 and Its Impact on the Ecosystem Services. *Land* **2021**, *10*, 1080. [CrossRef]

28. Abdullah, A.Y.M.; Masrur, A.; Adnan, M.S.G.; Baky, M.A.A.; Hassan, Q.K.; Dewan, A. Spatio-Temporal Patterns of Land Use/Land Cover Change in the Heterogeneous Coastal Region of Bangladesh between 1990 and 2017. *Remote Sens.* **2019**, *11*, 790. [CrossRef]
29. Kang, Y.; Guo, E.; Wang, Y.; Bao, Y.; Bao, Y.; Mandula, N. Monitoring Vegetation Change and Its Potential Drivers in Inner Mongolia from 2000 to 2019. *Remote Sens.* **2021**, *13*, 3357. [CrossRef]
30. Gandhi, G.M.; Parthiban, B.S.; Thummalu, N.; Christy, A. NDVI: Vegetation change detection using remote sensing and gis—A case study of Vellore District. *Procedia Comput. Sci.* **2015**, *57*, 1199–1210. [CrossRef]
31. Waseem, S.; Khayyam, U. Loss of vegetative cover and increased land surface temperature: A case study of Islamabad, Pakistan. *J. Clean. Prod.* **2019**, *234*, 972–983. [CrossRef]
32. Jełowicki, Ł.; Sosnowicz, K.; Ostrowski, W.; Osińska-Skotak, K.; Bakuła, K. Evaluation of Rapeseed Winter Crop Damage Using UAV-Based Multispectral Imagery. *Remote Sens.* **2020**, *12*, 2618. [CrossRef]
33. Keshta, A.E.; Riter, J.C.A.; Shaltout, K.H.; Baldwin, A.H.; Kearney, M.; Sharaf El-Din, A.; Eid, E.M. Loss of Coastal Wetlands in Lake Burullus, Egypt: A GIS and Remote-Sensing Study. *Sustainability* **2022**, *14*, 4980. [CrossRef]
34. Li, C.; Jia, X.; Zhu, R.; Mei, X.; Wang, D.; Zhang, X. Seasonal Spatiotemporal Changes in the NDVI and Its Driving Forces in Wuliangsu Lake Basin, Northern China from 1990 to 2020. *Remote Sens.* **2023**, *15*, 2965. [CrossRef]
35. Long, Q.; Wang, F.; Ge, W.; Jiao, F.; Han, J.; Chen, H.; Roig, F.A.; Abraham, E.M.; Xie, M.; Cai, L. Temporal and Spatial Change in Vegetation and Its Interaction with Climate Change in Argentina from 1982 to 2015. *Remote Sens.* **2023**, *15*, 1926. [CrossRef]
36. Dhillon, M.S.; Kübert-Flock, C.; Dahms, T.; Rummler, T.; Arnault, J.; Steffan-Dewenter, I.; Ullmann, T. Evaluation of MODIS, Landsat 8 and Sentinel-2 Data for Accurate Crop Yield Predictions: A Case Study Using STARFM NDVI in Bavaria, Germany. *Remote Sens.* **2023**, *15*, 1830. [CrossRef]
37. Han, W.; Chen, D.; Li, H.; Chang, Z.; Chen, J.; Ye, L.; Liu, S.; Wang, Z. Spatiotemporal Variation of NDVI in Anhui Province from 2001 to 2019 and Its Response to Climatic Factors. *Forests* **2022**, *13*, 1643. [CrossRef]
38. Jiang, F.; Deng, M.; Long, Y.; Sun, H. Spatial pattern and dynamic change of vegetation greenness from 2001 to 2020 in Tibet, China. *Front. Plant Sci.* **2022**, *13*, 892625. [CrossRef] [PubMed]
39. Li, M.; Yan, Q.; Li, G.; Yi, M.; Li, J. Spatio-Temporal Changes of Vegetation Cover and Its Influencing Factors in Northeast China from 2000 to 2021. *Remote Sens.* **2022**, *14*, 5720. [CrossRef]
40. Tong, S.; Zhang, J.; Bao, Y.; Lai, Q.; Lian, X.; Li, N.; Bao, Y. Analyzing vegetation dynamic trend on the Mongolian Plateau based on the hurst exponent and influencing factors from 1982–2013. *J. Geogr. Sci.* **2018**, *28*, 595–610. [CrossRef]
41. Saikia, A. NDVI variability in North East India. *Scott. Geogr. J.* **2009**, *125*, 195–213. [CrossRef]
42. Shibani, N.; Pandey, A.; Satyam, V.K.; Bhari, J.S.; Karimi, B.A.; Gupta, S.K. Study on the variation of NDVI, SAVI and EVI indices in Punjab State, India. *IOP Conf. Ser. Earth Environ. Sci.* **2023**, *1110*, 012070. [CrossRef]
43. Johnson, D.M.; Rosales, A.; Mueller, R.; Reynolds, C.; Frantz, R.; Anyamba, A.; Pak, E.; Tucker, C. USA Crop Yield Estimation with MODIS NDVI: Are Remotely Sensed Models Better than Simple Trend Analyses? *Remote Sens.* **2021**, *13*, 4227. [CrossRef]
44. Ghorbanian, A.; Mohammadzadeh, A.; Jamali, S. Linear and Non-Linear Vegetation Trend Analysis throughout Iran Using Two Decades of MODIS NDVI Imagery. *Remote Sens.* **2022**, *14*, 3683. [CrossRef]
45. Pimentel, D.; McNair, M.; Buck, L.; Pimentel, M.; Kamil, J. The Value of Forests to World Food Security. *Hum. Ecol.* **1997**, *25*, 91–120. [CrossRef]
46. Pearce, D.W. The economic value of forest ecosystems. *Ecosyst. Health* **2001**, *7*, 284–296. [CrossRef]
47. Grantham, H.S.; Duncan, A.; Evans, T.D.; Jones, K.R.; Beyer, H.L.; Schuster, R.; Walston, J.; Ray, J.C.; Robinson, J.G.; Callow, M.; et al. Anthropogenic modification of forests means only 40% of remaining forests have high ecosystem integrity. *Nat. Commun.* **2020**, *11*, 5978. [CrossRef]
48. Fearnside, P.M.; Guimarães, W.M. Carbon uptake by secondary forests in Brazilian Amazonia. *For. Ecol. Manag.* **1996**, *80*, 35–46. [CrossRef]
49. Landuyt, D.; De Lombaerde, E.; Perring, M.P.; Hertzog, L.R.; Ampoorter, E.; Maes, S.L.; De Frenne, P.; Ma, S.; Proesmans, W.; Blondeel, H.; et al. The functional role of temperate forest understorey vegetation in a changing world. *Glob. Chang. Biol.* **2019**, *25*, 3625–3641. [CrossRef]
50. Wieczynski, D.J.; Boyle, B.; Buzzard, V.; Duran, S.M.; Henderson, A.N.; Hulshof, C.M.; Kerkhoff, A.J.; McCarthy, M.C.; Michaletz, S.T.; Swenson, N.G.; et al. Climate shapes and shifts functional biodiversity in forests worldwide. *Proc. Natl. Acad. Sci. USA* **2019**, *116*, 587–592. [CrossRef]
51. Riccioli, F.; Marone, E.; Boncinelli, F.; Tattoni, C.; Rocchini, D.; Fratini, R. The recreational value of forests under different management systems. *New For.* **2019**, *50*, 345–360. [CrossRef]
52. Negrón-Juárez, R.I.; Holm, J.A.; Marra, D.M.; Rifai, S.W.; Riley, W.J.; Chambers, J.Q.; Koven, C.D.; Knox, R.G.; E McGroddy, M.; Di Vittorio, A.V.; et al. Vulnerability of Amazon forests to storm-driven tree mortality. *Environ. Res. Lett.* **2018**, *13*, 054021. [CrossRef]
53. Hutyra, L.R.; Munger, J.W.; Nobre, C.A.; Saleska, S.R.; Vieira, S.A.; Wofsy, S.C. Climatic variability and vegetation vulnerability in Amazonia. *Geophys. Res. Lett.* **2005**, *32*, L24712. [CrossRef]
54. Laurance, W.F.; Campbell, M.J.; Alamgir, M.; Mahmoud, M.I. Road expansion and the fate of Africa's tropical forests. *Front. Ecol. Evol.* **2017**, *5*, 75. [CrossRef]

55. Potapov, P.; Hansen, M.C.; Laestadius, L.; Turubanova, S.; Yaroshenko, A.; Thies, C.; Smith, W.; Zhuravleva, I.; Komarova, A.; Minnemeyer, S.; et al. The last frontiers of wilderness: Tracking loss of intact forest landscapes from 2000 to 2013. *Sci. Adv.* **2017**, *3*, e1600821. [CrossRef] [PubMed]
56. Van Khuc, Q.; Tran, B.Q.; Meyfroidt, P.; Paschke, M.W. Drivers of deforestation and forest degradation in Vietnam: An exploratory analysis at the national level. *For. Policy Econ.* **2018**, *90*, 128–141. [CrossRef]
57. Fan, L.; Wigneron, J.P.; Ciais, P.; Chave, J.; Brandt, M.; Sitch, S.; Yue, C.; Bastos, A.; Li, X.; Qin, Y.; et al. Siberian carbon sink reduced by forest disturbances. *Nat. Geosci.* **2023**, *16*, 56–62. [CrossRef]
58. Gorbunov, R.; Tabunshchik, V.; Gorbunova, T.; Safonova, M. Water Balance Components of Sub-Mediterranean Downy Oak Landscapes of Southeastern Crimea. *Forests* **2022**, *13*, 1370. [CrossRef]
59. Bokov, V.A. (Ed.) *Landscape and Geophysical Conditions for the Growth of Forests in the Southeastern Part of the Mountainous Crimea*; Tavria-Plus: Simferopol, Crimea, 2001; 133p.
60. Rudenko, L.G. (Ed.) *National Atlas of Ukraine*; Cartography: Kiev, Ukraine, 2007; 435p.
61. Banerjee, A.; Chen, R.E.; Meadows, M.; Singh, R.B.; Mal, S.; Sengupta, D. An Analysis of Long-Term Rainfall Trends and Variability in the Uttarakhand Himalaya Using Google Earth Engine. *Remote Sens.* **2020**, *12*, 709. [CrossRef]
62. Funk, C.; Peterson, P.; Landsfeld, M.; Pedreros, D.; Verdin, J.; Shukla, S.; Husak, G.; Rowland, J.; Harrison, L.; Hoell, A.; et al. The climate hazards infrared precipitation with stations—A new environmental record for monitoring extremes. *Sci. Data* **2015**, *2*, 150066. [CrossRef]
63. Marchi, M.; Castellanos-Acuna, D.; Hamann, A.; Wang, T.; Ray, D.; Menzel, A. ClimateEU, scale-free climate normals, historical time series, and future projections for Europe. *Sci. Data* **2020**, *7*, 428. [CrossRef]
64. Muñoz Sabater, J. ERA5-Land Monthly Averaged Data from 1981 to Present. Copernicus Climate Change Service (C3S) Climate Data Store (CDS). Available online: https://cds.climate.copernicus.eu/cdsapp#!/dataset/10.24381/cds.68d2bb30?tab=overview (accessed on 7 July 2023).
65. Ndayisaba, F.; Guo, H.; Bao, A.; Guo, H.; Karamage, F.; Kayiranga, A. Understanding the Spatial Temporal Vegetation Dynamics in Rwanda. *Remote. Sens.* **2016**, *8*, 129. [CrossRef]
66. Gu, Z.; Duan, X.; Shi, Y.; Li, Y.; Pan, X. Spatiotemporal variation in vegetation coverage and its response to climatic factors in the Red River Basin, China. *Ecol. Indic.* **2018**, *93*, 54–64. [CrossRef]
67. Jiang, F.; Kutia, M.; Ma, K.; Chen, S.; Long, J.; Sun, H. Estimating the aboveground biomass of coniferous forest in Northeast China using spectral variables, land surface temperature and soil moisture. *Sci. Total Env.* **2021**, *785*, 147335. [CrossRef] [PubMed]
68. Sun, J.; Hou, G.; Liu, M.; Fu, G.; Zhan, T.Y.; Zhou, H.; Tsunekawa, A.; Haregeweyn, N. Effects of climatic and grazing changes on desertification of alpine grasslands, northern Tibet. *Ecol. Indic.* **2019**, *107*, 105647. [CrossRef]
69. Wang, J.; Zhao, J.; Zhou, P.; Li, K.; Cao, Z.; Zhang, H.; Han, Y.; Luo, Y.; Yuan, X. Study on the Spatial and Temporal Evolution of NDVI and Its Driving Mechanism Based on Geodetector and Hurst Indexes: A Case Study of the Tibet Autonomous Region. *Sustainability* **2023**, *15*, 5981. [CrossRef]
70. Fan, D.; Ni, L.; Jiang, X.; Fang, S.; Wu, H.; Zhang, X. Spatiotemporal analysis of vegetation changes along the belt and road initiative region from 1982 to 2015. *IEEE Access* **2020**, *8*, 122579–122588. [CrossRef]
71. Tabunshchik, V.A. The distribution of the values of the NDVI on the territory of the Razdolnensky district of the Republic of Crimea in January–June 2018. *Geopolit. Ecogeodynamics Reg.* **2019**, *5*, 225–242.
72. Lupyan, E.A.; Bartalev, S.A.; Krasheninnikova Yu, S.; Plotnikov, D.E.; Tolpin, V.A.; Uvarov, I.A. Analysis of winter crops development in the southern regions of the European part of Russia in spring of 2018 with use of remote monitoring. *Sovrem. Probl. Distantsionnogo Zondirovaniya Zemli Iz Kosmosa* **2019**, *15*, 272–276. [CrossRef]
73. Shadchinov, S.M. Dependence of the Normalized Difference Vegetation Index Level on the Spatial Structure of the Landscape and Its Time Variability Using the Example of the West End of the Crimean Peninsula (Tarkhankut Peninsula). *Izv. Atmos. Ocean. Phys.* **2021**, *57*, 1586–1595. [CrossRef]
74. Gorbunov, R. Productivity dynamics of oak forests of the Crimean Peninsula. *E3S Web Conf.* **2020**, *169*, 03007. [CrossRef]
75. Shinkarenko, S.; Solodovnikov, D.; Bartalev, S.; Vasilchenko, A.; Vypritskii, A. Dynamics of the reservoir's areas of the Crimean Peninsula. *Mod. Probl. Remote Sens. Earth Space* **2021**, *18*, 226–241. [CrossRef]
76. Hutchinson, G.E. Concluding Remarks. *Cold Spring Harb. Symp. Quant. Biol.* **1957**, *22*, 415–427. [CrossRef]
77. Yang, J.; Wan, Z.; Borjigin, S.; Zhang, D.; Yan, Y.; Chen, Y.; Gu, R.; Gao, Q. Changing Trends of NDVI and Their Responses to Climatic Variation in Different Types of Grassland in Inner Mongolia from 1982 to 2011. *Sustainability* **2019**, *11*, 3256. [CrossRef]
78. Li, P.; Wang, J.; Liu, M.; Xue, Z.; Bagherzadeh, A.; Liu, M. Spatio-temporal variation characteristics of NDVI and its response to climate on the Loess Plateau from 1985 to 2015. *Catena* **2021**, *203*, 105331. [CrossRef]
79. Jordan, C.F. Derivation of Leaf-Area Index from Quality of Light on the Forest Floor. *Ecology* **1969**, *50*, 663–666. [CrossRef]
80. Crippen, R.E. Calculating the vegetation index faster. *Remote. Sens. Environ.* **1990**, *34*, 71–73. [CrossRef]
81. McDaniel, K.C.; Haas, R.H. Assessing mesquite-grass vegetation condition from Landsat. *Photogramm. Eng. Remote Sens.* **1982**, *48*, 441–450.
82. Richardson, A.J.; Everitt, J.H. Using spectral vegetation indices to estimate rangeland productivity. *Geocarto Int.* **1992**, *7*, 63–69. [CrossRef]
83. Huete, A.; Jackson, R.; Post, D. Spectral response of a plant canopy with different soil backgrounds. *Remote. Sens. Environ.* **1985**, *17*, 37–53. [CrossRef]

84. Elvidge, C.D.; Lyon, R.J. Influence of rock-soil spectral variation on the assessment of green biomass. *Remote. Sens. Environ.* **1985**, *17*, 265–279. [CrossRef]
85. Baret, F.; Guyot, G.; Major, D. TSAVI: A Vegetation Index Which Minimizes Soil Brightness Effects on LAI And APAR Estimation. In Proceedings of the 12th Canadian Symposium on Remote Sensing Geoscience and Remote Sensing Symposium, Vancouver, BC, Canada, 10–14 July 1989; Volume 3, pp. 1355–1358. [CrossRef]
86. Qi, J.; Chehbouni, A.; Huete, A.; Kerr, Y.; Sorooshian, S. A modified soil adjusted vegetation index. *Remote. Sens. Environ.* **1994**, *48*, 119–126. [CrossRef]
87. Huete, A.; Didan, K.; Miura, T.; Rodriguez, E.; Gao, X.; Ferreira, L. Overview of the radiometric and biophysical performance of the MODIS vegetation indices. *Remote. Sens. Environ.* **2002**, *83*, 195–213. [CrossRef]
88. Kauth, R.J.; Thomas, G.S. The tasseled Cap—A Graphic Description of the Spectral-Temporal Development of Agricultural Crops as Seen by LANDSAT. In Proceedings of the Symposium on Machine Processing of Remotely Sensed Data, West Lafayette, IN, USA, 29 June–1 July 1976; pp. 4B-41–4B-51.
89. Matsushita, B.; Yang, W.; Chen, J.; Onda, Y.; Qiu, G. Sensitivity of the Enhanced Vegetation Index (EVI) and Normalized Difference Vegetation Index (NDVI) to Topographic Effects: A Case Study in High-density Cypress Forest. *Sensors* **2007**, *7*, 2636–2651. [CrossRef] [PubMed]
90. Martín-Ortega, P.; García-Montero, L.G.; Sibelet, N. Temporal Patterns in Illumination Conditions and Its Effect on Vegetation Indices Using Landsat on Google Earth Engine. *Remote Sens.* **2020**, *12*, 211. [CrossRef]
91. Gorbunov, R. *Functioning and Dynamics of Regional Geoecosystems in the Conditions of Climate Change (on the Example of the Crimean Peninsula)*; KMK Publisher: Moscow, Russia, 2023.
92. Skorokhod, E.Y.; Churilova, T.Y.; Efimova, T.V.; Moiseeva, N.A.; Suslin, V.V. Bio-Optical Characteristics of the Black Sea Coastal Waters near Sevastopol: Assessment of the MODIS and VIIRS Products Accuracy. *Phys. Oceanogr.* **2021**, *28*, 216. [CrossRef]
93. Wu, Z.; Lei, S.; Bian, Z.; Huang, J.; Zhang, Y. Study of the desertification index based on the albedo-MSAVI feature space for semi-arid steppe region. *Environ. Earth Sci.* **2019**, *78*, 232. [CrossRef]
94. Höll, M.; Kantz, H.; Zhou, Y. Detrended fluctuation analysis and the difference between external drifts and intrinsic diffusionlike nonstationarity. *Phys. Rev. E* **2016**, *94*, 42201. [CrossRef] [PubMed]
95. Kantelhardt, J.W. Fractal and Multifractal Time Series. *arXiv* **2008**, arXiv:0804.0747. [CrossRef]
96. Crevecoeur, F.; Bollens, B.; Detrembleur, C.; Lejeune, T. Towards a "gold-standard" approach to address the presence of long-range auto-correlation in physiological time series. *J. Neurosci. Methods* **2010**, *192*, 163–172. [CrossRef]

Disclaimer/Publisher's Note: The statements, opinions and data contained in all publications are solely those of the individual author(s) and contributor(s) and not of MDPI and/or the editor(s). MDPI and/or the editor(s) disclaim responsibility for any injury to people or property resulting from any ideas, methods, instructions or products referred to in the content.

Article

Quantitatively Computing the Influence of Vegetation Changes on Surface Discharge in the Middle-Upper Reaches of the Huaihe River, China

Yuxin Wang [1], Zhipei Liu [1], Baowei Qian [1], Zongyu He [2] and Guangxing Ji [1,*]

[1] College of Resources and Environmental Sciences, Henan Agricultural University, Zhengzhou 450046, China
[2] Zhengzhou Architecture Design Institute, Zhengzhou 450052, China
* Correspondence: guangxingji@henau.edu.cn

Abstract: Changes in meteorology, hydrology, and vegetation will have significant impacts on the ecological environment of a basin, and the middle-upper reach of Huaihe River (MUHR) is one of the key regions for vegetation restoration in China. However, less studies have quantitatively accounted for the contribution of vegetation changes to land surface discharge in the MUHR. To quantitatively evaluate the influence of vegetation changes on land surface discharge in the MUHR, the Bernaola–Galavan (B–G) segmentation algorithm was utilized to recognize the mutation year of the Normalized Difference Vegetation Index (NDVI) time sequence data. Next, the functional relationship between the underlying surface parameter and the NDVI was quantitatively analyzed, and an adjusted Budyko formula was constructed. Finally, the effects of vegetation changes, climate factors, and mankind activities on the surface discharge in the MUHR were computed using the adjusted Budyko formula and elastic coefficient method. The results showed the following: (1) the surface runoff and precipitation from 1982 to 2015 in the MUHR presented a falling trend, yet the NDVI and potential evaporation presented an upward trend; (2) 2004 was the mutation year of the NDVI time series data, and the underlying surface parameter showed a significant linear regression relationship with the NDVI ($p < 0.05$); (3) the vegetation variation played a major role in the runoff variation during the changing period (2005–2015) in the MUHR. Precipitation, potential evaporation, and human activities accounted for −0.32%, −15.11%, and 18.24% of the surface runoff variation, respectively.

Keywords: surface discharge variation; vegetation variation; attribution analysis; Budyko hypothesis

1. Introduction

The formation and changes of surface discharge are mainly influenced by the underlying surface, climate factors, and mankind activities [1,2]. As a momentous component of the underlying surface, vegetation plays a role in water storage and water conservation and has significant effects on surface runoff in the basin [3–6]. Vegetation affects the changes in surface runoff through hydrological processes such as transpiration, interception, and water storage. In the past few decades, China's vegetation coverage has shown a fluctuating upward trend driven by the joint effects of multiple elements such as climate change and human activities [7,8]. It has been found by analyzing the remote sensing vegetation index that China has become the largest contributor to global vegetation greening, with a net change in leaf area index values of 1.35 million km^2 from 2000 to 2017, with a change rate of 17.8% [9]. This result has a great practical meaning for agricultural irrigation and basin water resource allocation and administration processes in quantitatively estimating the impacts of vegetation change on surface discharge changes.

The middle-upper reaches of the Huaihe River (MUHR) are situated in the eastern part of China, and are key regions for vegetation restoration in China. In recent years, the vegetation coverage in the Huaihe River has shown an increasing trend [10–12], and the

Normalized Difference Vegetation Index (NDVI) in the Huaihe River has generally shown a steady upward trend from 1999 to 2018 [13]. Under the comprehensive consequences of climate change, vegetation variation, and human activities, the discharge in the MUHR has decreased in recent decades [14,15]. Therefore, in the immediately following years, many researchers quantitatively analyzed the influence of different factors on the runoff changes in the MUHR [16–19]. Zhang et al. [13] found that the increase in NDVI of 10% resulted in an average decrease of 8.3% for the runoff in the Huang-Huai-Hai Basin. Through an SWIM (soil and water integrated model), distributed hydrological model, and statistical method, Gao et al. [20] quantitatively analyzed the coefficient of sensitivity of discharge to climate elements and computed the contribution proportion of climate elements to runoff, and concluded that the influence of climate change on the upstream discharge is mostly due to the impact of precipitation. Liu et al. [21] computed the elasticity coefficients of the runoff depth with the precipitation, potential evaporation, and underlying surface parameters based on the Budyko water and heat balance theory. They quantitatively analyzed the contributions of climate variation and human activities to the discharge changes in the MUHR and concluded that the underlying surface parameter variation is the key element leading to discharge reductions. Based on the Budyko hypothesis and differentiation formula, Ye et al. [22] explored the impacts of climate variation and human activities on the discharge characteristics at multiple timescales in the MUHR. The results proved that human activities were the main factors causing the decreased discharge. Sun et al. [23] utilized the simultaneous solution technique for multiple control factors based on a sensitivity test and the Budyko equation to separate the contributions of climate and human activities to the annual discharge changes in the Huaihe River Basin, and found that the physical mechanisms controlling the ET and discharge changes in the Huaihe River Basin have distinct spatial differences and interdecadal variations. Shi et al. [24] explored the spatial and temporal evolution of the vegetation cover in a tributary of the Huai River (Yihe) and its relationship with the surface runoff, and found that the relationship between the NDVI and surface discharge in the basin was dominated by a non-significant positive correlation. However, few studies have quantitatively analyzed the contribution rates of vegetation changes to the surface runoff changes in the MUHR.

Therefore, the aim of this research is to quantitatively calculate the impact of vegetation changes on discharge changes in the MUHR through the following four steps: (1) the change trends of the meteorological and hydrological data and NDVI data are analyzed. (2) the B–G segmentation algorithm is utilized to recognize the mutation year of the NDVI time sequence data; (3) the functional equation between the underlying surface parameter (w) and NDVI is quantified, and the modified Budyko formula is constructed; (4) the contribution ratio of the vegetation variation on the discharge variation in the MUHR is computed using the modified Budyko formula. This study contributes to further understanding the influence of vegetation variations on hydrology processes, and has a guiding significance for economic development and ecological environment governance in the MUHR.

2. Study Area and Data

The middle-upper reaches of the Huaihe River (MUHR, above Wujiadu Station) are located in the natural climate boundary zone between north and south China (Figure 1), with a longitude of 111°55′–118°4′ east and a latitude of 30°55′–34°55′ north. The middle reaches contain mountains, while the upstream reaches contain hills, with most of the vegetation being deciduous broad-leaved trees. The drainage area covers 121,300 km^2, occupying 40.1% of the Huaihe River Basin. The annual discharge rate from 1958 to 2016 was 266×10^8 m^3, comprising 58.7% of the discharge in Huaihe River Basin, making it is a momentous runoff-producing region for the basin. Influenced by anthropic factors and climatic variation, it is very easy for flood and drought disasters to form, affecting many cities along the line.

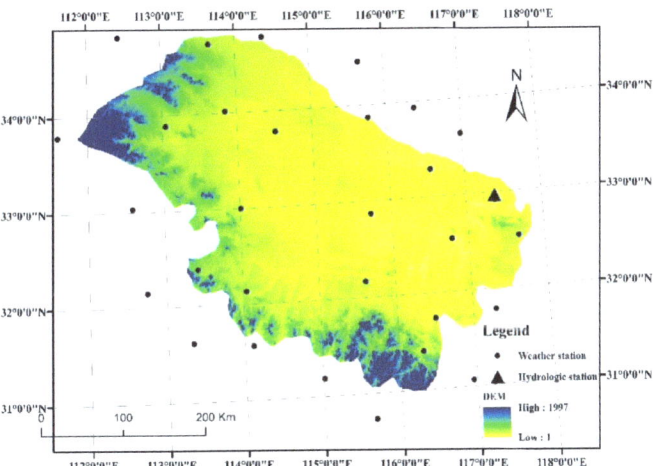

Figure 1. Location of the study area.

The NDVI information for the period of 1982–2015 were obtained from the NASA NDVI dataset (https://ecocast.arc.nasa.gov/data/pub/gimms/3g.v1/) (accessed on 1 January 2021). The runoff information for the Wujiadu Hydrological Station (1982–2015) were obtained from the Huaihe River Water Conservancy Commission (http://www.hrc.gov.cn/) (accessed on 1 January 2022). The meteorological station data for the areas in and around the MUHR (1982–2015) were acquired from the National Meteorological Service (http://www.cma.gov.cn/) (accessed on 1 January 2020).

3. Methods
3.1. Bernaola–Galavan (B–G) Segmentation Algorithm

Given a segmentation point i, the time sequence dataset (X) with a number N is divided into two subsequences, X_1 and X_2, respectively. The mean, standard deviation, and length values of X_1 and X_2 are U_1 and U_2, S_1 and S_2, and N_1 and N_2, respectively. Therefore, the formula for calculating the combined deviation S_D [25] is:

$$S_D = \left[(S_1^2 + S_2^2)/(N_1 + N_2 - 2)\right]^{1/2} (1/N_1 + 1/N_2)^{1/2} \tag{1}$$

A *t*-test was applied to measure the variability of the mean values of X_1 and X_2, and the calculation formula for statistic *T(i)* is:

$$T(i) = (U_1 - U_2)/S_D \tag{2}$$

When the variability of the mean values of X_1 and X_2 reaches the maximum, the *t*-test statistic also reaches the maximum (T_{max}), and the formula for calculating the significance probability $P(T_{max})$ corresponding to T_{max} [26] is:

$$P(T_{max}) = Prob(T \leq T_{max}) \tag{3}$$

$$P(T_{max}) \approx \left[\left(1 - I_{v/(V+T_{max}^2)}(\delta v, \delta)\right)\right]^{\gamma} \tag{4}$$

In the formula, $\gamma = 4.19\ln N - 11.54$, $\delta = 0.40$, N is the sample of the time series x(t), and $v = N - 2$, $I_x(a,b)$ is the incomplete β function. If $P(T_{max}) \geq P_0$, the sequence will be segmented; If $P(T_{max}) < P_0$, it cannot be divided. The range of P_0 is generally [0.5,0.95], and P_0 was set to 0.84 in this study [27].

If $P(T_{max}) \geq P_0$, one needs to calculate $P(T_{max})$ the two new subsequences to detect all mutation points. In addition, in order to ensure the effectiveness of the statistics, if the length of the subsequence $\leq l_0$, the subsequence will not be segmented; l_0 was set to 25 in this study [27].

3.2. Budyko Hypothesis

The Budyko hypothesis was founded based on the water balance equation:

$$R = P - ET - \Delta S \tag{5}$$

In the formula, R, P, and ET represent the runoff depth, precipitation, and actual evaporation in the watershed; ΔS is the change in water storage. When analyzing long time scales, the ΔS is negligible.

On the watershed scale, the precipitation can be directly obtained through the spatial interpolation of the rainfall station observation data, and the actual evaporation rates were computed using the Choudhury–Yang equation [28,29].

$$ET = \frac{P \times ET0}{(P^w + ET_0^w)^{1/w}} \tag{6}$$

Here, w reflects the characteristic parameters of the underlying surface; its value is dependent on the soil type, terrain condition, and landcover type, and it is applied to characterize the influence of human factors. ET_0 is the potential evaporation (mm), which can be computed using the Penman–Monteith formula:

$$ET_0 = \frac{0.408\Delta(R_n - G) + \gamma \frac{900}{T+273} U_2(e_a - e_d)}{\Delta + \gamma(1 + 0.34U_2)} \tag{7}$$

Combined with Formulas (5) and (6), Formula (5) can be converted:

$$R = P - \frac{P \times ET_0}{(P^w + ET_0^w)^{1/w}} \tag{8}$$

Li et al. [30] analyzed the functional relation between the NDVI and Budyko parameter (w) values of 26 major river basins around the world and discovered that there was a good linear functional equation between them:

$$w = a*NDVI + b \tag{9}$$

Combined with Formulas (8) and (9), Formula (5) can be changed into Equation (10):

$$R = P - \frac{P \times ET_0}{\left(P^{a*NDVI+b} + ET_0^{a*NDVI+b}\right)^{1/(a*NDVI+b)}} \tag{10}$$

The elasticity coefficients of runoff to the P (ε_P), ET_0 (ε_{ET0}), w (ε_w), and NDVI (ε_{NDVI}) can be computed using Formulas (11)–(14) [31,32].

$$\varepsilon_P = \frac{\left(1+\left(\frac{ET_0}{P}\right)^w\right)^{1/w+1} - \left(\frac{ET_0}{P}\right)^{w+1}}{\left(1+\left(\frac{ET_0}{P}\right)^w\right)\left[\left(1+\left(\frac{ET_0}{P}\right)^w\right)^{1/w} - \left(\frac{ET_0}{P}\right)\right]} \tag{11}$$

$$\varepsilon_{ET0} = \frac{1}{\left(1+\left(\frac{ET_0}{P}\right)^w\right)\left[1 - \left(1+\left(\frac{ET_0}{P}\right)^{-w}\right)^{1/w}\right]} \tag{12}$$

$$\varepsilon_w = \frac{\ln\left(1+\left(\frac{ET_0}{P}\right)^w\right) + \left(\frac{ET_0}{P}\right)^w \ln\left(1+\left(\frac{ET_0}{P}\right)^{-w}\right)}{w\left(1+\left(\frac{ET_0}{P}\right)^w\right)\left[1-\left(1+\left(\frac{ET_0}{P}\right)^{-w}\right)^{1/w}\right]} \quad (13)$$

$$\varepsilon_{NDVI} = \varepsilon_w \frac{a*NDVI}{a*NDVI+b} \quad (14)$$

The time sequence data are assigned into two stages: the base period (T_1) and the changing period (T_2). Thus, the change values of the P (ΔP), ET_0 (ΔET_0), w (Δw), and NDVI ($\Delta NDVI$) from T_1 to T_2 can be computed as follows:

$$\Delta P = P2 - P1 \quad (15)$$

$$\Delta ET_0 = ET_{02} - ET_{01} \quad (16)$$

$$\Delta w = w_2 - w_1 \quad (17)$$

$$\Delta NDVI = NDVI_2 - NDVI_1 \quad (18)$$

In the formula, P_1 and P_2 represent the mean precipitation in the T_1 and T_2 periods; ET_{01} and ET_{02} represent the mean potential evaporation in the T_1 and T_2 periods; w_1 and w_2 represent the characteristic parameters of the underlying surface in the T_1 and T_2 periods; and $NDVI_1$ and $NDVI_2$ represent the vegetation coverage in the T_1 and T_2 periods, respectively.

$$\Delta R_P = \varepsilon_P \frac{R}{P} \times \Delta P \quad (19)$$

$$\Delta R_{ET_0} = \varepsilon_{ET0} \frac{R}{ET_0} \times \Delta ET_0 \quad (20)$$

$$\Delta R_w = \varepsilon_w \frac{R}{w} \times \Delta w \quad (21)$$

$$\Delta R_{NDVI} = \varepsilon_{NDVI} \frac{R}{NDVI} \times \Delta NDVI \quad (22)$$

In the formulas, ΔR_P, ΔR_{ET0}, ΔR_w, and ΔR_{NDVI} respectively represent the surface runoff variation values caused by P, ET_0, w, and NDVI variations from T_1 to T_2.

Except for climate and vegetation changes, other factors affecting runoff changes, such as water conservancy projects, domestic water use for urban residents, and agricultural irrigation water, are classified as human activities in this study. Therefore, the amount of runoff change caused by the human activities from T_1 period to T_2 period can be expressed as:

$$\Delta R_{hum} = \Delta R_w - \Delta R_{NDVI} \quad (23)$$

Therefore, the aggregate of the runoff variation induced by various factors is:

$$\Delta R = \Delta R_P + \Delta R_{ET0} + \Delta R_{NVDI} + \Delta R_{hum} \quad (24)$$

Therefore, the contribution proportion of the P (ηR_p), ET_0 (ηR_{ET0}), human activities (ηR_H), and vegetation changes (ηR_{NDVI}) to the runoff variation in the MUHR can be computed using the following formulas:

$$\eta_{R_P} = \Delta R_P / \Delta R \times 100\% \quad (25)$$

$$\eta_{R_{ET0}} = \Delta R_{ET0} / \Delta R \times 100\% \quad (26)$$

$$\eta_{R_{NDVI}} = \Delta R_{NDVI} / \Delta R \times 100\% \quad (27)$$

$$\eta_{R_H} = \Delta R_{hum} / \Delta R \times 100\% \quad (28)$$

4. Results and Analysis

4.1. Trend Analysis

Figure 2a displays the annual variation tendency of the mean annual runoff depth values in the MUHR from 1982 to 2015. It can be recognized from Figure 2a that the average annual runoff depth displays a non-significant downward trend ($p > 0.05$). During the research period, the range of yearly runoff depths in the MUHR was 50–520 mm, and the maximum was reached in 2003. Figure 2b displays the yearly variation tendencies for the average annual NDVI changes in the MUHR. As we can see from Figure 2b, the NDVI was increasing. The range of the mean annual NDVI values of the MUHR was 0.46 to 0.60. In general, the gradient of the NDVI was 0.0026/a ($p < 0.05$), indicating that the vegetation recovery in the MUHR was significant.

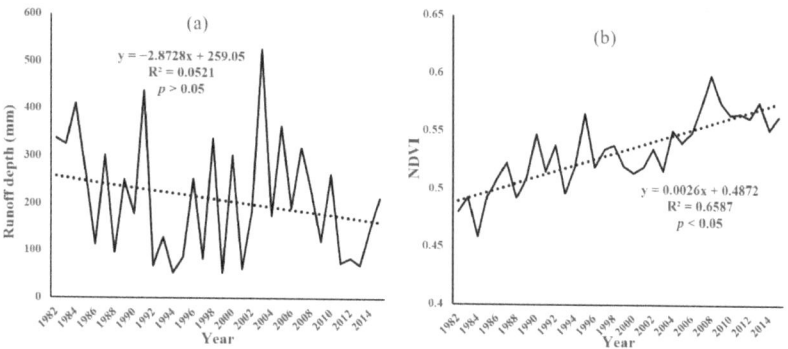

Figure 2. Annual variation tendencies of the runoff depth (**a**) and NDVI (**b**) values in in MUHR.

Figure 3a reveals the change tendencies for annual precipitation and potential evaporation from 1982 to 2015 in the MUHR. The annual rainfall shows a fluctuating and non-significant downward trend and its interannual change is dramatic, with a gradient of -1.173 mm/a ($p > 0.05$). During the study period, the annual average precipitation in the MUHR ranged from 600 to 1300 mm. The average annual potential evaporation in the MUHR were in the ranges of 950–1750, with a non-significant upward trend ($p > 0.05$) and a gradient of 0.0147 mm/a. Figure 3b reveals the change tendencies for actual evaporation from 1982 to 2015 in the MUHR. The actual evaporation show a non-significant upward trend ($p > 0.05$), with gradients of 1.6998 mm/a. During the study period, the average actual evaporation in the MUHR were in the ranges of 550–850 mm.

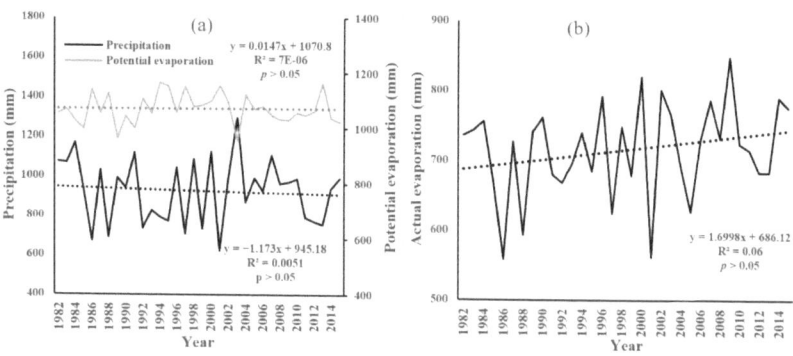

Figure 3. Change tendencies of the precipitation, potential evaporation, and actual evaporation in the MUHR.

4.2. Mutation Analysis of the NDVI

The B–G algorithm was used to distinguish the abrupt year of the NDVI time sequence data (Figure 4). It is worth noting that P_0 is a set threshold, and its value range is [0.5–0.95]. In this study, the value of P_0 is set to 0.84, and for the purpose of assuring the validity of the statistics, this research defines the value of subseries not less than 25, i.e., $l_0 \geq 25$. If the length of the subsequence is too short and the amount of data is too small, there is too much error in testing the mutation points. The result of the B–G algorithm revealed that the abrupt year of the NDVI time series data was around 2004. According to Formula (4), the probability at the maximum value of the t-test statistics $P(T_{max})$ was calculated. If $P(T_{max}) > P_0$, the mutation is considered to be significant. The calculation results show that the probability range of originality at the maximum value of the t-test statistics (2004) is 0.84128 > 0.84, proving the reliability of the result that the NDVI time series data mutated in 2004.

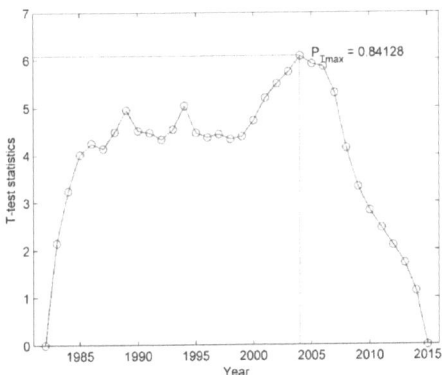

Figure 4. The B–G segmentation algorithm results for the NDVI data in the MUHR from 1982 to 2015.

4.3. Quantitative Analysis of Vegetation Variation Based on Streamflow Variation

To quantify the relationship between the NDVI and w, the w values in the MUHR from 1982 to 2015 were calculated using Equation (8). Next, the 10-year moving averages of w and NDVI were obtained in this study. Finally, the relationship between the NDVI and w in the MUHR was obtained (Figure 5). Figure 5 shows that the 10-year running average w has a linear function relationship with the 10-year running average NDVI.

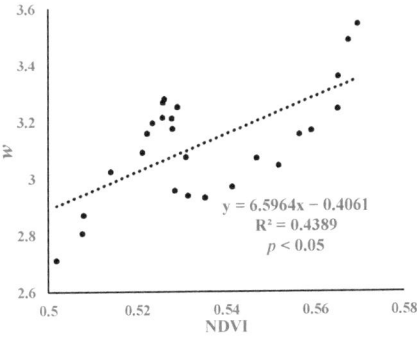

Figure 5. The relationship between the 10-year running average w and NDVI values.

In accordance with consequences of the B–G algorithm mutation analysis, the research phase was assigned to the base period (T_1: 1982–2004) and changing period (T_2: 2005–2015). In accordance with the mean potential evaporation, mean precipitation, and mean runoff depth values in the T_1 and T_2 periods, the characteristic parameters of the underlying surface (w) in the T_1 and T_2 periods were computed. The elasticity coefficients of the runoff to the P (ε_P), ET0 (ε_{ET0}), w (ε_w), and NDVI (ε_{NDVI}) can be computed using Formulas (11)–(14). The consequences are presented in Tables 1 and 2.

Table 1. The eigenvalues of the climate, hydrology, and NDVI variables in the MUHR.

Periods	ET_0/mm	P/mm	R/mm	w	NDVI
T_1	1076.66	924.59	218.49	2.74	0.52
T_2	1059.36	924.77	188.45	3.09	0.57
Δ	−17.30	0.18	−30.04	0.35	0.05

Table 2. Contribution rate analysis of the discharge variation in the MUHR.

ε_P	ε_{E0}	ε_w	ε_{NDVI}	ΔR_P	ΔR_{ET0}	ΔR_{NDVI}	ΔR_{hum}	ηR_P	ηR_{ET0}	ηR_{NDVI}	ηR_{hum}
2.43	−1.43	−1.45	−1.64	0.10	4.84	−31.09	−5.84	−0.32%	−15.11%	97.19%	18.24%

It can be seen from Table 1 that the ET_0 of the MUHR in the T_2 period has decreased to 17.30 mm, contrasting with the T_1 period. The precipitation of the T_2 period has slightly increased by 0.18 mm, contrasting with the T_1 period. The runoff depth in the T_2 period has reduced by 30.04 mm, contrasting with the T_1 period, but the NDVI value in the changing period presents an increasing trend, with an increase of 0.05.

It can be recognized from Table 2 that the elasticity coefficients of the runoff depths based on the ET_0, w, and NDVI are 2.43, −1.43, −1.45, and −1.64, respectively. The variation in the runoff depth caused by the precipitation, potential evaporation, human activities, and vegetation changes can be computed using Formulas (11)–(14), and equal 0.10 mm, 4.84 mm, −31.09 mm, and −5.84 mm, respectively. The variation in runoff depths caused by the significant growth of vegetation is the largest, accounting for 97.19%; that is, the vegetation variation is the major factor resulting in discharge changes in the MUHR. The precipitation, potential evaporation, and human activities account for −0.32%, −15.11%, and 18.24% of the surface runoff variation in the MUHR, respectively.

The findings of this paper are similar to those by scholars such as Zhang et al. [13] and Shi et al. [24], all of whom indicated that improved vegetation cover conditions in the watershed domain have a weakening effect on the runoff, but there are still some differences in the calculated contribution values, which may be due to several factors, including (1) the use of data from different time scales and (2) the use of different hydrological models and Budyko's assumption formula.

Many studies have confirmed that the vegetation changes caused by large-scale afforestation activities in the watershed can significantly affect the runoff [33–35]. Vegetation changes can affect runoff changes in many ways. (1) The higher the vegetation coverage rate, the stronger the ability to conserve water, and the surface runoff will be reduced. (2) The increase in vegetation leaf area can increase the evapotranspiration of plant leaves, and as water is discharged into the atmosphere through the respiration of leaf stomata, the soil water content will also decrease, affecting the surface runoff. (3) The improvement of the vegetation coverage degree efficiently improves the interception of rainfall, reducing the precipitation reaching the ground and affecting the change in runoff.

5. Conclusions

In this study, we first explored the change trends of meteorological and hydrological data and NDVI data and identified the mutation year of the NDVI time series data using the B–G segmentation algorithm. Next, the functional equation between the underlying surface characteristic parameter (w) and the NDVI was quantified, and the modified Budyko formula was structured. Finally, the impact degree of the vegetation change on the discharge change in the MUHR was computed using the modified Budyko formula.

The following conclusions were drawn. There is a significant linear functional relation between the NDVI and underlying surface parameters (w). The vegetation variation played a major role in the runoff variation during the changing period (2005–2015) in the MUHR, with a contribution rate of 97.19%. The precipitation, potential evaporation, and human activities accounted for −0.32%, −15.11%, and 18.24% of the surface runoff variation, respectively.

6. Discussions

The implementation of a series of soil and water conservation measures in the MUHR (especially the reforestation and grass restoration project in 1999) has significantly changed the rainfall–runoff relationship in the area. The underlying surface parameters (w) in the Budyko equation reflect the combined effect of the soil properties, topographic factors, and vegetation cover. The soil properties and topography are relatively stable parameters, while the vegetation factors become the main factors affecting w. The NDVI and w showed a strong synergistic trend, indicating that the vegetation restoration had a significant impact on w (Figure 5). The contribution analysis of the vegetation restoration to the runoff changes further verified that the increase in vegetation cover caused runoff attenuation in the MUHR.

There were several deficiencies in the attribution of different factors to discharge changes based on Budyko's theoretical assumptions. (1) Meteorological data for the individual dates were missing due to the limitation of the observation conditions. (2) This study assumed that each factor was relatively independent. This assumption ignores the interactions and connections between each factor [36–38], which will have an uncertain impact on the research results. (3) In addition to vegetation restoration, various soil and water conservation engineering measures such as the construction of terraces and sand dams also cause changes in w, and the effects of various water conservation measures on w cannot be quantitatively analyzed at present. (4) The change in water storage in the water balance equation based on Budyko's assumption is 0 on the multi-year average scale, which obviously ignores the interception for runoff by various soil and water conservation engineering measures and water conservancy projects.

In the future, we will consider building a distributed hydrological hydrothermal coupling model combined with higher-resolution remote sensing vegetation index data to analyze how vegetation affects the hydrological process in the MUHR [39–41]. This is an effective way to clarify the influence mechanism of vegetation variations on the water–heat relation. In addition, in this study, we did not distinguish the impact of reservoir construction projects on the discharge change, which will be further discussed in the subsequent study.

Author Contributions: Author Contributions: Conceptualization, G.J.; methodology, Y.W. and G.J.; software, Z.H.; validation, G.J.; formal analysis, Y.W., Z.L. and B.Q.; data curation, Y.W. and G.J.; writing—original draft preparation, Y.W., Z.L. and B.Q.; writing—review and editing, Y.W. and G.J.; visualization, Z.H.; project administration, G.J.; funding acquisition, G.J. All authors have read and agreed to the published version of the manuscript.

Funding: This research was funded by the National Key R&D Program of China (2021YFD1700900), the Think Tanks Key Research Projects of the Colleges and Universities of Henan Province (2022ZKYJ07), and the Special Fund for Top Talents of Henan Agricultural University (30501031).

Institutional Review Board Statement: Not applicable.

Informed Consent Statement: Not applicable.

Data Availability Statement: Not applicable.

Conflicts of Interest: The authors declare no conflict of interest.

References

1. Ji, G.; Lai, Z.; Xia, H.; Liu, H.; Wang, Z. Future Runoff Variation and Flood Disaster Prediction of the Yellow River Basin Based on CA-Markov and SWAT. *Land* **2021**, *10*, 421. [CrossRef]
2. Ji, G.; Wu, L.; Wang, L.; Yan, D.; Lai, Z. Attribution Analysis of Seasonal Runoff in the Source Region of the Yellow River Using Seasonal Budyko Hypothesis. *Land* **2021**, *10*, 542. [CrossRef]
3. Piao, S.L.; Wang, X.H.; Ciais, P.; Zhu, B.; Liu, J. Changes in satellite-derived vegetation growth trend in temperate and boreal Eurasia from 1982 to 2006. *Glob. Chang. Biol.* **2011**, *17*, 3228–3239. [CrossRef]
4. Jackson, R.B.; Jobbágy, E.G.; Avissar, R.; Roy, S.; Barrett, D.; Cook, C.; Farley, K.; Maitre, D.; Mccarl, B.; Cook, C. Trading water for carbon with biological carbon sequestration. *Science* **2005**, *310*, 1944–1947. [CrossRef]
5. Ji, G.; Song, H.; Wei, H.; Wu, L. Attribution Analysis of Climate and Anthropic Factors on Runoff and Vegetation Changes in the Source Area of the Yangtze River from 1982 to 2016. *Land* **2021**, *10*, 612. [CrossRef]
6. Yan, D.; Lai, Z.; Ji, G. Using Budyko-Type Equations for Separating the Impacts of Climate and Vegetation Change on Runoff in the Source Area of the Yellow River. *Water* **2020**, *12*, 3418. [CrossRef]
7. Wei, X.; Sun, G.; Liu, S.; Jiang, H.; Zhou, G.; Dai, L. The forest-streamflow relationship in China: A 40-year retrospect. *J. Am. Water Resour. Assoc.* **2008**, *44*, 1076–1085. [CrossRef]
8. Lü, Y.; Zhang, L.; Feng, X.; Zeng, Y.; Fu, B.; Yao, X.; Li, J.; Wu, B. Recent ecological transitions in China: Greening, browning, and influential factors. *Sci. Rep.* **2015**, *5*, 8732. [CrossRef] [PubMed]
9. Chen, C.; Park, T.J.; Wang, X.; Piao, S.; Xu, B.; Chaturvedi, R.; Fuchs, R.; Brovkin, V.; Ciais, P.; Fensholt, R.; et al. China and India lead in greening of the world through land-use management. *Nat. Sustain.* **2019**, *2*, 122–129. [CrossRef]
10. Liu, Z.; Wang, H.; Li, N.; Zhu, J.; Pan, Z.; Qin, F. Spatial and Temporal Characteristics and Driving Forces of Vegetation Changes in the Huaihe River Basin from 2003 to 2018. *Sustainability* **2020**, *12*, 2198. [CrossRef]
11. Liu, F.; Qin, T.; Girma, A.; Wang, H.; Weng, B.; Yu, Z.; Wang, Z. Dynamics of Land-Use and Vegetation Change Using NDVI and Transfer Matrix: A Case Study of the Huaihe River Basin. *Pol. J. Environ. Stud.* **2018**, *28*, 213–223. [CrossRef] [PubMed]
12. Gao, Z.; Liu, S.; Qin, X.; Xu, L. Spatial-temporal evolution of the Normalized Vegetation Index (NDVI) in the Huai River Basin from 1999 to 2018. *Pearl River* **2022**, *43*, 1–9. (In Chinese)
13. Zhang, J.; Zhang, C.; Bao, Z.; Li, M.; Wang, G.; Guan, X.; Liu, C. Effects of vegetation cover change on runoff in Huang-Huai-Hai River Basin. *Adv. Water Sci.* **2021**, *32*, 813–823. (In Chinese)
14. An, G.; Hao, Z. Variation of Precipitation and Streamflow in the Upper and Middle Huaihe River Basin, China, from 1959–2009. *J. Coast. Res.* **2017**, *80*, 69–79. [CrossRef]
15. Zhu, Y.; Wang, W.; Liu, Y.; Wang, H. Runoff changes and their potential links with climate variability and anthropogenic activities: A case study in the upper Huaihe River Basin, China. *Hydrol. Res.* **2015**, *46*, 1019–1036. [CrossRef]
16. Yang, C.; Chen, H.; Gu, Z.; Wang, W.; Ju, J.; Chen, L.; Zhu, H. Analysis on Influencing Factors of Runoff Evolution in Typical Watershed of Upper Huaihe River—A Case Study of Bailianya Watershed. *Sci. Soil Water Conserv.* **2020**, *18*, 110–116. (In Chinese)
17. Zhang, S.; Yang, D.; Jayawardena, A.W.; Xu, X.; Yang, H. Hydrological change driven by human activities and climate variation and its spatial variability in Huaihe Basin, China. *Hydrol. Sci. J.* **2016**, *61*, 1370–1382. [CrossRef]
18. Ma, F.; Ye, A.; Gong, W.; Mao, Y.; Miao, C.; Di, Z. An Estimate of Human and Natural Contributions to Flood Changes of the Huai River. *Glob. Planet. Change* **2014**, *119*, 39–50. [CrossRef]
19. Gao, C.; Ruan, T. The influence of climate change and human activities on runoff in the middle reaches of the Huaihe River Basin, China. *J. Geogr. Sci.* **2018**, *28*, 79–92. [CrossRef]
20. Gao, C.; Lu, M.; Zhang, X.; Li, P. Response analysis of runoff to climate change in the upper reaches of Huaihe River Basin. *J. North China Univ. Water Resour. Electr. Power (Nat. Sci. Ed.)* **2016**, *37*, 28–32. (In Chinese)
21. Liu, X.; Chen, M.; Wang, Z.; Zhu, S.; Cui, C.; Zhou, T. Analysis of the attribution of runoff changes in the middle and upper reaches of the Huaihe River basin. *Yellow River* **2020**, *42*, 16–22. (In Chinese)
22. Ye, T.; Shi, P.; Zhong, H.; Qu, S.; Wu, H.; Shen, L. Attribution analysis of runoff change in the upper and middle reaches of Huaihe River based on Budyko hypothesis and differential equation. *J. Hohai Univ. (Nat. Sci.)* **2022**, *50*, 25–32. (In Chinese)
23. Sun, S.; Wang, J.; Zhou, S.; Wang, J.; Yan, G.; Wang, H.; Bi, Z. Effects of climate and watershed characteristics on surface hydrological processes in the Huaihe River Basin. *Acta Ecol. Sin.* **2022**, *42*, 3933–3946. (In Chinese)
24. Shi, Z.; Zhao, Q.; Wang, Y.; Zhang, L. Temporal and Spatial Evolution of Vegetation Cover and Its Relationship with Runoff in Yihe River Basin. *Res. Soil Water Conserv.* **2023**, *30*, 1–8.
25. Gong, Z.; Feng, G.; Wan, S.; Li, J. Detection of North China and Global Climate Change Based on Heuristic Segmentation Algorithm. *Acta Phys. Sin.* **2006**, *55*, 477–484. (In Chinese) [CrossRef]
26. Huang, C.; Du, M.; Li, P.; Guo, Y.; Wang, L. Study on Variation and Driving Force of Rainfall Concentration under Changing Environment. *Adv. Water Sci.* **2019**, *30*, 496–506. (In Chinese)
27. Feng, G.; Gong, Z.; Dong, W.; Li, J. Research on climate mutation detection based on heuristic segmentation algorithm. *Acta Phys. Sin.* **2005**, *54*, 5494–5499. (In Chinese) [CrossRef]
28. Choudhury, B. Evaluation of an empirical equation for annual evaporation using field observations and results from a biophysical model. *J. Hydrol.* **1999**, *216*, 99–110. [CrossRef]
29. Yang, H.; Yang, D.; Lei, Z.; Sun, F. New analytical derivation of the mean annual water- energy balance equation. *Water Resour. Res.* **2008**, *44*, W03410. [CrossRef]

30. Li, D.; Pan, M.; Cong, Z.; Zhang, L.; Wood, E. Vegetation control on water and energy balance within the Budyko framework. *Water Resour. Res.* **2013**, *49*, 969–976. [CrossRef]
31. Ji, G.; Huang, J.; Guo, Y.; Yan, D. Quantitatively Calculating the Contribution of Vegetation Variation to Runoff in the Middle Reaches of Yellow River Using an Adjusted Budyko Formula. *Land* **2022**, *11*, 535. [CrossRef]
32. Xu, X.; Yang, D.; Yang, H.; Lei, H. Attribution analysis based on the Budyko hypothesis for detecting the dominant cause of runoff decline in Haihe basin. *J. Hydrol.* **2014**, *510*, 530–540. [CrossRef]
33. Yang, D.; Zhang, S.; Xu, X. Attribution Analysis of Runoff Change in the Yellow River Basin Based on Hydrothermal Coupling Equilibrium Equation. *Sci. Sin. (Technol.)* **2015**, *45*, 10241034. (In Chinese)
34. Zhao, F.; Zhang, L.; Xu, Z.; Scott, D.F. Evaluation of methods for estimating the effects of vegetation change and climate variability on streamflow. *Water Resour. Res.* **2010**, *46*, 742–750. [CrossRef]
35. Yang, Y.; Zeng, P. Effects of forest vegetation change on river runoff and sediment in China. *J. Beijing For. Univ.* **1994**, *16*, 3541. (In Chinese)
36. Wu, J.; Miao, C.; Wang, Y.; Duan, Q.; Zhang, X. Contribution analysis of the long-term changes in seasonal runoff on the Loess Plateau, China, using eight Budyko-based methods. *J. Hydrol.* **2017**, *545*, 263–275. [CrossRef]
37. Al-Safi, H.I.J.; Kazemi, H.; Sarukkalige, P.R. Comparative study of conceptual versus distributed hydrologic modelling to evaluate the impact of climate change on future runoff in unregulated catchments. *J. Water. Clim. Chang.* **2020**, *11*, 341–366. [CrossRef]
38. Jiang, C.; Xiong, L.; Wang, D.; Liu, P.; Guo, S.; Xu, C. Separating the impacts of climate change and human activities on runoff using the Budyko-type equations with time-varying parameters. *J. Hydrol.* **2015**, *522*, 326–338. [CrossRef]
39. Leuning, R.; Zhang, Y.Q.; Rajaud, A.; Cleugh, H.; Tu, K. A simple surface conductance model to estimate regional evaporation using MODIS leaf area index and the Penman-Monteith equation. *Water Resour. Res.* **2008**, *44*, 1–17. [CrossRef]
40. Liu, C.M.; Wang, Z.G.; Yang, S.T.; Sang, Y.; Liu, X.; Li, J. Hydro-informatic modeling system: Aiming at water cycle in land surface material and energy exchange processes. *Acta Geogr. Sin.* **2014**, *69*, 579–587. (In Chinese)
41. Zhang, D.; Liu, X.M.; Bai, P. Different influences of vegetation greening on regional water-energy balance under different climatic conditions. *Forests* **2018**, *9*, 412. [CrossRef]

Article

Response Mechanism of Annual Streamflow Decline to Vegetation Growth and Climate Change in the Han River Basin, China

Mengya Jia [1,2], Shixiong Hu [1,3,*], Xuyue Hu [1,2] and Yuannan Long [1,2]

[1] School of Hydraulic and Environmental Engineering, Changsha University of Science & Technology, Changsha 410114, China; mjia@live.esu.edu (M.J.); huxuyue62@163.com (X.H.); lynzhb@csust.edu.cn (Y.L.)
[2] Key Laboratory of Dongting Lake Aquatic Eco-Environmental Control and Restoration of Hunan Province, Changsha 410114, China
[3] Department of History & Geography, East Stroudsburg University, East Stroudsburg, PA 18301, USA
* Correspondence: shu@esu.edu

Abstract: Vegetation changes have a significant impact on the underlying surface of a watershed and alter hydrological processes. To clarify the synergistic evolution relationship between climate, vegetation, and hydrology, this study aims to reveal how vegetation restoration influences streamflow decline. This study first applied the trend-free pre-whitening Mann–Kendall (TFPW-MK) method to identify variation trends of various elements at Baihe and Shayang hydrologic stations from 1982 to 2015. Secondly, an extended Budyko equation was improved by fitting the linear relationship between annual *NDVI* and Budyko parameter (ω). Finally, based on the extended Budyko formula, the elastic coefficient method was applied to identify the influence of vegetation changes on runoff changes of the Baihe and Shayang stations from 1982 to 2015. The results displayed that (1) the annual *NDVI* of Baihe and Shayang hydrologic stations both presented an increasing trend, and streamflow presented an insignificant decrease trend. The mutation year of the annual runoff depth of Baihe and Shayang stations both occurred in 1990. (2) The annual *NDVI* had a significant and positive linear relationship with ω. (3) The streamflow decline of Baihe and Shayang stations is mainly influenced by precipitation variation and human activities. (4) Vegetation growth had a positive effect on the streamflow decline of Baihe and Shayang stations, with a contribution rate of 14.06% and 17.87%. This effect of vegetation growth on discharge attenuation should be given high priority.

Keywords: vegetation growth; response mechanism; streamflow decline; extended Budyko equation

Citation: Jia, M.; Hu, S.; Hu, X.; Long, Y. Response Mechanism of Annual Streamflow Decline to Vegetation Growth and Climate Change in the Han River Basin, China. *Forests* **2023**, *14*, 2132. https://doi.org/10.3390/f14112132

Academic Editor: Ge Sun

Received: 4 September 2023
Revised: 22 October 2023
Accepted: 23 October 2023
Published: 26 October 2023

Copyright: © 2023 by the authors. Licensee MDPI, Basel, Switzerland. This article is an open access article distributed under the terms and conditions of the Creative Commons Attribution (CC BY) license (https:// creativecommons.org/licenses/by/ 4.0/).

1. Introduction

The water cycle, as a key link connecting the ocean, land, and atmosphere, has significant implications for global climate and natural environment changes. Climatic and anthropogenic factors are two main factors that affect the water cycle, including precipitation, runoff, evaporation, and land use change [1–3]. Climate change directly affects hydrological cycle elements such as temperature, leading to terrestrial hydrological system change. Human activities have significantly affected runoff by altering the underlying surface conditions and canopy interception and evapotranspiration of the basin through many methods, such as off-river water intake, hydraulic engineering construction, and urban expansion [4–6]. By reason of the combined impact of climatic and anthropogenic factors, the hydrological cycle of the basin has undergone significant changes, resulting in frequent occurrence of extreme events such as extreme droughts and floods [7–10], posing serious challenges to water resource management, ecosystem balance in river basins, and agricultural irrigation and production [11–13]. Therefore, quantitative analysis of the effect of different factors on discharge variation is profit to clarify the evolution trend of watershed runoff, guiding managers to make correct decisions, and providing an important theoretical foundation for achieving water resource security and sustainable utilization.

Hanjiang River Basin (HJR) is a first-order tributary of the Yangtze River Basin. The upstream HJR is one of the water sources for China's South–North Water Transfer Central Project, and midstream and downstream HJR are one of China's major grain and cotton production bases. Because of the remarkable growth of water consumption by anthropic factors and the influence of global warming, the discharge of HJR was significantly reduced. Therefore, the analysis of the variation patterns and drivers of streamflow attenuation in HJR has become a hot research topic for scholars [14]. Du et al. [15] carried out the evolution law and driver of hydrological elements in the upper reaches of the HJR by mathematical statistical tests and hydrological modeling methods. Results indicated that there was a downward trend in precipitation and streamflow, with a sudden change year in 1990. Average annual runoff from 1991 to 2016 decreased by 8.12 billion cubic meters compared to 1962 to 1990, with approximately 43.2% affected by climate change and 56.8% affected by human activities. Zou et al. [16] explored the effect of precipitation changes and anthropic factors on runoff by the double cumulative curve method. Results indicated that the effect of anthropic factors on the reduction of runoff in HJR gradually increased during the research period. During the research period, the influence of human factors on streamflow attenuation in Shiquan station and Xiantao station increased by 18.56% and 11.15% in Phase III (2003–2018) compared to Phase II (1991–2002), respectively.

In order to protect and expand forests, reduce land desertification, and prevent soil erosion, the Chinese government has vigorously implemented key projects such as "returning farmland to forests and grasslands" and the "Three–North Shelterbelt Program". From 2000 to 2017, the vegetation leaf area index increased significantly, accounting for 25% of the global net increase in leaf area. The implementation of a series of ecological restoration measures has caused rapid changes in vegetation in the Han River Basin. Zhan et al. [17] analyzed the characteristics and driving forces of vegetation change in midstream and downstream of HJR from 2001 to 2015. NDVI displayed a growth trend, and anthropic factors had a facilitating impact on vegetation recovery. Vegetation changes have altered hydrological processes such as precipitation interception, evapotranspiration, and soil infiltration, and have had a significant impact on watershed water cycle and runoff. The implementation of a series of ecological restoration measures in the HJR has a significant impact on its precipitation–runoff relationship. In recent years, some scholars have begun to analyze the impact of vegetation change on the HJR. Peng et al. [18] found that changes in the underlying surface are the main driver contributing to streamflow attenuation of HJR, and vegetation change is an important reason for the overall decrease in runoff of HJR. However, previous studies have mostly qualitatively analyzed the positive or negative impact of vegetation changes on runoff in the HJR, few researchers quantitatively assessed the contribution of vegetation growth on streamflow attenuation.

As a consequence, this paper aims to reveal how vegetation restoration influences streamflow variation by following contents: (1) identifying the changing characteristics of meteorological, hydrological, and annual *NDVI* data of HJR; (2) constructing a corrected Budyko equation; and (3) quantifying the impact of vegetation growth on discharge attenuation in HJR by corrected Budyko equation and elasticity analysis methods. This study helps to clarify vegetation spatiotemporal characteristics of HJR and its influence on runoff changes, which is instrumental in sustainable economic development and ecological environment protection of HJR.

2. Research Area and Data

Hanjiang River originates from the southern foothills of the Qinling Mountains. The entire area of the HJR is 1577 km, with a watershed area of 15.3×10^4 km^2, with the range of 106°11′–114°14′ E and 30°08′–34°20′ N. The mainstream of the Hanjiang River runs from east to west, passing between the Qinling Mountains and Daba Mountains, and the river gradient is large. The mountainous area within the watershed accounts for approximately 70% of the total watershed area. The average annual runoff of the entire Hanjiang River basin is 57.7 billion m^3. The streamflow supply of the Hanjiang River mainly comes from atmospheric precipitation, and the distribution of surface water resources is basically consistent with the distribution of precipitation.

HJR belongs to the East Asian subtropical monsoon climate region, and its climate has obvious seasonal characteristics. The vegetation types in the watershed are mainly subtropical evergreen broad-leaved forest, evergreen broad-leaved forest, and deciduous broad-leaved mixed forest, with high vegetation coverage. The precipitation in the basin is mainly influenced by the warm and humid air currents in the southeast and southwest. The distribution of precipitation in the basin is uneven within the year, with an average annual precipitation of 873 mm, an average annual temperature of 12.0–16.0 °C, an extreme maximum temperature of 42.7 °C, and an extreme minimum temperature of −17.6 °C. The average wind speed for many years is between 1.0–3.0 m/s.

There are 14 major reservoirs built in the main stream of the Hanjiang River, including Danjiangkou, Shiquan, Ankang, etc. There are six major large reservoirs built in the tributaries, including Linhekou and Doulingzi. Key water diversion projects in the basin include the Water Diversion from the Han to Wei River Project, the Qingquangou Water Diversion Project, and the Water Diversion from the Yangtze River to Hanjiang River Project [19–21].

In this paper, Baihe and Shayang hydrological station of HJR are selected for this study. The discharge data of Baihe hydrological station and Shayang hydrological station from 1982 to 2015 are selected from the China Hydrological Yearbook and Yangtze River Water Resources Commission. Meteorological station data from 1982 to 2015 are downloaded from the China Meteorological Administration, and these stations' potential evaporation can be computed by the Penman–Monteith equation. Annual NDVI (1982–2015) data are drawn from the GIMMS NDVI3g V1.0 dataset by the method of calculating the average value, and the spatial resolution of NDVI data is 0.05 degrees. Annual NDVI (1982–2015) data are drawn from the GIMMS *NDVI*3g V1.0 dataset by the method of calculating the average value. We use the tool "Zonal Statistics as Table" in ARCGIS to calculate the average value.

3. Methodology

3.1. Trend-Free Pre-Whitening Mann–Kendall Method

The Mann–Kendall (MK) trend test is a nonparametric rank test method, which does not require the original data to obey a normal distribution, nor is it affected by a few outliers and missing data, and is widely used in the field of hydrological statistics [22,23]. However, the non-parametric rank test method requires the original sequence to be independent, and if the original sequence has autocorrelation, it will significantly amplify the trend of the sequence. In this paper, we use the trend-free pre-whitening (TFPW) MK method to preprocess the original data series. TFPW-MK trend testing method is a pre-removal type MK trend testing method proposed for the autocorrelation problem of the sequence to be tested [24]. Through the trend-free pre-whitening method, we eliminate the influence of dominant trends on the estimation of autocorrelation coefficients in the original data series, and can effectively reduce the impact of autocorrelation on the test results in the sequence and avoid distortion of the test results [25]. Therefore, in this paper, we made use of the TFPW-MK trend test method to identify the variation trend characteristics of runoff depth (R), precipitation (Pre), potential evapotranspiration (ET0) and NDVI.

3.2. Mutational Analysis Methods

The Pettitt catastrophe test is a nonparametric catastrophe point detection method, which is simple in calculation and less affected by outliers and is widely used in the catastrophe analysis of hydrometeorological elements. The specific calculation process can be found in reference [26]. The Pettitt catastrophe test $K_{t,N}$ method defines the statistic to obtain the most significant possible mutation point and uses the statistic to determine whether the mutation point meets the given significance level. $p < 0.05$ indicates statistically significant.

$$K_{t,N} = \max|U_{t,N}|, (1 \leq t \leq N) \tag{1}$$

$$P = 2\exp\left\{-6K_{t,N}^2/\left(N^3 + N^2\right)\right\} \tag{2}$$

Among them, U_t is the statistical variable of the sample time series data, and N is the number of sample sequences.

The accumulated deviation value mutation test approach is an intuitive curve judgment approach. The cumulative deviation curve of the variables is plotted with time as the horizontal coordinate for determining the continuous evolution characteristics of time series data, and the extreme point of the curve corresponds to the time when the sudden change occurs. The advantages of this method are not only the ease of calculation, but also the clarity of the time when the mutation occurs.

3.3. Extended Budyko Equation

There are a great many methods that can identify the effect of different factors on discharge variation, such as hydrological modeling [27], statistical analysis [28], and elasticity coefficient based on the Budyko hypothesis [29,30]. (1) The hydrological modeling method has high analysis accuracy and strong physical mechanism, which can simulate the hydrological process of precipitation, evapotranspiration, runoff and other factors. However, the calibration of model parameters is difficult and highly influenced by subjective factors of human beings, which increases the uncertainty of the results. Wen et al. [31] made use of the SWAT model to identify the effect of different factors on runoff variation of the Jinghe River and found that anthropogenic factors were major drivers influencing streamflow attenuation. (2) The statistical analysis method mainly conducts attribution analysis of streamflow variation by analyzing the correlation and trend characteristics of long-term series hydro-meteorological data, including comparative method of the slope changing ratio of cumulative quantity (SCRCQ), etc. Wang et al. [32] decoupled the influence of precipitation, evapotranspiration, and anthropogenic factors on discharge variation in Songhua by SCRCQ. (3) The Budyko method is easier to obtain input data and simpler calculation than hydrological modeling [33] and has become a commonly used method for scholars to study the driver of long-term runoff changes [34–37]. Through the derivation and derivation of a series of equations, the attribution of watershed runoff is quantitatively calculated. Zhao et al. [38] explored the driving factors of runoff in the middle reaches of the Yellow River by Budyko hydrothermal coupling equilibrium model. Except for the Beiluo River and Yanhe River, the main driver for streamflow changes was human activities.

$$R = Pre - ET \tag{3}$$

R is runoff depth and Pre is precipitation. In light of this theory of Budyko, there is a functional relationship between the actual evaporation (ET) and drought index ($AI = ET0/Pre$) at the multi-year scale of the basin.

$$ET = P - f\left(\frac{ET0}{Pre}\right) \tag{4}$$

$$R = Pre - f\left(\frac{ET0}{Pre}\right) \tag{5}$$

$ET0$ is potential evaporation. There are many equations for $f\left(\frac{ET0}{Pre}\right)$, and Fu's equation based on the Budyko framework is Equation (4) [39].

$$f\left(\frac{ET0}{Pre}\right) = 1 + \frac{ET0}{Pre} - \left(1 + \left(\frac{ET0}{Pre}\right)^\omega\right)^{1/\omega} \tag{6}$$

where ω is a variable that characterizes watershed property information and represents the impact of environmental changes such as climate, topography, vegetation type, and soil on the Budyko curve in the watershed, reflecting the macroscopic manifestation of the micro-interaction between various environmental factors. The soil properties will undergo corresponding changes with the restoration of vegetation. The topography undergoes corresponding changes with the impact of the erosion process, such as gully head retrogressive erosion, erosion gully undercutting, etc. However, changes in land use and surface cover conditions have a much greater impact on hydrological processes than soil properties and topography. In particular, with the implementation of the large-scale conversion of farmland to forest and grass project in 1999, vegetation coverage became the most critical factor for underlying surface conditions and significantly affected the hydrological process of the watershed.

$$R = Pre\left(\left(1 + \left(\frac{ET0}{Pre}\right)^{\omega}\right)^{1/\omega} - \left(\frac{ET0}{Pre}\right)\right) \quad (7)$$

Therefore, vegetation as the most critical factor for underlying surface conditions has been introduced into the interpretation of water and heat coupling control parameters in many studies in recent years. Li et al. [40] found that the underlying surface parameters (ω) in Budyko's hypothesis have a linear relationship with NDVI.

$$\omega = a*NDVI + b \quad (8)$$

a and b are constants for a given basin and need to be computed, and the constant (a) characterizes the sensitivity degree of ω to vegetation changes.

Combining Equations (7) and (8), Equation (9) describing the relationship between runoff depth and climate and vegetation indices is obtained, so the influence of vegetation growth on discharge attenuation of the HJR can be estimated quantitatively from NDVI data obtained by remote sensing.

$$R = Pre\left(\left(1 + \left(\frac{ET0}{Pre}\right)^{(a*NDVI+b)}\right)^{1/(a*NDVI+b)} - \left(\frac{ET0}{Pre}\right)\right) \quad (9)$$

3.4. Elasticity Coefficient Method

The elasticity coefficients of runoff depth (R) to ET0, Pre, and NDVI can be calculated.

$$\varepsilon_{ET0} = \frac{\Delta R}{\Delta ET0}\frac{ET0}{R} \quad (10)$$

$$\varepsilon_{Pre} = \frac{\Delta R}{\Delta Pre}\frac{Pre}{R} \quad (11)$$

$$\varepsilon_{NDVI} = \frac{\Delta R}{\Delta NDVI}\frac{NDVI}{R} \quad (12)$$

where $\frac{\Delta R}{\Delta ET0}$, $\frac{\Delta R}{\Delta Pre}$, and $\frac{\Delta R}{\Delta NDVI}$ can be obtained by partial differentiation of Equation (9), respectively.

$$\Delta R_{ET0} = \varepsilon_{ET0}\frac{R}{ET0}\Delta ET0 \quad (13)$$

$$\Delta R_{Pre} = \varepsilon_{Pre}\frac{R}{Pre}\Delta Pre \quad (14)$$

$$\Delta R_{NDVI} = \varepsilon_{NDVI}\frac{R}{NDVI}\Delta NDVI \quad (15)$$

$$\Delta R_{hum} = \varepsilon_{\omega}\frac{R}{\omega}\Delta\omega - \Delta R_{NDVI} \quad (16)$$

where ΔR_{ET0}, ΔR_{Pre}, ΔR_{NDVI} and ΔR_{hum} represent the changes in runoff depth due to changes in the $ET0$, Pre, $NDVI$ and underlying surface parameters, respectively. ε_{ET0}, ε_{Pre}, ε_{NDVI} and ε_ω represent the elastic coefficient of streamflow to $ET0$, Pre, $NDVI$ and ω.

$$\Delta R = \Delta R_{ET0} + \Delta R_{Pre} + \Delta R_{NDVI} + \Delta R_{hum} \tag{17}$$

$$\eta_{R_{ET0}} = \Delta R_{ET0}/\Delta R \times 100\% \tag{18}$$

$$\eta_{R_{Pre}} = \Delta R_{Pre}/\Delta R \times 100\% \tag{19}$$

$$\eta_{R_{NDVI}} = \Delta R_{NDVI}/\Delta R \times 100\% \tag{20}$$

$$\eta_{R_{hum}} = \Delta R_{hum}/\Delta R \times 100\% \tag{21}$$

$\eta_{R_{ET0}}$, $\eta_{R_{Pre}}$, $\eta_{R_{NDVI}}$ and $\eta_{R_{hum}}$ represent the contribution rates of $ET0$, Pre, $NDVI$ and anthropic factors on runoff.

4. Results and Analysis

4.1. Trend Analysis

In this paper, we made use of the TFPW-MK trend test method to identify the variation trend characteristics of runoff depth (R), precipitation (Pre), potential evapotranspiration ($ET0$) and $NDVI$ at the Baihe and Shayang hydrological stations (Table 1). The significance test of the TFPW-MK method needs to be judged based on the z value. If the absolute value of z is greater than 1.96 and less than 2.576, it indicates that the results are significant at the 0.05 level. If the absolute value of z is greater than 2.576, it indicates that the result is significant at the 0.01 level.

Table 1. TFPW-MK trend test results of meteorological and hydrological elements on Baihe and Shayang hydrological stations.

Hydrographic Station	Variable	β	Z	Level of Significance
Baihe	R	−3.044	−0.697	—
	Pre	0.552	0.201	—
	ET0	1.627	1.844	—
	NDVI	0.001	2.897	0.01
Shayang	R	−2.930	−1.193	—
	Pre	−0.988	−0.232	—
	ET0	1.758	1.844	—
	NDVI	0.001	1.627	—

For Baihe hydrological station, the Z value of R from 1982 to 2015 was −0.697, which indicated that the annual R at Baihe hydrological station showed an insignificant downward trend overall. The z value of Pre and ET0 at Baihe hydrological station from 1982 to 2015 was 0.201 and 1.844, which indicated that the annual Pre and ET0 at Baihe hydrological station both showed an insignificant upward trend overall. The Z value of NDVI at Baihe hydrological station from 1982 to 2015 was 2.897, which indicated that the annual NDVI showed a significant upward trend overall. For the Shayang hydrological station, the Z values of R and Pre from 1982 to 2015 were −1.193 and −0.232, which indicated that the annual R and Pre both showed an insignificant downward trend overall. The Z value of ET0 and NDVI at Shayng hydrological station from 1982 to 2015 was 1.844 and 1.627, which indicated that the annual ET0 and NDVI both showed an insignificant upward trend overall.

4.2. Abrupt Analysis

Mutation analysis for annual runoff depth data (1982–2015) of the Baihe and Shayang stations was conducted using the Pettitt mutation test method (Figure 1). From Figure 1, the statistics of the Pettitt mutation test method for annual runoff depth data (1982–2015) of the Baihe and Shayang stations both showed an overall trend of increasing from 1982 to 1990 and then decreasing from 1991 to 2015. And the statistics in 1990 both were the maximum value and exceeded the critical value of $\alpha = 0.05$. The results indicated that the sudden change year for annual runoff depth data (1982–2015) of the Baihe and Shayang stations both were around 1990.

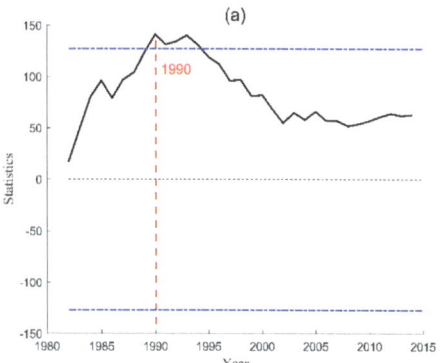 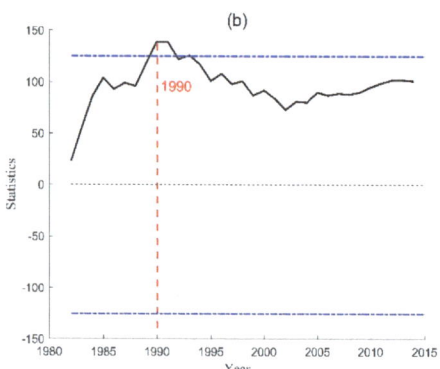

Figure 1. Pettitt mutation test results of runoff depth in Baihe (**a**) and Shayang (**b**) during 1982–2015.

The accumulated deviation value mutation test approach was used to test annual runoff depth data (1982–2015) of the Baihe and Shayang stations to confirm the accuracy of the mutation results of the Pettitt mutation test method (Figure 2). Figure 2 presented that the accumulated deviation value of annual runoff depth data (1982–2015) of the Baihe and Shayang stations both showed an overall trend of increasing from 1982 to 1990 and then decreasing from 1991 to 2015, which implied that annual runoff depth data (1982–2015) of these two hydrological stations both mutated in 1990. Therefore, the Pettitt and cumulative deviation value mutation test methods both verified that annual runoff depth data (1982–2015) of Baihe and Shayang stations mutated in 1990.

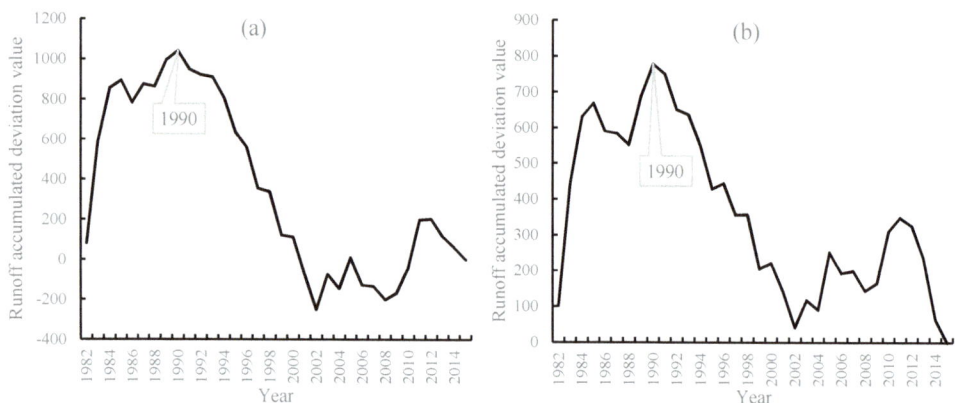

Figure 2. Runoff depth accumulated deviation value mutation analysis in Baihe (**a**) and Shayang (**b**) from 1982 to 2015.

4.3. Influence Assessment of Revegetation on Streamflow

In an effort to further investigate the effect of revegetation on streamflow attenuation in HJR, we should obtain the linear relationship equations of NDVI and ω. Next, for obtaining the fitting equations of NDVI and ω, the 7-year moving average NDVI and ω of Baihe and Shayang hydrological stations are computed. Finally, the regression coefficients for Baihe ($a = 6.51$ and $b = -3.59$) and Shayang ($a = 14.23$ and $b = -9.10$) are obtained by least squares fitting, respectively (Figure 3). The determination coefficient (R^2) of the fitting equation are 0.39 and 0.59, and the equations both are significant ($p < 0.01$). Comparing the fitting equations of NDVI and ω in other study areas, these determination coefficients range from 0.35 to 0.55 [30,41,42]. Determination coefficients in the figures are close to their numerical value. Therefore, the fitting equations of NDVI and ω can be applied to calculating the contribution of vegetation change to runoff attenuation.

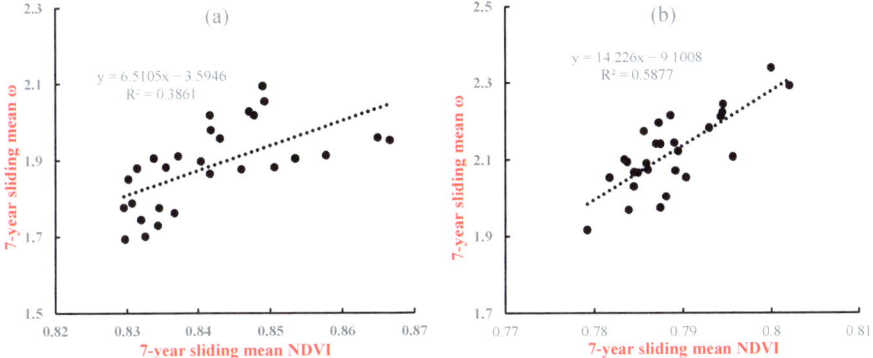

Figure 3. Functional Equations for 7-year sliding mean NDVI and 7-year sliding mean ω in Baihe (**a**) and Shayang (**b**) stations.

In accordance with the mutation analysis results, 1990 was taken as the abrupt year for splitting the annual runoff depth of Baihe and Shayang into two sub-periods (P1: 1982–1990 and P2: 1991–2015), and values of hydro-meteorological elements and NDVI of HJR were computed (Table 2). We applied the T-test method to test the significance of the difference between the two sub-periods (P1: 1982–1990 and P2: 1991–2015).

Table 2. Values of hydro-meteorological elements and NDVI in HJR.

Station	Periods	ET0/mm	Pre/mm	R/mm	ω	NDVI
Baihe	P1 (1982–1990)	847.52	919.51	493.75	1.69	0.833
	P2 (1991–2015)	898.16	813.61	336.39	1.94	0.846
	Difference	50.64	−105.90	−157.36	0.25	0.013
	P	0.002	0.086	0.006	0.001	0.099
Shayang	P1 (1982–1990)	895.15	882.37	391.37	1.92	0.783
	P2 (1991–2015)	940.64	806.06	273.82	2.20	0.791
	Difference	45.49	−76.30	−117.56	0.28	0.008
	P	0.005	0.127	0.003	0.058	0.294

From Table 2, potential evaporation and ω of the Biahe station both significantly increased by 50.64 mm and 0.25 from P1 to P2 ($p < 0.05$). Runoff depth significantly decreased by 157.36 mm, from 493.75 mm (P1) to 336.39 mm (P2) ($p < 0.05$). The change in precipitation showed no significant decrease, decreasing by 105.90 mm from 919.51 mm (P1) to 813.61 mm (P2). The change in NDVI showed no significant increase, increasing by 0.013 from 0.833 (P1) to 0.836 mm (P2).

For Shayang station, potential evaporation significantly increased by 45.49 mm from 895.15 mm (P1) to 940.64 mm (P2) ($p < 0.05$). Runoff depth significantly decreased by 117.56 mm, from 391.37 mm (P1) to 273.82 mm (P2) ($p < 0.05$). The change in precipitation showed no significant decrease, decreasing by 76.30 mm from 882.37 mm (P1) to 806.06 mm (P2). The change in NDVI and ω both showed no significant increase, increasing by 0.008 and 0.28 from P1 to P2.

The elasticity coefficient of the runoff depth of the Baihe and Shayang stations on each driving factor was calculated. And then the amount of runoff changes due to changes in P, $ET0$, ω, and $NDVI$ could be computed, as shown in Table 3.

Table 3. Attributing analysis of discharge variation in HJR.

Station	ε_{ET0}	ε_{Pre}	ε_ω	ε_{NDVI}	ΔR_{ET0}	ΔR_{Pre}	ΔR_{hum}	ΔR_{NDVI}	ηR_{ET0}	ηR_{pre}	ηR_{hum}	ηR_{NDVI}
Baihe	−0.64	1.64	−1.32	−3.83	−13.88	−78.09	−44.15	−22.26	8.76%	49.30%	27.88%	14.06%
Shayang	−0.84	1.84	−1.39	−7.36	−12.59	−51.89	−33.27	−21.29	10.57%	43.55%	28.01%	17.87%

From T_1 to T_2, the increase of 50.64 mm in $ET0$ led to a decrease of 13.88 mm in the runoff depth of Baihe station. The decrease of 105.90 mm in Pre caused a decrease of 78.09 mm in runoff depth. The increase of 0.25 in ω and 0.013 in $NDVI$ caused a decrease of 44.15 mm and 22.26 mm in the runoff depth of Baihe station. The contributions of $ET0$, Pre, human activities and $NDVI$ to the runoff depth of Baihe station were 8.76%, 49.30%, 27.88% and 14.06%, respectively.

For Shayang station, the increase of 50.64 mm in $ET0$, 0.25 in ω, and 0.013 in $NDVI$ led to a decrease of 12.59 mm, 33.27 mm, and 21.29 mm in runoff depth. The decrease of 105.90 mm in Pre caused a decrease of 51.89 mm in runoff depth. The contributions of $ET0$, Pre, human activities and $NDVI$ to the runoff depth of Shayang station were 10.57%, 43.55%, 28.01% and 17.87%, respectively. In summary, streamflow decline in the HJR is mainly influenced by precipitation variation and human activities. Moreover, vegetation growth had a positive effect on streamflow decline.

With the implementation of the large-scale conversion of farmland to forest and grass project in 1999, vegetation coverage became the most critical factor for underlying surface conditions and significantly affected the hydrological process of the watershed. From Figure 3, we found that ω had a positive functional relationship with NDVI in the HJB, indicating that revegetation has a significant impact on ω. In addition, according to the sensitivity analysis results, the elasticity coefficients of runoff dep®(R) to NDVI were negative, indicating that R will decrease with increasing NDVI.

Although there are different research methods on the impact of vegetation change on runoff, there is generally a consistent conclusion that an increase in vegetation coverage is an important factor leading to a decrease in surface runoff [29,30]. This study presented that increasing vegetation coverage had a reducing effect on the runoff of the Baihe and Shayang stations, which is consistent with previous research findings [43,44], so this effect of vegetation growth on discharge attenuation should be given high priority. The government should plan vegetation restoration reasonably to reduce the reduction of water resources caused by unreasonable vegetation restoration.

5. Conclusions and Discussion

5.1. Conclusions

Runoff changes in the HJB have a significant impact on hydrological processes and the ecosystem of the Yangtze River. Since the implementation of the ecological restoration project, vegetation coverage of HJB has presented a growth trend as a whole. Vegetation changes can affect the underlying surface conditions of a watershed, and alter hydrological processes. To clarify the synergistic evolution relationship between climate, vegetation, and hydrology, this study aims to reveal how vegetation restoration influences streamflow decline.

The results displayed that (1) *NDVI* presented an increasing trend, and streamflow of Baihe and Shayang hydrologic station presented an insignificant decrease trend. The mutation year of the annual runoff depth of Baihe and Shayang stations both occurred in 1990. (2) The Budyko parameter has a positive linear relationship with *NDVI* ($p < 0.01$). (3) Streamflow decline in the HJR is mainly influenced by precipitation variation and human activities. Moreover, vegetation growth had a positive effect on the streamflow decline of Baihe and Shayang stations, with contribution rates of 14.06% and 17.87%. This effect of vegetation growth on discharge attenuation should be given high priority.

5.2. Discussions

This study quantitatively calculates the contribution rate of vegetation restoration to runoff changes in the Han River Basin by introducing NDVI into the Budyko formula, revealing how vegetation restoration affects the discharge decrease in HJR, filling the gap in attribution analysis of runoff changes in the HJR. Despite the strict control of data and models in this study, there are uncertainties. First of all, the determination coefficients of the fitting equations are low, so other factors need to be introduced to improve the determination coefficient of the fitting equation. Secondly, the method used in this study is only applicable to areas where vegetation change has a significant impact on the underlying surface.

In addition, the issue of interactive effects of the influence of climatic conditions and anthropic factors. Current studies have established a framework in which climate variability and human activities are independent factors, whereas, in reality, there is a strong interaction between the two in the eco-hydrological systems [44–47]. Therefore, how to couple the objectivity of statistical methods and the physical mechanism of hydrological modeling methods, while reducing uncertainty and improving accuracy, is an important direction for future research [48,49]. Based on the Budyko hypothesis, the variation of water storage in a watershed on a multi-year average scale in the water balance equation is 0, which clearly overlooks the interception of watershed runoff by soil and water conservation. Therefore, quantitative research on the contribution of different soil and water conservation to runoff remains a challenge and needs to be addressed in the attribution analysis of water and sediment changes. Further exploration of attribution analysis theory and model algorithms will also become the focus of future research.

Author Contributions: Conceptualization, S.H.; Methodology, M.J.; Software, M.J.; Formal analysis, M.J.; Data curation, M.J. and Y.L.; Writing—original draft, M.J.; Writing—review & editing, M.J., S.H., X.H. and Y.L.; Project administration, S.H. and X.H.; Funding acquisition, S.H. and Y.L. All authors have read and agreed to the published version of the manuscript.

Funding: This research was funded by the project of Philosophy and Social Science Plan in Henan Province (Grant No. 2022BJJ070) and the key project of Philosophy and Social Science Research in Colleges and Universities in Henan Province (Grant No. 2022-JCZD-15).

Data Availability Statement: Not applicable.

Conflicts of Interest: The authors declare no conflict of interest.

References

1. Zhou, S.; Wang, Y.; Guo, A.; Zhang, R.; Liu, Q. Future spatiotemporal changes in water resources in the Yellow River Basin. *J. Hydroelectr. Power* **2018**, *37*, 28–39. (In Chinese)
2. Ji, G.; Song, H.; Wei, H.; Wu, L. Attribution Analysis of Climate and Anthropic Factors on Runoff and Vegetation Changes in the Source Area of the Yangtze River from 1982 to 2016. *Land* **2021**, *10*, 612. [CrossRef]
3. Abbott, B.; Bishop, K.; Zarnetske, J.; Minaudo, C.; Chapin, F.; Krause, S.; Hannah, D.; Conner, L.; Ellison, D.; Godsey, S.; et al. Human domination of the global water cycle absent from depictions and perceptions. *Nat. Geosci.* **2019**, *12*, 533–540. [CrossRef]
4. Yang, X.; Wu, W.; Zheng, C.; Wang, Q. Identification of attribution of runoff changes in the Yi River Basin based on the Budyko hypothesis. *Res. Soil Water Conserv.* **2023**, *30*, 100–106. (In Chinese)
5. Yang, S.; Jiang, R.; Xie, J.; Wang, Y.; Wang, Y. Research on the Characteristics and Attributions of Runoff Changes in the Upper Wei River. *J. Water Resour. Water Eng.* **2019**, *30*, 37–42. (In Chinese)
6. Ji, G.; Wu, L.; Wang, L.; Yan, D.; Lai, Z. Attribution Analysis of Seasonal Runoff in the Source Region of the Yellow River Using Seasonal Budyko Hypothesis. *Land* **2021**, *10*, 542. [CrossRef]

7. Li, Y.; Li, H.; Huang, J.; Liu, C. An approximation method for evaluating flash flooding mitigation of sponge city strategies-A case study of Central Geelong. *J. Clean. Prod.* **2020**, *257*, 120525. [CrossRef]
8. Ji, G.; Lai, Z.; Xia, H.; Liu, H.; Wang, Z. Future Runoff Variation and Flood Disaster Prediction of the Yellow River Basin Based on CA-Markov and SWAT. *Land* **2021**, *10*, 421. [CrossRef]
9. Aissia, M.A.B.; Chebana, F.; Ouarda, T.B.M.J.; Roy, L.; Bruneau, P.; Barbet, M. Dependence evolution of hydrological characteristics, applied to floods in a climate change context in Quebec. *J. Hydrol.* **2014**, *519*, 148–163. [CrossRef]
10. Ji, G.; Lai, Z.; Yan, D.; Wu, L.; Wang, Z. Spatiotemporal patterns of future meteorological drought in the Yellow River Basin based on SPEI under RCP scenarios. *Int. J. Clim. Change Strateg. Manag.* **2022**, *14*, 39–53. [CrossRef]
11. Greve, P.; Orlowsky, B.; Mueller, B.; Sheffield, J.; Reichstein, M.; Seneviratne, S. Global assessment of trends in wetting and drying over land. *Nat. Geosci.* **2014**, *7*, 716–721. [CrossRef]
12. Ge, J.; Pitman, A.J.; Guo, W.; Zan, B.; Fu, C. China's tree-planting could falter in a warming world. *Nature* **2019**, *573*, 474.
13. Qin, Y.; Batzoglou, J.; Siebert, S.; Huning, L.; AghaKouchak, A.; Mankin, J.; Hong, C.; Tong, D.; Davis, S.; Mueller, N. Agricultural risks from changing snowmelt. *Nat. Clim. Change* **2020**, *10*, 459–465. [CrossRef]
14. Deng, L.; Guo, S.; Tian, J.; Wang, H.; Wang, J. Runoff variation and attribution analysis in the upper Han River basin. *South-to-North Water Transf. Water Sci. Technol.* **2023**, *21*, 761–769. (In Chinese)
15. Du, T.; Cao, L.; Ouyang, S.; Ping, Y.R.; Bao, B. Evolution Law and Attribution Analysis of Hydrological Elements in the Upper Han River Basin. *Rep. Yangtze River Acad. Sci.* **2023**, 1–7.
16. Zou, L.; Zhang, Y.; Chen, T.; Liu, H.Y. Research on the Characteristics of Precipitation and Runoff Evolution in the Han River Basin. *Hydrology* **2023**, *43*, 103–109.
17. Zhan, Y.; Fan, J.; Meng, T.; Li, Z.; Yan, Y.; Huang, J.; Chen, D.; Sui, L. Analysis on vegetation cover changes and the driving factors in the mid-lower reaches of Han River Basin between 2001 and 2015. *Open Geosci.* **2021**, *13*, 675–689. [CrossRef]
18. Peng, T.; Mei, Z.Y.; Dong, X.H.; Wang, J.B.; Liu, J.; Chang, W.J.; Wang, G.X. Attribution of runoff changes in the Han River Basin based on the Budyko hypothesis. *South North Water Divers. Water Conserv. Technol. (Chin. Engl.)* **2021**, *19*, 1114–1124.
19. Chen, H. *Research on Parameter Transplantation Method and Hydrological Hydrodynamic Coupling Model in the Han River Basin*; Dalian University of Technology: Dalian, China, 2020. (In Chinese)
20. Huo, L. *Evolution Analysis of Water Resources System in the Upper Reaches of Hanjiang River under Climate Change and Early Warning Research on Carrying Capacity*; Huazhong University of Science and Technology: Wuhan, China, 2020. (In Chinese)
21. Li, F.; Deng, Z.; Deng, R.; Peng, C. Research on Ecological Flow Guarantee Measures for the Main Stream of the Han River. *People's Yangtze River* **2021**, *52*, 50–54. (In Chinese)
22. Tang, B.; Tong, L.; Kang, S.; Zhang, L. Impacts of climate variability on reference evapotranspiration over 58 years in the Haihe river basin of north China. *Agric. Water Manag.* **2011**, *98*, 1660–1670. [CrossRef]
23. Luo, M.; Liu, T.; Frankl, A.; Duan, Y.C.; Meng, F.H.; Bao, A.M.; Kurban, A.; De Maeyer, P. Defining spatiotemporal characteristics of climate change trends from downscaled GCMs ensembles: How climate change reacts in Xinjiang, China. *Int. J. Climatol.* **2018**, *38*, 2538–2553. [CrossRef]
24. Yue, S.; Wang, C.Y. Applicability of prewhitening to eliminate the influence of serial correlation on the Mann-Kendall test. *Water Resour. Res.* **2002**, *38*, 4-1–4-7. [CrossRef]
25. Xv, J.; Yang, D.; Lei, Z.; Xv, J.; Yang, D.; Lei, Z.; Li, C.; Peng, J. Examination of long-term trends in precipitation and runoff in the Yangtze River Basin. *Yangtze River* **2006**, *85*, 63–67. (In Chinese)
26. Pettitt, A.N. A Non-Parametric Approach to the Change-Point Problem. *J. R. Stat. Soc.* **1979**, *28*, 126–135. [CrossRef]
27. Nie, N.; Zhang, W.; Liu, M.; Chen, H.; Zhao, D. Separating the impacts of climate variability, land-use change and large reservoir operations on streamflow in the Yangtze River basin, China, using a hydrological modeling approach. *Int. J. Dig. Earth* **2021**, *14*, 231–249. [CrossRef]
28. Wang, S.; Yan, Y.; Yan, M.; Zhao, X. Quantitative estimation of the impact of precipitation and human activities on runoff change of the Huangfuchuan River basin. *J. Geogr. Sci.* **2012**, *22*, 906–918. [CrossRef]
29. Liu, Y.; Chen, W.; Li, L.; Huang, J.; Wang, X.; Guo, Y.; Ji, G. Assessing the contribution of vegetation variation to streamflow variation in the Lancang River Basin, China. *Front. Ecol. Evol.* **2023**, *10*, 1058055. [CrossRef]
30. Ji, G.; Yue, S.; Zhang, J.; Huang, J.; Guo, Y.; Chen, W. Assessing the Impact of Vegetation Variation, Climate and Human Factors on the Streamflow Variation of Yarlung Zangbo River with the Corrected Budyko Equation. *Forests* **2023**, *14*, 1312. [CrossRef]
31. Wen, D.; Cui, B.; Liu, Z.; Zhang, K. Relative effects of human activities and climate change on the river runoff in an arid basin in northwest China. *Hydrol. Process.* **2015**, *28*, 4854–4864.
32. Wang, S.; Wang, Y.; Ran, L.; Su, T. Climatic and anthropogenic impacts on runoff changes in the Songhua River basin over the last 56years (1955–2010), Northeastern China. *Catena* **2015**, *127*, 258–269. [CrossRef]
33. Ji, G.; Huang, J.; Guo, Y.; Yan, D. Quantitatively Calculating the Contribution of Vegetation Variation to Runoff in the Middle Reaches of Yellow River Using an Adjusted Budyko Formula. *Land* **2022**, *11*, 535. [CrossRef]
34. Donohue, R.J.; Roderick, M.L.; Mcvicara, T.R. Can dynamic vegetation information improve the accuracy of Budyko's hydrological model? *J. Hydrol.* **2010**, *390*, 23–34. [CrossRef]
35. Caracciolo, D.; Pumo, D.; Viola, F. Budyko's based method for annual runoff characterization across different climatic areas: An application to United States. *Water Resour. Manag.* **2018**, *32*, 3189–3202. [CrossRef]

36. Zhang, X.; Dong, Q.; Cheng, L.; Xia, J. A Budyko-based framework for quantifying the impacts of aridity index and other factors on annual runoff. *J. Hydrol.* **2019**, *579*, 124224. [CrossRef]
37. Yang, X.; Wang, C.; Du, J.; Qiu, S.; Liu, J. Dynamic evolution of attribution analysis of runoff based on the complementary Budyko equation in the source area of Lancang river. *Front. Earth Sci.* **2023**, *11*, 1160520. [CrossRef]
38. Zhao, G.J.; Tian, P.; Mu, X.M.; Jiao, J.Y.; Wang, F.; Gao, P. Quantifying the impact of climate variability and human activities on streamflow in the middle reaches of the Yellow River basin, China. *J. Hydrol.* **2014**, *519*, 387–398. [CrossRef]
39. Fu, B. On the calculation of the evaporation from land surface. *Chin. J. Atmos. Sci.* **1981**, *5*, 23–31. (In Chinese)
40. Li, D.; Pan, M.; Cong, Z.; Zhang, L.; Wood, E. Vegetation control on water and energy balance within the Budyko framework. *Water Resour. Res.* **2013**, *49*, 969–976. [CrossRef]
41. Yue, S.; Huang, J.; Zhang, Y.; Chen, W.; Guo, Y.; Cheng, M.; Ji, G. Quantitative Evaluation of the Impact of Vegetation Restoration and Climate Variation on Runoff Attenuation in the Luan River Basin Based on the Extended Budyko Model. *Land* **2023**, *12*, 1626. [CrossRef]
42. Gao, H.; Li, Q.; Xiong, G.; Li, B.; Zhang, J.; Meng, Q. Quantitative assessment of hydrological response to vegetation change in the upper reaches of Luanhe River with the modified Budyko framework. *Front. Ecol. Evol.* **2023**, *11*, 1178231. [CrossRef]
43. Calder, I.R.; Smyle, J.; Aylward, B. Debate over flood-proofing effects of planting forests. *Nature* **2007**, *450*, 945. [CrossRef] [PubMed]
44. Wang, Y.; Liu, Z.; Qian, B.; He, Z.; Ji, G. Quantitatively Computing the Influence of Vegetation Changes on Surface Discharge in the Middle-Upper Reaches of the Huaihe River, China. *Forests* **2022**, *13*, 2000. [CrossRef]
45. Wu, J.; Miao, C.; Wang, Y.; Duan, Q.; Zhang, X. Contribution analysis of the long-term changes in seasonal runoff on the Loess Plateau, China, using eight Budyko-based methods. *J. Hydrol.* **2017**, *545*, 263–275. [CrossRef]
46. Al-Safi, H.I.J.; Kazemi, H.; Sarukkalige, P.R. Comparative study of conceptual versus distributed hydrologic modelling to evaluate the impact of climate change on future runoff in unregulated catchments. *J. Water. Clim. Chang.* **2020**, *11*, 341–366. [CrossRef]
47. Jiang, C.; Xiong, L.; Wang, D.; Liu, P.; Guo, S.; Xu, C. Separating the impacts of climate change and human activities on runoff using the Budyko-type equations with time-varying parameters. *J. Hydrol.* **2015**, *522*, 326–338. [CrossRef]
48. Leuning, R.; Zhang, Y.Q.; Rajaud, A.; Cleugh, H.; Tu, K. A simple surface conductance model to estimate regional evaporation using MODIS leaf area index and the Penman-Monteith equation. *Water Resour. Res.* **2008**, *44*, 1–17. [CrossRef]
49. Liu, C.M.; Wang, Z.G.; Yang, S.T.; Sang, Y.; Liu, X.; Li, J. Hydro-informatic modeling system: Aiming at water cycle in land surface material and energy exchange processes. *Acta Geogr. Sin.* **2014**, *69*, 579–587. (In Chinese)

Disclaimer/Publisher's Note: The statements, opinions and data contained in all publications are solely those of the individual author(s) and contributor(s) and not of MDPI and/or the editor(s). MDPI and/or the editor(s) disclaim responsibility for any injury to people or property resulting from any ideas, methods, instructions or products referred to in the content.

Article

Assessing the Impact of Vegetation Variation, Climate and Human Factors on the Streamflow Variation of Yarlung Zangbo River with the Corrected Budyko Equation

Guangxing Ji, Shuaijun Yue, Jincai Zhang, Junchang Huang, Yulong Guo and Weiqiang Chen *

College of Resources and Environmental Sciences, Henan Agricultural University, Zhengzhou 450046, China; guangxingji@henau.edu.cn (G.J.)
* Correspondence: chwqgis@163.com

Abstract: The Yarlung Zangbo River (YZR) is the largest river on the Qinghai Tibet Plateau, and changes in its meteorology, hydrology and vegetation will have a significant impact on the ecological environment of the basin. In order to deepen our understanding of the relationship of climate–vegetation–hydrological processes in YZR, the purpose of this study is to explore how vegetation growth in the YZR affects its runoff changes. We first identified the abrupt year of discharge in the YZR using a heuristic segmentation algorithm and cumulative anomaly mutation test approach. After that, the functional equation for NDVI and the Budyko parameter (n) was computed. Finally, the NDVI was introduced into the Budyko equation to evaluate the impact of vegetation changes on the streamflow in the YZR. Results showed that: (1) NDVI and discharge in the YZR both presented an increasing trend, and the mutation year of annual runoff in Nuxia station occurred in 1997. (2) n had a significant negative correlation with NDVI in the YZR ($p < 0.01$). (3) The contributions of Pr, ET_0, $NDVI$, and n on streamflow change in the S2 period (1998–2015) were 5.26%, 1.14%, 43.04%, and 50.06%. The results of this study can provide scientific guidance and support for the evaluation of the effects of ecological restoration measures, as well as the management and planning of water resources in the YZR.

Keywords: streamflow change; NDVI; climate change; human activities; attribution analysis

Citation: Ji, G.; Yue, S.; Zhang, J.; Huang, J.; Guo, Y.; Chen, W. Assessing the Impact of Vegetation Variation, Climate and Human Factors on the Streamflow Variation of Yarlung Zangbo River with the Corrected Budyko Equation. *Forests* **2023**, *14*, 1312. https://doi.org/10.3390/f14071312

Academic Editor: Timothy A. Martin

Received: 16 May 2023
Revised: 19 June 2023
Accepted: 23 June 2023
Published: 26 June 2023

Copyright: © 2023 by the authors. Licensee MDPI, Basel, Switzerland. This article is an open access article distributed under the terms and conditions of the Creative Commons Attribution (CC BY) license (https://creativecommons.org/licenses/by/4.0/).

1. Introduction

Climate change characterized by global warming has significantly changed the hydrological cycle, which has had a very profound impact on natural ecosystems and the development of human society. For example, changes in the hydrological cycle process will increase the frequency of extreme hydrological events [1,2], change the spatial and temporal distribution of water resources, damage the ecological environment, and exacerbate the current situation of unbalanced and inadequate regional development [3,4]. In addition, since the Industrial Revolution, human activities have increasingly interfered with hydrological processes in basins, significantly changing hydrological cycle elements in terms of time and space, quantity, and quality [5,6]. Runoff is an important resource related to natural environmental changes and human social progress, and directly affects agricultural irrigation and production, ecological protection and restoration, and economic development and stability [7–9]. Therefore, clarifying and understanding the runoff evolution rules and its influencing factors under the context of changing environments can provide suggestions and guidance for regional ecological environment development and human production activities.

As the "water tower of Asia", the Qinghai Tibet Plateau is the birthplace of major rivers in Asia (the Yellow River, Yangtze River, Nu River, Indus River, Yarlung Zangbo River, Lancang River, etc.) [10] and is also the ecological security barrier in Asia and the world [11]. The Yarlung Zangbo River (YZR) is located in the southern part of the Qinghai–Tibet Plateau, with an average altitude of 4000 m. It is an important water vapor channel for the warm and humid Indian Ocean flow to enter the hinterland of the Qinghai–Tibet Plateau [12]. The national ecological function protection zone at the headstream provides an important ecological safety barrier for the economic center of Tibet and the Grand Canyon region in the downstream. The one river and two rivers region in the middle reaches of the basin (YZR, Nianchu River, and Lhasa River) is an important commodity grain base and population gathering place in Tibet [13]. In addition, the YZR is an important international river, and its water resources changes are related to the economic development and social stability of the Southeast Asian countries.

In recent years, many scholars have analyzed the characteristics of runoff changes and influencing factors in the YZR [14,15]. Liu et al. [16] found that both climate and land use changes can lead to runoff change trends in the YZR. Wang et al. [17] used the trend analysis and wavelet analysis methods to analyze the annual runoff change trend in Yangcun and Nuxia stations on the YZR from 1970 to 2012. The results showed that runoff presented an insignificant growth trend. Li et al. [18] used the heuristic segmentation algorithm to identify the mutation year of the runoff, and then used Mann–Kendall nonparametric test to analyze the temporal variation characteristics of runoff, and used the concentration degree and concentration period method to study the intra-annual change law of runoff based on the monthly runoff data of Nuxia station in the YZR from 1961 to 2015. Based on meteorological data, monthly scale runoff data, and land use data, Liu et al. [19] used improved hydrological models to clarify the impact of climate and underlying surface changes on runoff from 1991 to 2010. The results showed that during the period from 1991 to 2010, the contribution rates of climate and underlying surface changes to runoff change varied significantly between different periods, and climate change contributed more to runoff change than underlying surface changes. From a spatial perspective, the contribution rate of climate change to watershed runoff production was larger in the upstream and middle reaches, and smaller in the northeast of the downstream. Yang et al. quantitatively analyzed the impact of climate change on the runoff of the upper Yarlung Zangbo River. The results showed that the annual runoff and evapotranspiration of the upper reaches of the Yarlung Zangbo River showed a significant increasing trend from 1981 to 2010, and the increase of precipitation was the main factor for the increase in runoff [20].

Since 1999, the Chinese government has implemented a number of large-scale ecological restoration projects, such as the returning farmland to forest/grassland project, the Three-North Shelterbelt Program project [9], which has significantly increased the vegetation coverage across China [21]. The elevation of the YZR varies greatly. The cold and dry conditions in the upper reaches have changed into warm and humid conditions in the lower reaches, resulting in high vegetation diversity in the basin. In recent years, many scholars have analyzed the characteristics of vegetation changes and influencing factors in the YZR [22–24]. Lv et al. [25] analyzed the temporal and spatial changes in vegetation cover in the YZR, and found that there was an obvious positive correlation between NDVI and precipitation. Sun et al. [26] investigated how vegetation growth changed in the YZR, and determined the driving mechanisms. Wang et al. [27] analyzed the variation characteristics of vegetation cover from 1985 to 2018 in the YZR, and found that NDVI showed an overall growth trend. Meng et al. [28] found that the vegetation change in the YZR was greatly affected by precipitation and temperature, of which the impact of temperature was slightly stronger than precipitation, and the impact of climate change on the vegetation in the basin had a lagging effect. Cui et al. [29] assessed the sensitivity of vegetation change in the YZR to temperature and precipitation, and found that NDVI had a positive relationship with temperature and precipitation changes, and annual NDVI was more sensitive to temperature than to precipitation. Zuo et al. [30] analyzed the impact mechanism of climate change on vegetation dynamics in the YZR. There was a significant positive correlation between NDVI and precipitation and drought in the upstream and middle reaches of the basin, while there was a significant negative correlation between

NDVI and temperature in the southeast of the middle and lower reaches. The vegetation changes in the YZR could change the underlying surface characteristics and the energy balance, and then significantly affect the river runoff of the basin [31,32]. Nevertheless, few research works have computed the contribution of vegetation variation on the streamflow of the YZR.

The debate on the relationship between "vegetation and water" dates back to at least the mid-19th century [33]. Some research works displayed how vegetation growth had a negative effect on discharge increase [34]. However, some research works demonstrated that vegetation growth had few effects on discharge [35,36] and even a positive effect on discharge increase [37,38]. In order to deepen our understanding of the relationships of the climate–vegetation–hydrological process in the YZR, the purpose of this study is to explore how vegetation growth in the YZR affects runoff changes. The contents of this paper include: (1) Analyzing the temporal variation characteristics of hydro-meteorological elements in the YZR, and identifying the abrupt year of runoff at Nuxia hydrological station; (2) Analyzing the temporal variation characteristics of NDVI and the Budyko parameter (n), and quantifying the functional equation for NDVI and n. (3) Assessing the impact of vegetation variation, human activities, and climate change on runoff change. The results of this study can provide scientific guidance and support for the evaluation of the effects of ecological restoration measures, as well as the management and planning of water resources in the YZR.

2. Research Area and Data

The YZR is the largest river on the Qinghai–Tibet Plateau. It originates from the Jamayangzong Glacier, leaves China in Bashika and enters India in Assam, where it is renamed the Brahmaputra River. The total length of the river is 2057 km, covering an area of about 240,000 km^2, accounting for about 20% of the Tibet Autonomous Region, with an average altitude of 4500 m. The terrain is high in the west and low in the east. Influenced by the warm and humid air flow in the Indian Ocean, the climatic conditions in the upstream and downstream of the basin vary greatly. The temperature varies significantly with altitude, with a gradual increasing trend from upstream to downstream. The annual average temperature in the high-altitude region where the source is located is about 0 to 3 °C, and in the middle reaches it is about 5 to 9 °C, while the monthly average maximum temperature in Lhasa is about 10 to 17 °C. The precipitation tends to increase gradually from upstream to downstream, with an annual precipitation of about 420 mm in Shigatse, while the annual precipitation in Bashika can reach 5000 mm. The vertical distribution of precipitation is seasonally uneven. The precipitation in the rainy season (June to September) accounts for 65% to 80% of the annual precipitation, while the precipitation in the dry season (October to April of the next year) is sparse. The average annual runoff is 166.1 billion m^3, and the distribution of runoff is uneven within the year. The water resources in the wet season account for 70%, while the water resources in the dry season only account for less than 20%. The upstream vegetation of the basin is mainly composed of alpine grasslands, alpine meadows, and some alpine vegetation, while the middle reaches are mainly covered by shrubs and meadows. The downstream vegetation is mainly composed of coniferous forests, broad-leaved forests, and some alpine vegetation.

Nuxia hydrological station is located in the main stream of the YZR and controls more than 80% of the basin area. It is one of the important hydrological stations in the Tibetan Plateau. The runoff data (1982–2015) of Nuxia station were obtained from the Tibet autonomous region hydrology and water resources survey (http://www.xzsw.com.cn/ accessed on 1 March 2023). The multi-year meteorological station data (1982–2015) in and around the YZR were obtained from the China Meteorological Administration. First of all, we computed the daily potential evapotranspiration at meteorological stations in and around the YZR using the Penman–Monteith equation, and then computed monthly precipitation and potential evaporation. Finally, we utilized the Kriging method in the ArcGIS to interpolate the monthly precipitation and potential evaporation, and we computed annual scale precipitation and potential evapotranspiration through adding monthly scale data. NDVI for the period of 1982–2015 was obtained from the GIMMS NDVI3g V1.0 dataset, and its temporal resolution and spatial resolution were day and 0.05 degree. Monthly and annual NDVI were computed using the maximum synthesis method.

3. Research Methods

3.1. Methods for Detecting Trends and Variability

A simple linear regression method was used for analyzing the temporal variation characteristics of hydro-meteorological elements in the YZR. The linear regression method has simple algorithms, fast operation speed, and is good at obtaining linear relationships and changing trends in data, with strong explanatory power. Therefore, this method is often used to analyze the temporal variation characteristics of data [39,40].

The heuristic segmentation algorithm treats mutation detection as a segmentation problem that can divide a non-smooth sequence into smooth subsequences with different mean values. The year of mutation was determined by identifying the maximum mean difference of each subsequence, depending on whether the mutation point met different statistical significances [41]. For traditional detection methods, some of the shortcomings that arise when dealing with nonlinear and non-stationary time series data could be compensated [42]. The heuristic segmentation method uses multiple iterations in the segmentation process, which greatly improved the computational efficiency and had good practicality [43,44].

To determine the accuracy of the heuristic segmentation to detect mutation points, this study will use the cumulative abnormal mutation test method to diagnose the mutation years of the same time series data to corroborate the mutation results of the heuristic segmentation method. The advantage of these two methods is that they are not only easy to calculate, but are also clear about the time of mutation occurrence.

3.2. Corrected Budyko Equation

According to the degree of description of the rainfall runoff generation system, the attribution analysis methods for watershed runoff changes are classified into three categories: (1) Empirical relationship method. Considering the watershed as a whole, based on the causal relationship between rainfall input and runoff output, a hydrological statistical relationship was established for the base period. The simulated runoff computed the hydrological statistical relationship based on rainfall conditions during the measurement period and compared it with the measured runoff during that period to obtain the contribution of human activities to water and sediment changes. This method was based on observation and experience, is also known as the hydrological statistical method, and is intuitive, concise, and has a certain degree of accuracy. The double cumulative curve method for regression after the dependent and independent variables have been accumulated year by year belongs to this category. (2) Semi empirical formula method. Based on the relevant theories of meteorology, hydrology, probability theory, etc., we established a relationship equation that was mutually causal and continuous and had a certain physical basis. Then, different conditional variables were introduced and equations were used to estimate the impact of climate factors and human activities on runoff changes, also known as the conceptual model method and semi quantitative model method. For example, the runoff elasticity coefficient method based on Budyko's water heat balance theory belongs to this category. This type of method considers the impact of climate factors (potential evapotranspiration and rainfall) on runoff, making the rainfall and runoff processes clearer. (3) Hydrological modeling method based on physical processes. The surface was divided into units according to different geographical element types, and then we reproduced physical processes such as rainfall, interception, infiltration, and runoff by considering the input and output of each unit in the vertical and horizontal directions. Finally, by changing the input conditions, the contribution of geographical elements to runoff changes can be estimated. For example, the SWAT (Soil and Water Assessment Tool) model, the WEPP

(Water Erosion Prediction Project) model, and the Xin'anjiang model all belong to this category.

There are presuppositions for the Budyko formula, which was applied for quantitatively computing the contribution of different factors to runoff. They are: (1) human factor, climatic factor, and vegetation are independent; (2) the runoff change in the base period is only affected by climatic factor; and (3) except for runoff changes caused by precipitation, potential evapotranspiration, and vegetation changes, all other factors that affect runoff changes are unanimously considered to be human factors [45–48].

$$R = Pr - ET \tag{1}$$

R, Pr and ET, respectively, denote runoff depth, precipitation, and actual evaporation of the basin.

$$ET = \frac{Pr \times ET_0}{(Pr^n + ET_0^n)^{1/n}} \tag{2}$$

ET_0 denotes potential evapotranspiration (mm) and is computed using the Penman–Monteith equation. Budyko parameter n reflects the comprehensive impact of soil, terrain, and anthropogenic factors. The soil and terrain are not prone to changes in a short period of time, so anthropogenic factors have become the main factor affecting the Budyko parameter (n). Therefore, Budyko parameter n is utilized for characterizing anthropogenic factors. Anthropogenic factors influence streamflow in YZR through numerous means, including water conservancy project, tree planting, and afforestation, etc., Li et al. [49] proved that Budyko parameter n has a significant correlation with $NDVI$ in a basin.

$$n = a * NDVI + b \tag{3}$$

$$R = Pr - \frac{Pr \times ET_0}{\left(Pr^{a*NDVI+b} + ET_0^{a*NDVI+b}\right)^{1/(a*NDVI+b)}} \tag{4}$$

ε_P is elasticity coefficient of R for Pr, ε_{ET0} is elasticity coefficient of R for ET_0, ε_n is elasticity coefficient of R for n, and ε_{NDVI} is elasticity coefficient of R for $NDVI$, and they are computed with the following equations [50].

$$\varepsilon_P = \frac{\left(1+\left(\frac{ET_0}{Pr}\right)^n\right)^{1/n+1} - \left(\frac{ET_0}{Pr}\right)^{n+1}}{\left(1+\left(\frac{ET_0}{Pr}\right)^n\right)\left[\left(1+\left(\frac{ET_0}{Pr}\right)^n\right)^{1/n} - \left(\frac{ET_0}{Pr}\right)\right]} \tag{5}$$

$$\varepsilon_{ET0} = \frac{1}{\left(1+\left(\frac{ET_0}{Pr}\right)^n\right)\left[1-\left(1+\left(\frac{ET_0}{Pr}\right)^{-n}\right)^{1/n}\right]} \tag{6}$$

$$\varepsilon_n = \frac{\ln\left(1+\left(\frac{ET_0}{Pr}\right)^n\right) + \left(\frac{ET_0}{Pr}\right)^n \ln\left(1+\left(\frac{ET_0}{Pr}\right)^{-n}\right)}{n\left(1+\left(\frac{ET_0}{Pr}\right)^n\right)\left[1-\left(1+\left(\frac{ET_0}{Pr}\right)^{-n}\right)^{1/n}\right]} \tag{7}$$

$$\varepsilon_{NDVI} = \varepsilon_n \frac{a*NDVI}{a*NDVI+b} \tag{8}$$

According to mutation analysis result, the study period was separated into S_1 and S_2. Thus, the change values of precipitation (ΔPr), potential evapotranspiration (ΔET_0), underlying surface characteristic parameter (Δn), and NDVI ($\Delta NDVI$) from S_1 to S_2 were computed. ΔR_{Pr}, ΔR_{ET0}, ΔR_n, and ΔR_{NDVI}, respectively, represent streamflow variation values caused by variation values of precipitation, potential evapotranspiration, n, and NDVI from S_1 to S_2.

$$\Delta R_{Pr} = \varepsilon_P \frac{R}{Pr} \times \Delta Pr \tag{9}$$

$$\Delta R_{ET0} = \varepsilon_{ET0} \frac{R}{ET_0} \times \Delta ET_0 \tag{10}$$

$$\Delta R_n = \varepsilon_n \frac{R}{n} \times \Delta n \tag{11}$$

$$\Delta R_{NDVI} = \varepsilon_{NDVI} \frac{R}{NDVI} \times \Delta NDVI \tag{12}$$

$$\Delta R_{hum} = \Delta R_n - \Delta R_{NDVI} \tag{13}$$

ηR_{pr}, ηR_{ET0}, ηR_{NDVI} and ηR_H, respectively, denote the contribution rate of Pr, potential evapotranspiration, $NDVI$, and anthropogenic factor on streamflow, and are computed with the following equations.

$$\Delta R = \Delta R_{Pr} + \Delta R_{ET0} + \Delta R_{NVDI} + \Delta R_{hum} \tag{14}$$

$$\eta R_{Pr} = \Delta R_{Pr}/\Delta R \times 100\% \tag{15}$$

$$\eta R_{ET0} = \Delta R_{ET0}/\Delta R \times 100\% \tag{16}$$

$$\eta R_{NDVI} = \Delta R_{NDVI}/\Delta R \times 100\% \tag{17}$$

$$\eta R_H = \Delta R_{hum}/\Delta R \times 100\% \tag{18}$$

4. Results

4.1. Trends Analysis of Climate Factors

The simple linear regression method was employed for analyzing the temporal variation characteristics of precipitation and potential evapotranspiration during 1982–2015 (Figure 1). The slope of annual Pr from 1982 to 2015 was 0.2084 mm/a, indicating an increasing trend ($p > 0.05$) of annual Pr. The fluctuation range of annual Pr in the YZR was 411.00–631.30 mm, and the maximum and minimum values occurred in 1991 and 1982. Figure 1b demonstrated that the ET_0 of the YZR fluctuated and decreased, with a fluctuation range of 901.78–1023.79. The maximum and minimum values appeared in 2009 and 2000, and the slope of ET_0 from 1982 to 2015 was -0.3873/a ($p < 0.05$).

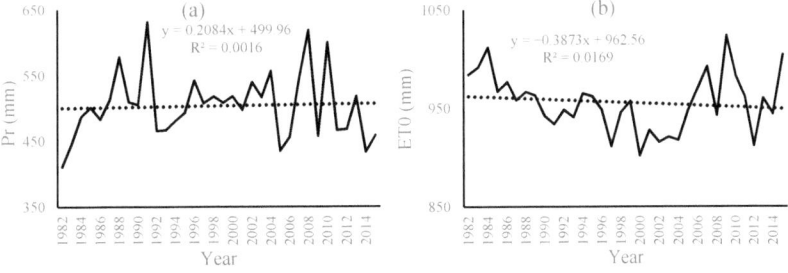

Figure 1. Temporal varying characteristics of annual average precipitation (**a**) and reference evaporation (**b**) in the YZR.

4.2. Trend Analysis and Abrupt Analysis of Runoff Depth

Figure 2 displayed the temporal varying characteristics of runoff depth in the YZR. Figure 2 demonstrates that the runoff depth of the YZR climbed up and then declined, showing an overall growth trend, with a fluctuation range of 202.30–458.78. The maximum and minimum values appeared in 2000 and 1983, and the slope of the runoff depth from 1982 to 2015 was 2.0218 mm/a ($p < 0.05$).

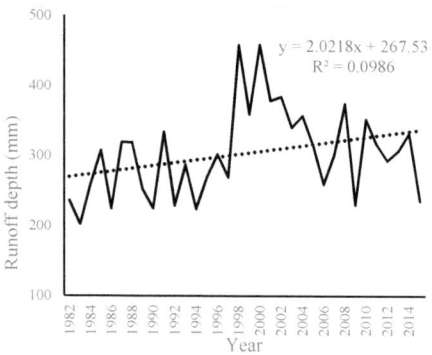

Figure 2. Temporal varying characteristics of runoff depth in the YZR.

A heuristic segmentation algorithm was applied for distinguishing the abrupt years of annual discharge data through two steps (Figure 3): (1) Calculating the T-test statistics for each year to measure the variability of the mean values of two subsequences. The year with the largest T-test statistic (T_{max}) may be the year of mutation. The largest T-test statistics were about equal to 3.6 and occurred in 1997, implying that the mutation year of annual runoff in Nuxia station may have occurred in 1997. (2) Calculating the significance probability $P(T_{max})$ corresponding to the largest T-test statistic (T_{max}). The greater the $P(T_{max})$, the better the significance. The basic parameter range was set between 0.5 and 0.95, and it is generally considered plausible to take the value between this range, and in this paper, in order to distinguish it from other points as a distinction, we took a different value according to its actual situation, i.e., P0 took the value of 0.84. Therefore, $P(T_{max})$ was 0.673 and was greater than 0.67 (P0), which proved that annual runoff of Nuxia station mutated in 1997.

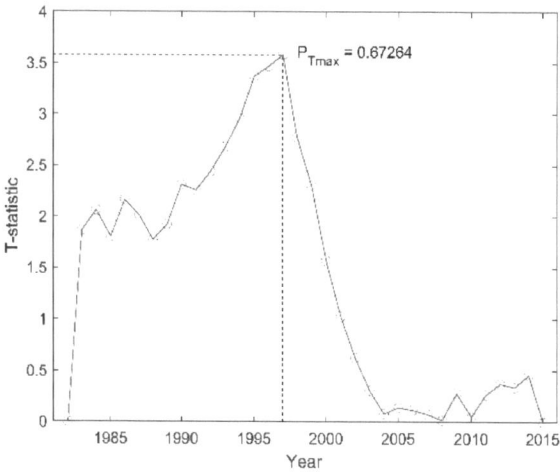

Figure 3. Heuristic segmentation algorithm abrupt test result of runoff in YZR from 1982 to 2015.

In order to verify the sudden change result of the heuristic segmentation algorithm, the cumulative anomaly mutation test approach was applied to diagnose the sudden change year of the same time series data (Figure 4). Figure 4 demonstrates that the cumulative anomaly values of runoff in the YZR showed an overall decline trend from 1982 to 1997, and showed an overall growth trend from 1997 to 2015. As a consequence, the turning point of the runoff cumulative anomaly value was 1997, which proved that the annual runoff of Nuxia station mutated in 1997. Therefore, based on the results of the two methods, we believe that the annual runoff of Nuxia station mutated in 1997, and the result of this study is consistent with the research result of Li et al. [18].

Figure 4. Runoff accumulated anomaly mutation analysis in YZR from 1982 to 2015.

4.3. Functional Equation for NDVI and n

If the elements ET_0, Pr, and R of the Budyko equation can be determined, the Budyko parameter (n) can be solved. For computing the effect of vegetation change on discharge in the YZR, we first explored the temporal varying characteristics of NDVI and n in the YZR (Figure 5). There were significant fluctuations in NDVI and Budyko parameter (n), with a fluctuation range of 0.382–0.456 and 1.069–1.527. The mean value of n decreased from 1.396 in 1982–1998 to 1.253 in 1998–2015, with a relative change rate of 10.266%.

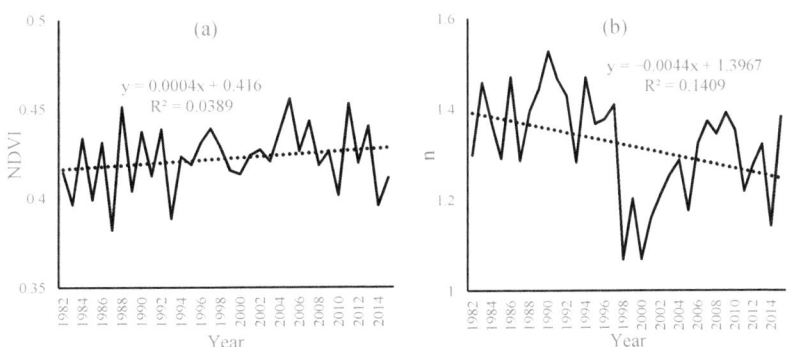

Figure 5. Temporal varying characteristics of NDVI (**a**) and n (**b**) in YZR.

Li et al. [51] found that the change rate of Budyko parameter (n) was well correlated with the vegetation index of the middle Yellow River basin; Zhang et al. [52] established a model for the lower bedding surface parameter n, which can be estimated directly using vegetation change. The 9-year sliding mean values of NDVI and Budyko parameter (n) were computed. Next, we drew a scatter diagram of the 9-year sliding mean values of NDVI and Budyko parameter (n) (Figure 6). In the end, the functional equation for NDVI and Budyko parameter (*n*) was computed (a = −9.9744 and b = 5.55), and its determination coefficient was 0.4604 ($p < 0.01$), implying that *n* had a significant negative correlation with NDVI.

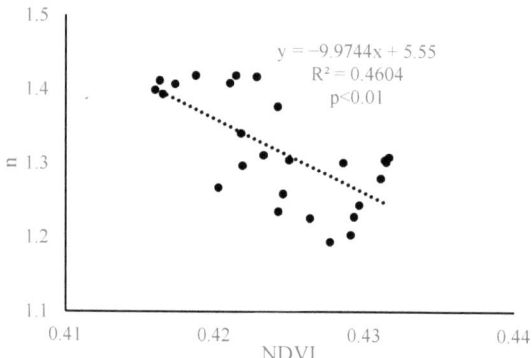

Figure 6. Functional equations for NDVI and n.

4.4. Influence Assessment of NDVI on Streamflow

According to the results of the heuristic segmentation algorithm and cumulative anomaly mutation test approach, the study period was separated into S1 (1982–1997) and S2 (1998–2015), and values of hydro-meteorological elements and NDVI of the YZR in different periods were obtained (Table 1). ET_0 in the YZR decreased by 9.22 mm from S1 (960.66 mm) to S2 (951.45 mm), and *Pr* and *R* increased by 4.74 mm and 69.96 mm from S1 to S2. NDVI increased by 0.007 from S1 (0.418) to S2 (0.425). *n* decreased by 0.143 from 1.396 (S1) to 1.253 (S2).

Table 1. Values of hydro-meteorological elements and NDVI in YZR.

Periods	ET_0/mm	R/mm	Pr/mm	n	NDVI
S1 (1982–1997)	960.66	265.87	501.10	1.396	0.418
S2 (1998–2015)	951.45	335.83	505.84	1.253	0.425

According to Table 1, the differences in *Pr*, ET_0, *NDVI*, and *n* between S1 and S2 were computed, and then the elastic coefficients of *R* on *Pr*, ET_0, *NDVI*, and *n* were computed using Formulas (5)–(8). Then, the amounts of runoff caused by *Pr*, ET_0, *NDVI*, and *n* were calculated with Formulas (9)–(13) (Table 2). The elasticity coefficients of *R* for *Pr* (ε_P) and *NDVI* (ε_{NDVI}) were 1.27 and 6.18, indicating that a 1% decrease in annual *Pr* and *NDVI* would lead to a decrease of 1.27% and 6.18% in runoff depth. The elasticity coefficients of *R* for ET_0 (ε_{ET0}) and Budyko parameter *n* (ε_n) were −0.27 and −1.96, indicating that a 1% decrease in annual ET_0 and Budyko parameter *n* would lead to an increase of 1.27% and 6.18% in runoff depth. Finally, based on Formulas (15)–(18), the contributions of *Pr*, ET_0, *NDVI*, and *n* on streamflow at the Nuxia Hydrological Station were computed (Table 2). The contributions of *Pr*, ET_0, *NDVI*, and *n* on the streamflow change in the S2 period (1998–2015) were 5.26%, 1.14%, 43.04%, and 50.06%. In summary, the increase of runoff in the YZR is mainly influenced by *NDVI* growth and underlying surface change. Some research works demonstrated that the vegetation growth had a positive effect on discharge increase [37,38], and the areas where *NDVI* growth led to increased runoff were mostly located in large watersheds with complex terrain [53]. The Yarlung Zangbo River is a large watershed with a complex terrain. Therefore, the conclusion of this study is similar to existing studies.

Table 2. Attribution analysis of discharge variation in YZR.

ε_P	ε_{E0}	ε_n	ε_{NDVI}	ΔR_P	ΔR_{ET0}	ΔR_{NDVI}	ΔR_{hum}	ηR_P	ηR_{ET0}	ηR_{NDVI}	ηR_{hum}
1.27	−0.27	−1.96	6.18	3.62	0.78	29.65	34.83	5.26%	1.14%	43.04%	50.06%

5. Conclusions and Discussion

5.1. Conclusions

The Yarlung Zangbo River (YZR) is the largest river on the Qinghai–Tibet Plateau. In order to deepen our understanding of the relationship of climate–vegetation–hydrological processes in the YZR, we first identified the abrupt year of discharge in the YZR using a heuristic segmentation algorithm and cumulative anomaly mutation test approach. After that, the functional equation for NDVI and n was computed. In the end, NDVI was introduced into the Budyko equation to evaluate the impact of vegetation change on streamflow in the YZR. It turned out that: (1) NDVI and discharge in the YZR both presented an increasing trend, and the mutation year of annual runoff in Nuxia station occurred in 1997. (2) n had a significant negative correlation with NDVI in the YZR ($p < 0.01$). (3) The contributions of Pr, $ET0$, $NDVI$, and n on streamflow change in the S2 period (1998–2015) were 5.26%, 1.14%, 43.04%, and 50.06%.

5.2. Discussion

Since 1999, the Chinese government has implemented a number of large-scale ecological restoration projects, such as the returning farmland to forest/grassland project, the Three-North Shelterbelt Program project, which has significantly increased the vegetation coverage across China. The elevation of the YZR varies greatly. The cold and dry conditions in the upper reaches change into warm and humid conditions in the lower reaches, resulting in high vegetation diversity in the basin. The implementation of a series of soil and water conservation measures in the YZR has significantly changed the rainfall–runoff relationship in the area. The underlying surface of the watershed reflects the comprehensive impact of soil, terrain, and vegetation. The soil and terrain are not prone to changes in a short period of time, so vegetation change has become the main factor affecting the changes in the underlying surface of the watershed. Liu et al. and Han et al. both found that the vegetation cover in the YZR had an overall increasing trend [54,55]. Any increase or decrease in vegetation will in turn lead to changes in runoff. The debate on the relationship between "vegetation and water" dates back to at least the mid-19th century [33]. Some research works have demonstrated that vegetation growth has a negative effect on discharge increase [34]. However, some research works have shown that the vegetation growth has few effect on discharge [35,36] and even a positive effect on discharge increase [37,38].

From Figure 6, we found that NDVI in the YZR showed a strong correlation with the Budyko parameter (n), indicating that vegetation restoration has a significant impact on the Budyko parameter (n). Subsequent attribution analysis indicated that the change in vegetation cover led to an increase in runoff in the YZR. Some research works demonstrated that vegetation growth had a positive effect on discharge increase, and the areas where NDVI growth led to increased runoff were mostly located in large watersheds with complex terrain [53]. The Yarlung Zangbo River is a large watershed with a complex terrain. Therefore, the conclusion of this study is similar to those of existing studies. The impact of vegetation reconstruction on water resources should be given significant attention. The government should plan reasonable vegetation restoration according to the actual conditions of the basin. Yang et al. [56] analyzed the impact of different vertical structures on water yield using a simulated rainfall experiment in the field, and found that the vertical structure of vegetation is an important factor influencing water yield. Therefore, in the subsequent vegetation restoration process, the vegetation structure in the YZR should be optimized.

Although this study strictly controlled data, several uncertainties existed. First of all, meteorological data of individual dates at some stations were missing. Next, precipitation and potential evapotranspiration data were obtained through interpolation, using data from national meteorological stations. Moreover, the study ignored the reciprocal effects between climatic factor, vegetation, and human factor [57–59], so we should systematically quantify the influence of climatic conditions and anthropic factor interactions on eco-hydrological systems [60–62]. In the future, we will try to build a distributed hydrological hydrothermal coupling model for computing how vegetation affects the hydrological process in the YZR [63–65]. In addition, the attribution analysis of runoff in the study area only considered four factors: precipitation, potential evapotranspiration, vegetation, and human activities. In fact, many glaciers are distributed in the YZR. With the increase in temperature, the impact of glacier melting on the runoff change of the YZR is increasing. In a follow-up study, the contribution rate of glacier melting to runoff change will be quantitatively analyzed [15,66].

Author Contributions: Conceptualization, G.J. and W.C.; Methodology, G.J. and S.Y.; Software, G.J., S.Y., J.Z. and J.H.; Validation, S.Y., J.H. and Y.G.; Formal analysis, S.Y. and J.Z.; Data curation, S.Y.; Writing—original draft, G.J. and W.C.; Writing—review and editing, G.J. and W.C.; Visualization, J.Z. and Y.G.; Project administration, W.C.; Funding acquisition, W.C. All authors have read and agreed to the published version of the manuscript.

Funding: This research was funded by the National Key R&D Program of China (2021YFD1700900), the Research Project of Henan Federation of Social Sciences (2023-ZZJH-189), the Research Project of Henan Federation of Social Sciences (SKL-2022-2249) and the special fund for top talents in Henan Agricultural University (30501031).

Data Availability Statement: Not applicable.

Conflicts of Interest: The authors declare no conflict of interest.

References

1. Ji, G.; Lai, Z.; Xia, H.; Liu, H.; Wang, Z. Future Runoff Variation and Flood Disaster Prediction of the Yellow River Basin Based on CA-Markov and SWAT. *Land* **2021**, *10*, 421. [CrossRef]
2. Ji, G.; Lai, Z.; Yan, D.; Wu, L.; Wang, Z. Spatiotemporal patterns of future meteorological drought in the Yellow River Basin based on SPEI under RCP scenarios. *Int. J. Clim. Chang. Strateg. Manag.* **2022**, *14*, 39–53. [CrossRef]
3. Greve, P.; Orlowsky, B.; Mueller, B.; Sheffield, J.; Reichstein, M.; Seneviratne, S. Global assessment of trends in wetting and drying over land. *Nat. Geosci.* **2014**, *7*, 716–721. [CrossRef]
4. Li, Y.; Li, H.; Huang, J.; Liu, C. An approximation method for evaluating flash flooding mitigation of sponge city strategies-A case study of Central Geelong. *J. Clean. Prod.* **2020**, *257*, 120525. [CrossRef]
5. Abbott, B.; Bishop, K.; Zarnetske, J.; Minaudo, C.; Chapin, F.; Krause, S.; Hannah, D.; Conner, L.; Ellison, D.; Godsey, S.; et al. Human domination of the global water cycle absent from depictions and perceptions. *Nat. Geosci.* **2019**, *12*, 533–540. [CrossRef]
6. Ji, G.; Wu, L.; Wang, L.; Yan, D.; Lai, Z. Attribution Analysis of Seasonal Runoff in the Source Region of the Yellow River Using Seasonal Budyko Hypothesis. *Land* **2021**, *10*, 542. [CrossRef]
7. Piao, S.; Ciais, P.; Huang, Y.; Shen, Z.; Peng, S.; Li, J.; Zhou, L.; Liu, H.; Ma, Y.; Ding, Y.; et al. The impacts of climate change on water resources and agriculture in China. *Nature* **2010**, *467*, 43–51. [CrossRef]
8. Qin, Y.; Batzoglou, J.; Siebert, S.; Huning, L.; AghaKouchak, A.; Mankin, J.; Hong, C.; Tong, D.; Davis, S.; Mueller, N. Agricultural risks from changing snowmelt. *Nat. Clim. Chang.* **2020**, *10*, 459–465. [CrossRef]
9. Ji, G.; Song, H.; Wei, H.; Wu, L. Attribution Analysis of Climate and Anthropic Factors on Runoff and Vegetation Changes in the Source Area of the Yangtze River from 1982 to 2016. *Land* **2021**, *10*, 612. [CrossRef]
10. Yi, Y.; Wu, S.; Zhao, D.; Zheng, D.; Pan, T. Modeled effects of climate change on actual evapotranspiration in different eco-geographical regions in the Tibetan Plateau. *J. Geogr. Sci.* **2013**, *23*, 195–207.
11. Piao, S.; Zhang, X.; Wang, T.; Liang, E.; Wang, S.; Zhu, J.; Niu, B. Responses and feedback of the Tibetan Plateau's alpine ecosystem to climate change. *Chin. Sci. Bull.* **2019**, *64*, 2842–2855. (In Chinese) [CrossRef]
12. Li, Y.; Su, F.; Chen, D.; Tang, H. Atmospheric Water Transport to the Endorheic Tibetan Plateau and its Effect on the Hydrological Status in the Region. *J. Geophys. Res. Atmos.* **2019**, *124*, 12864–12881. [CrossRef]
13. Tang, Q.; Lan, C.; Su, F.; Fang, H.; Zhang, S.; Han, D.; Liu, X.; He, L.; Xu, X.; Tang, Y.; et al. Streamflow change on the Qinghai-Tibet Plateau and its impacts. *Chin. Sci. Bull.* **2019**, *64*, 2807–2821. (In Chinese)
14. Zhao, Z.; Fu, Q.; Gao, C.; Zhang, X.; Xu, Y. Simulation of monthly runoff considering flow components in Yarlung Zangbo River. *J. China Hydrol.* **2017**, *37*, 26–30. (In Chinese)
15. Xin, J.; Sun, X.; Liu, L.; Li, H.; Liu, X.; Li, X.; Cheng, L.; Xu, Z. Quantifying the contribution of climate and underlying surface changes to alpine runoff alterations associated with glacier melting. *Hydrol. Process.* **2021**, *35*, e14069. [CrossRef]

16. Liu, Z.; Yao, Z.; Huang, H.; Wu, S.; Liu, G. Land use and climate changes and their impacts on runoff in the Yarlung Zangbo River Basin, China. *Land Degrad. Dev.* **2014**, *25*, 203–215. [CrossRef]
17. Wang, X.; Qin, G.; Li, H. Analysis on characteristics and variation trend of annual runoff of mainstream of Yarlung Tsangpo River. *Yangtze River* **2016**, *47*, 23–26. (In Chinese)
18. Li, H.; Niu, Q.; Wang, X.; Liu, L.; Xu, Z. Variation Characteristics of Runoff in the Yarlung Zangbo River Basin from 1961 to 2015. *J. Soil Water Conserv.* **2021**, *35*, 110–115. (In Chinese)
19. Liu, J.; Ren, Y.; Zhang, W.; Tao, H.; Yi, L. Study on the influence of climate and underlying surface change on runoff in the Yarlung Zangbo River basin. *J. Glaciol. Geocryol.* **2021**, *43*, 275–287. (In Chinese)
20. Yang, D.; Wang, Y.; Tang, L.; Yan, D.; Cui, T. Analysis of runoff changes and their causes under climate changes in upper Yarlung Zangbo River basin. *J. Hydroelectr. Eng.* **2023**, *42*, 42–49. (In Chinese)
21. Feng, X.; Fu, B.; Piao, S.; Zeng, Z.; Lü, Y.; Zeng, Y.; Li, Y.; Jiang, X.; Wu, B. Revegetation in China's Loess Plateau is approaching sustainable water resource limits. *Nat. Clim. Chang.* **2016**, *6*, 1019–1022. [CrossRef]
22. Liu, L.; Niu, Q.; Heng, J.; Li, H.; Xu, Z. Transition Characteristics of the Dry-Wet Regime and Vegetation Dynamic Responses over the Yarlung Zangbo River Basin, Southeast Qinghai-Tibet Plateau. *Remote Sens.* **2019**, *11*, 1254. [CrossRef]
23. Liu, X.; Xu, Z.; Peng, D. Spatio-Temporal Patterns of Vegetation in the Yarlung Zangbo River, China during 1998–2014. *Sustainability* **2019**, *11*, 4334. [CrossRef]
24. Fu, H.; Zhao, W.; Zhan, Q.; Yang, M.; Xiong, D.; Yu, D. Temporal Information Extraction for Afforestation in the Middle Section of the Yarlung Zangbo River Using Time-Series Landsat Images Based on Google Earth Engine. *Remote Sens.* **2021**, *13*, 4785. [CrossRef]
25. Lv, Y.; Dong, G.; Yang, S.; Zhou, Q.; Cai, M. Spatio-Temporal Variation in NDVI in the Yarlung Zangbo River Basin and Its Relationship with Precipitation and Elevation. *Resour. Sci.* **2014**, *36*, 603–0611. (In Chinese)
26. Sun, W.; Wang, Y.; Fu, Y.; Xue, B.; Wang, G.; Yu, J.; Zuo, D.; Xu, Z. Spatial heterogeneity of changes in vegetation growth and their driving forces based on satellite observations of the Yarlung Zangbo River Basin in the Tibetan Plateau. *J. Hydrol.* **2019**, *574*, 324–332. [CrossRef]
27. Wang, S.; Wang, S.; Fan, F. Change patterns of NDVI (1985—2018) in the Yarlung Zangbo River Basin of China based on time series segmentation algorithm. *Acta Ecol. Sin.* **2020**, *40*, 6863–6871. (In Chinese)
28. Meng, Q.; Liu, Y.; Ju, Q.; Liu, J.; Wang, G.; Jin, J.; Guan, T.; Liu, C.; Bao, Z. Vegetation change and its response to climate change in the Yarlung Zangbo River basin in the past 18 years. *South-North Water Transf. Water Sci. Technol.* **2021**, *19*, 539–550. (In Chinese)
29. Cui, L.; Pang, B.; Zhao, G.; Ban, C.; Ren, M.; Peng, D.; Zuo, D.; Zhu, Z. Assessing the Sensitivity of Vegetation Cover to Climate Change in the Yarlung Zangbo River Basin Using Machine Learning Algorithms. *Remote Sens.* **2022**, *14*, 1556. [CrossRef]
30. Zuo, D.; Han, Y.; Xu, Z.; Li, P. Impact mechanism of climate change on vegetation dynamics in the Yarlung Zangbo River Basin. *Water Resour. Prot.* **2022**, *38*, 1–8. (In Chinese)
31. Piao, S.L.; Wang, X.H.; Ciais, P.; Zhu, B.; Liu, J. Changes in satellite-derived vegetation growth trend in temperate and boreal Eurasia from 1982 to 2006. *Glob. Change Biol.* **2011**, *17*, 3228–3239. [CrossRef]
32. Jackson, R.B.; Jobbágy, E.G.; Avissar, R.; Roy, S.; Barrett, D.; Cook, C.; Farley, K.; Maitre, D.; Mccarl, B.; Cook, C. Trading water for carbon with biological carbon sequestration. *Science* **2005**, *310*, 1944–1947. [CrossRef]
33. Calder, I.R.; Smyle, J.; Aylward, B. Debate over flood-proofing effects of planting forests. *Nature* **2007**, *450*, 945. [CrossRef]
34. Wang, Y.; Liu, Z.; Qian, B.; He, Z.; Ji, G. Quantitatively Computing the Influence of Vegetation Changes on Surface Discharge in the Middle-Upper Reaches of the Huaihe River, China. *Forests* **2022**, *13*, 2000. [CrossRef]
35. Buttle, J.M.; Metcalfe, R.A. Boreal forest disturbance and streamflow response, northeastern Ontario. *Can. J. Fish. Aquat. Sci.* **2000**, *57*, 5–18. [CrossRef]
36. Ceballos-Barbancho, A.; Morán-Tejeda, E.; Luengo-Ugidos, M.Á.; Llorente-Pinto, J.M. Water resources and environmental change in a Mediterranean environment: The south-west sector of the Duero river basin (Spain). *J. Hydrol.* **2008**, *351*, 126–138. [CrossRef]
37. Zhou, G.; Wei, X.; Luo, Y.; Zhang, M.; Li, Y.; Qiao, Y.; Liu, H.; Wang, C. Forest recovery and river discharge at the regional scale of Guangdong Province, China. *Water. Resour. Res.* **2010**, *46*, W09503. [CrossRef]
38. Wang, S.; Fu, B.; He, C.; Sun, G.; Gao, G. A comparative analysis of forest cover and catchment water yield relationships in northern China. *For. Ecol. Manag.* **2011**, *262*, 1189–1198. [CrossRef]
39. Xu, Z.; Ban, C.; Zhang, R. Evolution laws and attribution analysis in the Yarlung Zangbo River basin. *Adv. Water Sci.* **2022**, *33*, 519–530. (In Chinese)
40. Xia, H.; Wang, H.; Ji, G. Regional Inequality and Influencing Factors of Primary PM Emissions in the Yangtze River Delta, China. *Sustainability* **2019**, *11*, 2269. [CrossRef]
41. Liu, X.; Wu, Z.; Liu, Y.; Zhao, X.; Rui, Y.; Zhang, J. Spatial-temporal characteristics of precipitation from 1960 to 2015 in the Three Rivers' Headstream Region, Qinghai, China. *Acta Geogr. Sin.* **2019**, *74*, 1803–1820. (In Chinese)
42. Bernaola-Galván, P.; Ivanov, P.C.; Amaral, L.A.N.; Stanley, H.E. Scale Invariance in the nonstationarity of human heart rate. *Phys. Rev. Lett.* **2001**, *87*, 168105. [CrossRef]
43. Feng, G.; Gong, Z.; Dong, W.; Li, J.P. Abrupt climate change detection based on heuristic segmentation algorithm. *Acta Phys. Sin.* **2005**, *54*, 5494–5499. (In Chinese) [CrossRef]
44. Gong, Z.; Feng, G.; Wang, S.; Li, J.P. Analysis of features of climate change of Hubei area and the global climate change based on heuristic segmentation algorithm. *Acta Phys. Sin.* **2006**, *55*, 477–483. (In Chinese) [CrossRef]

45. Yang, H.; Yang, D.; Lei, Z.; Sun, F. New analytical derivation of the mean annual water- energy balance equation. *Water Resour. Res.* **2008**, *44*, W03410. [CrossRef]
46. Li, H.; Shi, C.; Zhang, Y.; Ning, T.; Sun, P.; Liu, X.; Ma, X.; Liu, W.; Collins, A.L. Using the Budyko hypothesis for detecting and attributing changes in runoff to climate and vegetation change in the soft sandstone area of the middle Yellow River basin, China. *Sci. Total Environ.* **2020**, *703*, 135588. [CrossRef]
47. Caracciolo, D.; Pumo, D.; Viola, F. Budyko's Based Method for Annual Runoff Characterization across Different Climatic Areas: An Application to United States. *Water. Resour. Manag.* **2018**, *32*, 3189–3202. [CrossRef]
48. Zhang, X.; Dong, Q.; Cheng, L.; Xia, J. A Budyko-based framework for quantifying the impacts of aridity index and other factors on annual runoff. *J. Hydrol.* **2019**, *579*, 124224. [CrossRef]
49. Li, D.; Pan, M.; Cong, Z.; Zhang, L.; Wood, E. Vegetation control on water and energy balance within the Budyko framework. *Water Resour. Res.* **2013**, *49*, 969–976. [CrossRef]
50. Ji, G.; Huang, J.; Guo, Y.; Yan, D. Quantitatively Calculating the Contribution of Vegetation Variation to Runoff in the Middle Reaches of Yellow River Using an Adjusted Budyko Formula. *Land* **2022**, *11*, 535. [CrossRef]
51. Li, Y.; Liu, C.; Zhang, D.; Liang, K.; Li, X.; Dong, G. Reduced Runoff Due to Anthropogenic Intervention in the Loess Plateau, China. *Water* **2016**, *8*, 458. [CrossRef]
52. Zhang, S.; Yang, H.; Yang, D.; Jayawardena, A. Quantifying the effect of vegetation change on the regional water balance within the Budyko Framework. *Geophys. Res. Lett.* **2015**, *43*, 1140–1148. [CrossRef]
53. Zhou, G.Y.; Wei, X.H.; Chen, X.Z.; Zhou, P.; Liu, X.D.; Xiao, Y.; Sun, G.; Scott, D.F.; Zhou SY, D.; Han, L.S.; et al. Global pattern for the effect of climate and land cover on water yield. *Nat. Commun.* **2015**, *6*, 5918. [CrossRef]
54. Liu, L.; Niu, Q.; Heng, J.; Li, H.; Xu, Z. Characteristics of dry and wet conversion and dynamic vegetation response in Yarlung Zangbo River basin. *Trans. Chin. Soc. Agric. Eng.* **2020**, *36*, 175–184+338. (In Chinese)
55. Han, X.; Zuo, P.; Li, P.; Xu, Z.; Gao, X. Spatiotemporal variability of vegetation cover and its response to climate change in Yarlung Zangbo River Basin. *Adv. Sci. Technol. Water Resour.* **2021**, *41*, 16–23. (In Chinese)
56. Yang C, Yao W, Xiao P, Qin, D. Effects of vegetation cover structure on runoff and sediment yield and its regulation mechanism. *J. Hydraul. Eng.* **2019**, *50*, 1078–1085.
57. Yan, D.; Lai, Z.; Ji, G. Using Budyko-Type Equations for Separating the Impacts of Climate and Vegetation Change on Runoff in the Source Area of the Yellow River. *Water* **2020**, *12*, 3418. [CrossRef]
58. Gao, H.; Li, Q.; Xiong, G.; Li, B.; Zhang, J.; Meng, Q. Quantitative assessment of hydrological response to vegetation change in the upper reaches of Luanhe River with the modified Budyko framework. *Front. Ecol. Evol.* **2023**, *11*, 1178231. [CrossRef]
59. Liu, Y.; Chen, W.; Li, L.; Huang, J.; Wang, X.; Guo, Y.; Ji, G. Assessing the contribution of vegetation variation to streamflow variation in the Lancang River Basin, China. *Front. Ecol. Evol.* **2023**, *10*, 1058055. [CrossRef]
60. Wu, J.; Miao, C.; Wang, Y.; Duan, Q.; Zhang, X. Contribution analysis of the long-term changes in seasonal runoff on the Loess Plateau, China, using eight Budyko-based methods. *J. Hydrol.* **2017**, *545*, 263–275. [CrossRef]
61. Al-Safi, H.I.J.; Kazemi, H.; Sarukkalige, P.R. Comparative study of conceptual versus distributed hydrologic modelling to evaluate the impact of climate change on future runoff in unregulated catchments. *J. Water. Clim. Chang.* **2020**, *11*, 341–366. [CrossRef]
62. Jiang, C.; Xiong, L.; Wang, D.; Liu, P.; Guo, S.; Xu, C. Separating the impacts of climate change and human activities on runoff using the Budyko-type equations with time-varying parameters. *J. Hydrol.* **2015**, *522*, 326–338. [CrossRef]
63. Leuning, R.; Zhang, Y.Q.; Rajaud, A.; Cleugh, H.; Tu, K. A simple surface conductance model to estimate regional evaporation using MODIS leaf area index and the Penman-Monteith equation. *Water Resour. Res.* **2008**, *44*, 1–17. [CrossRef]
64. Liu, C.M.; Wang, Z.G.; Yang, S.T.; Sang, Y.; Liu, X.; Li, J. Hydro-informatic modeling system: Aiming at water cycle in land surface material and energy exchange processes. *Acta Geogr. Sin.* **2014**, *69*, 579–587. (In Chinese)
65. Zhang, D.; Liu, X.M.; Bai, P. Different influences of vegetation greening on regional water-energy balance under different climatic conditions. *Forests* **2018**, *9*, 412. [CrossRef]
66. Ji, H.; Peng, D.; Gu, Y.; Luo, X.; Pang, B.; Zhu, Z. Snowmelt Runoff in the Yarlung Zangbo River Basin and Runoff Change in the Future. *Remote Sens.* **2023**, *15*, 55. [CrossRef]

Disclaimer/Publisher's Note: The statements, opinions and data contained in all publications are solely those of the individual author(s) and contributor(s) and not of MDPI and/or the editor(s). MDPI and/or the editor(s) disclaim responsibility for any injury to people or property resulting from any ideas, methods, instructions or products referred to in the content.

MDPI AG
Grosspeteranlage 5
4052 Basel
Switzerland
Tel.: +41 61 683 77 34

Forests Editorial Office
E-mail: forests@mdpi.com
www.mdpi.com/journal/forests

Disclaimer/Publisher's Note: The title and front matter of this reprint are at the discretion of the Guest Editors. The publisher is not responsible for their content or any associated concerns. The statements, opinions and data contained in all individual articles are solely those of the individual Editors and contributors and not of MDPI. MDPI disclaims responsibility for any injury to people or property resulting from any ideas, methods, instructions or products referred to in the content.

www.ingramcontent.com/pod-product-compliance
Lightning Source LLC
LaVergne TN
LVHW072331090526
838202LV00019B/2397